THE FIGHT IN THE FIELDS

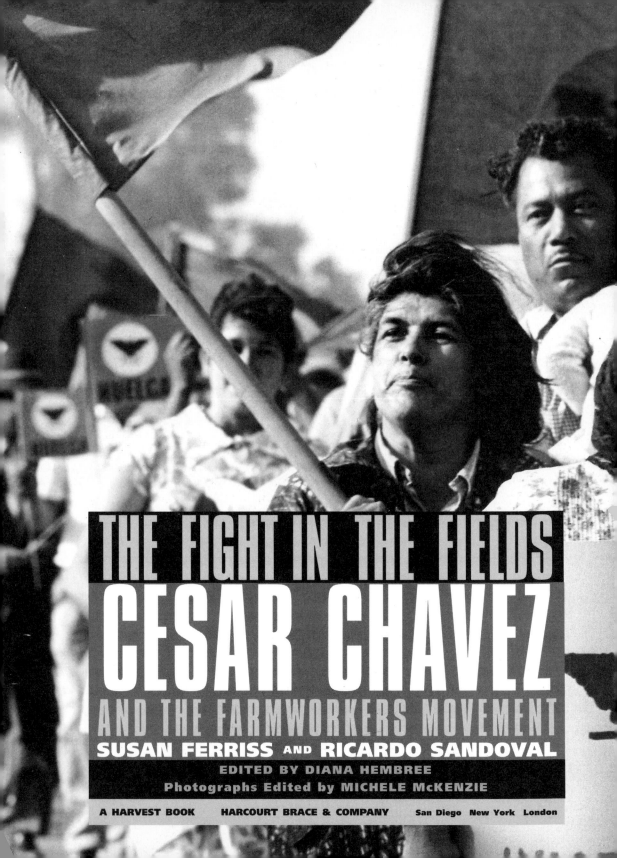

THE FIGHT IN THE FIELDS
CESAR CHAVEZ
AND THE FARMWORKERS MOVEMENT
SUSAN FERRISS AND RICARDO SANDOVAL

EDITED BY DIANA HEMBREE

Photographs Edited by MICHELE McKENZIE

A HARVEST BOOK   HARCOURT BRACE & COMPANY   San Diego New York London

Library of Congress Cataloging-in-Publication Data
Ferriss, Susan.
The fight in the fields: Cesar Chavez and the Farmworkers movement/by
Susan Ferriss and Ricardo Sandoval; edited by Diana Hembree.—1st ed.
p.   cm.
Includes bibliographical references and index.
ISBN 0-15-100239-8   ISBN 0-15-600598-0 (pbk.)
1. Chavez, Cesar, 1927–1995. 2. United Farm Workers—History. 3. Labor
leaders—United States—Biography. 4. Trade-unions—Migrant agricultural
laborers—United States—History. 5. Mexican American migrant agricultural
laborers—History. I. Sandoval, Ricardo. II. Hembree, Diana. III. Title.
HD6509.C48F47   1997
331.88′13′092—dc21   96-40375
[B]

Text set in Galliard and Frutiger Black
Display text set in Dante and OptiElizabeth
Designed by Camilla Filancia
Printed in the United States of America
First Harvest edition 1998

N M L K J I H G F

CONTENTS

▌N 1989, the two of us, Rick Tejada-Flores and Ray Telles—both filmmakers and old friends—came together with a shared set of concerns about telling stories on television that weren't being told. One of the most important stories, and one of the most neglected, was that of Cesar Chavez and the farmworkers movement. Although neither one of us came from the fields, we found that in many ways, the farmworkers' struggle had shaped who we were and how we viewed the world.

Ray Telles's deep interest in the farmworkers movement was rooted in his family: "My father and uncle, both electricians, went to Delano to help out in the early days of the union. As a college student I became involved in the Chicano movement and, like many, I came to recognize Cesar as a major force in determining its direction. Later, as a journalist covering the farmworkers, I was able to draw on this background to do stories that were both close to home and very important to my family. When I made films on the toll exacted on farmworkers by pesticides or on the UFW's battles with a hostile state government, the specific problems were new, but the stories were all too familiar."

Rick Tejada-Flores worked as a volunteer for the UFW, taking pictures for the union and later creating the union's first documentary film. "Those two years with the union were a defining experience in my life. Like so many people who came into contact with Chavez and the farmworkers, I was indelibly changed by the process. Although it was the start of my work as a filmmaker, the real lessons that I learned were about how society works and how it can be changed."

The UFW's story, which had touched both of our lives so profoundly, is the story of how ordinary people can become extraordinary, how the powerless can exert influence, and how the voiceless can be heard.

Cesar Chavez was the most important Latino leader this country has ever seen. Part of his greatness was that his vision reached out to touch millions of Americans— not just Mexican Americans, but ordinary people of all sorts. The connection that the farmworkers forged between the haves and the have-nots created a remarkable moment in American history—an era in which people who would not normally meet connected and worked together to correct terrible injustices.

The film and the book are an attempt to capture the intensity and focus of this remarkable movement and to help people learn its powerful lessons. With the death of Cesar Chavez in 1993, there is an urgency to preserve these memories while they are still fresh—but not because Chavez's death was the end of the story. The farmworkers movement is flourishing today, and all across the country there has been an outpouring of honors for the man who inspired it. But unless people remember what Cesar Chavez believed in and dedicated his life to, this belated respect will be an empty gesture.

It was clear from the start that no matter how good a film we made, there would be many stories and issues that we could not include. Hence this book, a rich biography of both Chavez and the farmworkers movement from its birth to its present renaissance. The book's editorial team—writers Susan Ferriss and Ricardo Sandoval, editor Diana Hembree, photo editor Michele McKenzie, and researcher Juan Avila Hernandez—all share our vision and concern. Their work has developed its own path and its own voice. As you read *The Fight in the Fields,* their words and images will open the doors to another world.

RICK TEJADA-FLORES
RAY TELLES
1996

ACKNOWLEDGMENTS

WRITING ABOUT farm labor has been a passion for us for many years. We are grateful to Ray Telles and Rick Tejada-Flores for inviting us to write this book in conjunction with their documentary film, *The Fight in the Fields,* and for offering us a wealth of resources to use. Many thanks as well to our hardworking colleagues on the book team and our editor at Harcourt Brace, Walter Bode, assistant editor Theo Lieber, managing editor Marianna Lee, copy editor Rachel Myers, production director David Hough, designer Camilla Filancia, and the rest of the editorial and production staff; Gary Soto, who wrote the foreword; and Sandra Dijkstra, our agent. We are also grateful to our editors and fellow reporters at the *San Francisco Examiner* and the *San Jose Mercury News* for their patience and enthusiasm, and to the Ferriss and Sandoval families for supporting us during this project and our journalism careers. Special thanks are in order for the Chavez family, the United Farm Workers, and Wayne State University. We are also grateful to Jacques Levy for allowing us to draw from his rich biography *Cesar Chavez: Autobiography of La Causa,* as well as his unpublished interviews with Chavez and other union figures. The distinguished work on the union's early days by authors John Gregory Dunne, Anne Loftis and Dick Meister, Peter Matthiessen, Ronald Taylor, and others should also be acknowledged. Finally, we would also like to thank the editors and writers who have encouraged us to pursue stories on farmworkers: Paul Wilner, Joe Livernois, Fred Hernandez, Thom Akeman, Jane Kay, Patricia Yollin, Eric Brazil, Greg Lewis, Jim Malone, and Mike May.

WHEN I WAS thirteen, I worked in the garden of an elderly woman named Mrs. Hall. My task, for which I was paid a dollar an hour, was to weed flower beds and rake leaves that had floated down from a stately eucalyptus. There was not much work, maybe two hours' worth, and since it took me about an hour to get to her house by foot and an hour to get back to mine, I made the most of it. I remember raking dropped oranges, which were puckered from the summer heat, raking them in such a way that from some distance I might have looked to be playing street hockey. I kicked dried oranges around the yard, feigning work, trying to appear busy because that particular day there was hardly a leaf out of place or a weed poking through the flower beds.

This was Fresno in 1965. This was me trying to find work, trying to be a good boy, when one morning jolly Mrs. Hall told me to put down my rake and join her for a discussion at the Fresno Free Public Library, for which she served as a "friend." The discussion, I would soon discover, involved the issue of farmworkers, particularly timely since Cesar Chavez and his union were kicking up a political storm in Delano. I slid into Mrs. Hall's car, a white Ford Comet, in part because I feared that if I said no she might not hire me back to kick oranges around her yard. At the library—the children's section—the people attending had separated themselves. This was not a round-table discussion, but a division of labor, if not race. At one end sat the Chicanos, and on the other sat what I could only identify, in my ignorant youth, as the white people. The discussion resembled a trial and, in fact, took place literally in the shadow of the old Fresno courthouse halfway down the block. The topic at hand was whether field conditions were as bad as the newly founded United Farm Workers union claimed. Introductions started quietly enough with some smiles, leg crossings, pencil scribblings, and chair scootings. Soon, though, hot tongues lashed out from both ends of the table, but it was the Chicanos who became boisterous. I suspect that if you worked for poor pay for so many years, you too might want to scream.

I didn't know what to do that summer afternoon. I stood by a huge globe and spun it slowly, letting my finger drag along the equator. I listened, then didn't listen as my mind wandered, but I was suddenly snapped out of dreamland when I was called upon for my testimony. I was asked my opinion of whether the farmworkers had a right to raise their voices. The spinning globe was still in traction, still turning dizzily, and after I sized up both sides—the whites, then the Chicanos, then back to the whites—I mumbled, "I don't know." Immediately, I felt lame, *estúpido*. With those three words—"I don't know"—I felt that I had let down the entire world. The turning globe slowed and my head lowered in shame.

What did I know? I left the library and worked for Mrs. Hall perhaps two or three more times, then called it quits, perhaps out of shame, perhaps because I was bored by getting paid for not doing work. Two years later, at age fifteen, I entered the fields to pick grapes and to chop cotton and beets. I knelt under the grapevines, knife in hand, the bunches dropping into my pan; I staggered along the cotton rows, wincing from the sun that glared down at the ghostly mirage of heat vapors. Later still, I took labor buses on the west side of Fresno—known as Chinatown—and roared away for my daily keep. Then I understood what my parents and grandparents, all farmworkers at some time in their lives, were arguing about. Finally, three years too late, I understood the debate in the library.

Thirty years have passed now, and Cesar Chavez is no longer with us. But we still have the union. We have intelligent and passionate reminders such as this book, a companion to the film *The Fight in the Fields,* whose title echoes that of Carey McWilliams's book *Factories in the Field.* What both this book and the film do is relate a story of labor in California and beyond, a story that reaffirms the spirit of a man who began a social movement so large that Cesar's presence, even in death, is indelible. Even though most Chicanos in the 1970s were urban and, for the most part, disconnected from agricultural life, Cesar inspired us all through his will for justice, a will that was nonviolent, single-minded, and, dare I say, saintly. Cesar marshaled his will in fertile valleys—San Joaquin, Imperial, Salinas, Napa, Rio Grande, among others. He sided with the poor. He argued for nonviolence. He prayed and fasted. He could utter these words with conviction: "What you do for the least of my brothers, you do for me." Who could argue that his efforts were not pious?

Cesar's legacy is evident among us. We have schools named after him, public buildings, parks, streets, and boulevards. There are murals with his image, *corridos* wheezing from accordions, scholarships in his name. While his memory is honored, there is occasionally petty controversy. Even in Fresno, my hometown, where you would think that he would be revered, controversy arose in 1994 when Kings Canyon Boulevard was renamed Cesar Chavez Boulevard, then switched back again, after protesters had their say with city officials. At the statewide level, there was talk of proclaiming a Cesar Chavez day in California. Every politician, no matter their ethnicity, district, or party, hurried away from the idea.

What was it, then, that Cesar insisted upon for his union? We could vaguely mouth words such as "justice," "dignity," "fairness," with a rhetoric that stirs the soul. Winery worker Salvador Mendoza summed it up when he said Cesar wanted "the *patrón* to share the riches he is able to earn." Such a clear proclamation, such a tidy remark. Cesar and the union wanted for the workers a living wage and basic amenities such as clean water, toilets in the field, decent housing, and health benefits, which are especially important in an industry where workers are susceptible to injury and sickness. He wanted to curtail the spraying of pesticides while the workers were in the field. (In and around Fresno there are more than a thousand wells closed because of pesticides.) Moreover, he wanted workers to be allowed to bargain collectively without the interference of large corporate growers, government, or police.

Journalists Susan Ferriss and Ricardo Sandoval offer a portrait of Cesar Chavez and the tumultuous times of the United Farm Workers. The writing is thorough and mature. The testimonies are intense, the research probing, and the strategy of photo and text is appropriate for those reading the history of the union for the first time. The images in the book are mesmerizing. For instance, in chapter eight, we read about the 1985 discovery of "Rancho las Cuevas," or "Ranch of the Caves," where workers slept in coffin-like holes dug out of a hillside, an eerie metaphor for premature death itself.

The specifics found in these pages would raise hairs on a fist. There is nothing timid in the text, nothing that resembles artful posturing in the array of photos. Despite passion in the text and the dramatic, unstaged photography, there is a lucid and deliberate organization which begins with the present struggles—scenes from the union's efforts in Watsonsville with added references to California's chic wine country of Napa Valley and Sonoma county. The initial pages acquaint us with a live, not dormant or dead, union. The narrative then moves back in time and scans Cesar's roots in Arizona, where his father farmed until he lost his property during the Great Depression. The narrative then picks up speed and we learn about Cesar's life: the racial prejudices he faced as a child; his flirtation as a teenager with *pachucoismo* in Delano and in San Jose's barrio Sal Si Puedes; his years in the fields as a worker; his first years as an organizer putting together from disarray the building blocks of a union that would become the United Farm Workers of America. Incident after incident touches our conscience, leaving us feeling that we are witnessing a lasting struggle in American labor politics.

Not since the civil rights movement has a national cause—or *causa*—been so heady, so noble and meritorious, as the United Farm Workers movement. This was a massive struggle to right wrongs, a struggle that has continued despite political interference from Nixon to Reagan to California's current administration. In the course of this movement, Cesar became—whether he accepted this status or not—a spiritual leader for all Chicanos. He had no choice; we needed him to remind us of our social responsibilities; today we need him more than ever, particularly in face of such challenges as California's much-argued Proposition 187, the initiative that aimed to strip immigrant children of such rights as health care and schooling.

Like many social and moral issues, the cause for the farmworkers was so remarkably simple that it became complicated. An example is when the growers felt the early economic sting of *la huelga*—"the strike." The growers had brought in Mexican nationals, all of them desperately poor, to attempt to harvest crops and, thus, break the strikes. What would Cesar do? How could he pit the Chicanos against Mexicans? His vision was large, but his tactics for appeasing groups were brilliantly soft, if not comforting. He brought Mexicans into the union, admittedly slowly; nevertheless, there was never the mean-spiritedness of separating Chicanos from Mexican nationals. Cesar advocated a patient coaxing, an education in the fields, for those Mexican workers.

The union has offered large corporate growers repeated opportunities to rethink their position, to make amends, to adequately compensate their workers. During each negotiation, each strike, each suit and countersuit, the union continued to offer growers the privilege of negotiating and undoing their wrongs. But their stubbornness, as one reads here, is tantamount to what can only be construed as greed—there is no other word for their behavior. They have been driven by greed even if it hurts people.

I'm thinking of the Fresno poet Luis Omar Salinas and his ironic poem "My Fifty-Plus Years Celebrate Spring," particularly the ending:

> I've heard it said
> hard work ennobles
> the spirit—
> If that is the case
> the road to heaven
> must be crowded
> beyond belief.

Heaven must be crowded with farmworkers, shoulder to shoulder with men and women who worked the crops. Perhaps they are holding their grape knives, hoes, or pruning shears, their eyes stinging with dust but clearly seeing that what they did with their lives was to provide for others. Growers could learn from such examples.

GARY SOTO

■ FOLLOWING PAGE: *Some strawberry workers call the crop they harvest* fruta del diablo—*"fruit of the devil"—because picking requires them to stoop so low. Workers arrange the berries in flats placed on miniature wheelbarrows.*

Come on, men. Be brave! The boss isn't anywhere near you. He's in his nice house. No one is going to hear you but us.

—NELLY PRUNEDA,
United Farm Workers
union organizer

INTRODUCTION

JUST BEFORE NOON on a summer day in 1996, two United Farm Workers union organizers pulled off a country highway and began driving cautiously down a dirt road outside Watsonville, a small town on California's lush central coast. A salty fog was just beginning to yield to a blue sky. The organizers' old Toyota was coated with the thick dust of many other trips just like this one, and a cloud of brown powder billowed behind the vehicle as it pulled into a strawberry farm. The ranch, as big farms are often called in California, was colossal: row after row of squat green plants marching as far as the eye could see, covering rolling hills, and stretching toward the nearby Monterey Bay. Since daybreak, farmworkers, wearing sun hats and swathed in scarves and gloves to protect themselves against pesticides and dirt, had been working, bending over constantly, their fingers flying to pick what some of them ruefully call *la fruta del diablo*—"the devil's fruit."

At the edge of the field, the organizers glanced around to make sure there were no unpleasant surprises in store for them, but everything was going smoothly. They were right on time, and as required by law, the supervisor allowed them on the property and the crew of harvesters started walking en masse toward a clearing, where they usually ate lunch. There, two portable toilets were available for their use—a simple amenity that might not have been there were it not for the United Farm Workers union's work over several decades.

"Hola, hola!" Lauro Barajas called out cheerfully, as he walked toward the clearing, holding a large piece of butcher paper for writing down wage proposals. Nelly Pruneda, Barajas's organizing partner, flashed a generous smile. Under a baseball hat covered with union buttons, she wore dangling pearl earrings and carefully applied makeup.

■ *UFW organizers in Watsonville, California, the "berry capital of the world," show strawberry workers proposals for wages and benefits under a union contract.*

"Any questions today?" she asked in Spanish, as the workers sat down with exhausted groans, stretching their legs out and opening lunch bags. She knew that some of the men had talked enthusiastically with organizers on other days.

The men smiled at her, but looked distant. "Come on, men," Pruneda teased. "Be brave! The boss isn't anywhere near you. He's in his nice house. No one is going to hear you but us." The wisecrack eased the tension and everyone laughed, including pickers who wore grower-supplied buttons declaring, *Libre sin Unión*—"Free without Unions." Observing the buttons, Barajas remarked, "The *patrón* is afraid. That's why he gives you buttons and why he is going to organize another barbecue for you. That's OK. Eat your tacos. I would, too. But that doesn't mean I can be easily bought off . . . Now, tell me, what do you think of this idea: six dollars an hour, fifty cents piece rate, and the employer pays eighty-five cents toward a medical plan?"

The union meeting had commenced. More than thirty years after Cesar Chavez founded the United Farm Workers, and more than three years after his death, organizing immigrant farmworkers remains one of the most delicate and difficult propositions for social change in America. But the challenges for young organizers like Barajas and Pruneda—and those challenges are many—are a far cry from what Chavez and the original UFW activists faced when they began their remarkable revolution in 1962.

That was when an epic struggle for labor rights and justice exploded in California, just as the shock waves from the civil rights movement in the South were spreading across the nation. The farmworkers' emergence shook up race relations in California and gave birth to a new chapter in the Southwest's Chicano rights movement. The farmworkers movement changed the lives of field laborers in ways that were unimaginable then, improving wages and conditions so much that today a slice of the farmworker population can actually call itself middle class.

Almost all the laws and protections farmworkers now have are the fruit of Cesar Chavez's legacy. When he unexpectedly died in his sleep in Arizona on April 22, 1993, he was sixty-six years old, and still fighting. He was continuing a ten-year-old boycott of grapes that had become more symbolic than practical—a cry of protest against the threat of pesticides, the poisoning of water and workers, and the loss of children who may have died from pesticide-related cancers. As California moves toward the twenty-first century, the name Cesar Estrada Chavez still summons fierce loyalty from those who admired him—and derisive gut reactions from many farmers and political conservatives.

"We think Cesar Chavez was basically a scoundrel," said Wes Bisgaard, manager of the Imperial Valley chapter of the California Farm Bureau Federation, the state's principal agricultural trade group. Bisgaard and the desert valley's farm community were beside themselves in 1995 after hearing that a local college in El Centro—prodded by student groups—was ready to name a new library after Chavez. "We're not appreciative of naming anything after him."

Even after his death, Cesar's legacy sparks controversy and splits communities, just as his union crusades did when he was alive. While many cities in California were quick to add his name to new libraries and schools, debates raged—and still rage—in some towns where Latinos had to insist on public recognition of Chavez. Demonstrations for and against the naming of public buildings and thoroughfares were particularly fierce in the San Joaquin Valley, cradle of the farmworkers movement.

Even in liberal San Francisco, a two-year controversy ensued about renaming a street after Cesar Chavez in the predominantly Latino Mission District. Those who backed the tribute believed covert racism spurred the minor, but loud, opposition to the renaming of the street—Who, they asked, would not agree to honor Chavez? The dispute was finally decided by city voters in 1995, and the Mission's Army Street was renamed. Yet even in 1996, some are still disgruntled and have been known to spray-paint slashes across Cesar Chavez Boulevard signs. "It Will Always Be Army Street," reads a local bumper sticker. "The growers still refuse to accept our heritage as legitimate and refuse to admit we have a right to be in this country. That's what Cesar was fighting for, and that's what we're fighting for, too," said an angry David Zavala, head of the Mexican-American Political Association chapter in the Imperial Valley, lashing out at opponents of the new Cesar Chavez Library in that area.

For many farmworkers, residual resentment from growers and some politicians convinces them that *la causa*—"the cause"—still has a mission. "Cesar cared a lot for us, until the last hour of his life. His concern was looking after his people," said Mary Magaña, a McFarland farmworker who has worked for years under a UFW contract at a San Joaquin Valley rose company. Magaña first met Chavez in 1962, when he was an obscure dreamer just starting to organize skeptical farmworkers. "We older people, we know what it was like before. We know the union is good."

THE PRODUCT OF a depression-era migrant childhood, Chavez was known for having a will as fierce as the growers were rigid. As a child and a young man, he had experienced some of the worst America had to offer to poor minorities. And when he started the movement in the 1960s, farmworkers had precious little with which to defend themselves. They didn't have the legal right to organize and vote for collective bargaining. They didn't have the right to have clean drinking water, access to portable toilets, lunch breaks, or short rest breaks during the workday. And they were not entitled to the minimum wage or unemployment insurance. Benefits such as health insurance, pensions, and paid vacations were dreams. Housing was horrible, and most migrant kids didn't have a chance of finishing high school—and nobody seemed to care.

But the movement forced people who lived in cities to consider what kind of hidden world produced the food they ate every day. Harsh stories flowed from the vineyards and orchards, and some Americans began to understand what it must be like to be a migrant worker, trapped in a cycle of poverty, working harder and dying younger than any other class of people in the United States. Farmworkers in California produced most of America's fruits and vegetables, yet they were among the least protected of industrial employees. And it had been that way for a century. Few attempts to reform the system had succeeded before Chavez; these attempts were often met with violent reprisal swiftly meted out in farm towns dependent on the state's "factories in the field."

Chavez came to believe the only path to equality would be through a farmworkers union. His goal seemed impossible: Farmworkers and domestics were the only two groups of private employees Congress excluded from the National Labor Relations Act of 1935, the law that guarantees most Americans the right to union organizing. The NLRA still excludes farmworkers; to date, only in California and Hawaii do they have the basic right under the law to organize, vote for a union, and demand that employers make a good-faith effort to bargain collectively.

Crafted by legislators under intense pressure from Chavez in 1975, California's Agricultural Labor Relations Act has proved imperfect. The law remains a historic victory, nonetheless, helping to pave the way to dignity for thousands by breaking the growers' near-absolute power over immigrant and ethnic-minority workers. Until then, growers could bargain among themselves to set pay and work rules, and they could refuse, with impunity, to even talk to workers' representatives. "The farmworker is an outsider, even though he may be a resident worker," Chavez said in the late 1960s. "He is an outsider economically, and he is an outsider racially."

The farmworkers' singular level of powerlessness was rooted in the earliest days of California's agricultural economy, which took shape following the Gold Rush. In the 1860s, bankers, railroad czars, and industrialists began to seize large tracts of California's land to grow food in mass quantities; there were handsome profits to be made from the East Coast markets and from the state's own expanding population. The new transcontinental railroad carried grain and produce to the East; the same ten thousand Chinese immigrants who had built those tracks were recruited to sow crops and work the harvests. After an economic crisis led to a white backlash

■ *A pensive Cesar Chavez in April of 1974, following a year of strikes met with violence.*

against the Chinese, and the Chinese Exclusion Act was passed, growers hired destitute whites until the next wave of immigrants arrived.

At the turn of the century, Japanese were recruited to work California's fields, including those of the new sugar-beet industry. The Japanese were praised for their horticultural skills, but when they dared join a new group of Mexican laborers to form associations that demanded wage increases—as they did in Ventura County's sugar-beet industry in 1903—they were roundly denounced as undesirables. An editorial in the *Oxnard Courier* speculated that such unions were worthless because they were "in the hands of people whose experience has been only to obey a master rather than think and manage for themselves. . . ." The mainstream American labor movement, for its part, appeared to agree, and did nothing to assist the farmworkers because they were not white. American Federation of Labor president Samuel Gompers, hailed as the workingman's hero, was no friend to the sugar-beet workers of Ventura; he refused to recognize their fledgling union because its members included Japanese.

■ *Five thousand people in Fresno protested a city council proposal to change the name of Cesar Chavez Boulevard back to its original name.*

When the numbers of Japanese immigrants in the fields declined, growers next turned to East Indians, Filipinos, and Mexicans, and had various laws manipulated to keep the flow of immigrants coming. A class of middlemen known as labor contractors evolved and, for decades, routinely traveled through Mexico and sent word out to the Mexican interior that peones were wanted in *el norte*. They came by rail, truck, horse, or foot. During the Great Depression, desperately poor whites and blacks and Mexican Americans supplanted immigrants in the fields. But when World War II broke out, the U.S. government brought in Mexican braceros—guest workers—to harvest crops. When the bracero program ended, Mexicans kept coming and growers continued to recruit them. Today, California agribusiness, which at $24.5 billion a year is the state's largest industry, remains dependent on Mexicans. Growers acknowledge this. Yet sometimes a patronizing streak still creeps in when they talk about it privately. In the mid-1980s, one grower advised another who was new to managing strawberries not to worry: "When the little brown guys show up, you put the berries in the box. When they leave, you stack up the crates and call it a season."

THE DEPENDENCE OF California agribusiness on Mexican workers is no small matter, considering the staggering array of produce they harvest. The Golden State produces almost all of the country's artichokes, avocados, apricots, olives, nectarines, prunes, and processed tomatoes; and most of its fresh fruits and vegetables, including broccoli, lettuce, grapes, lemons, strawberries, melons, and peaches. The markets for these foods are global: In Tokyo, there's a demand for perfect-looking California broccoli, and in London, baskets of Watsonville strawberries are for sale. Some of the biggest agribusiness corporations—including wine and grape growers— are so well-connected in Washington that they were granted more than $100 million in federal subsidies in 1995 to help advertise their products abroad. Predictions were that by the year 2000 these crops would no longer be harvested by hand. They were wrong; much of this work still rests on the backs of as many as seven hundred thousand seasonal workers.

The world of the farmworker is undoubtedly more professional and just than before Chavez's revolution, but it is still a parallel universe: For one thing, it is more ethnically segregated and inaccessible to outsiders than it was in the 1960s. Housing is still poor or even out of reach for an underclass of workers, some of whom tack together lean-tos to sleep in at night after a full day of picking. Thirty years ago, about half of all California's farmworkers had been born in the United States, but today more than 90 percent are thought to be foreign-born. Mexican village and familial connections are often the best ticket into entry-level jobs, and Spanish is the primary language. Chicano or Mexican foremen and contractors usually do the hiring, whether from a predawn gathering place for day laborers or from inside an air-conditioned company office.

The fields have a culture of their own, with a Spanish and "Spanglish" lexicon that only the most observant non-Hispanic growers understand. *Un raitero* is a driver who sells rides to the fields, or *los files. La pisca* is the harvest, *una wineria* is a winery, *las canerias* are where vegetables and fruits are frozen or canned. The Immigration and Naturalization Service and its agents are *la migra.* Farmworkers with permanent residency call their green cards *micas,* and growers are *los rancheros,* or *patrones,* "bosses." White people, generally grouped together as "Anglos," are also called *gavachos.*

Farmworkers are *campesinos,* and often refer to themselves by what they pick; they are *freseros,* "strawberry workers"; *lechugeros,* "lettuce cutters"; or *uveros,* "grape pickers." At the bottom of the hierarchy are *los ilegales,* the illegal immigrants, or *los sin-papeles,* literally, "those without papers." The United Farm Workers union is *la unión de campesinos* or *la unión de Chavez,* and ardent union supporters are sometimes known as *Chavistas.*

Delfina Corcoles is a *Chavista* who wasn't born yet when the cause began in the small town of Delano in the San Joaquin Valley. But in Watsonville, where an enormous berry industry has multiplied over hills and hollows in recent decades, Corcoles said she was ready to join the movement in the summer of 1996. "I can't be bought by a taco," said the tough-talking thirty-year-old, who's been harvesting strawberries for eleven years. "I've been wanting a union for a long, long time." She complained about ever-shifting wage rates, no real raises in more than a decade, and a frightening lack of job security.

Jesus Lopez Avalos feels the same. "The reason we are united with the union is so the *patrones* will pay us just wages and a medical plan that will cover our families," said the burly Mexican immigrant, looking out from under the shade of a broad-brimmed straw hat. "Even though I've been working for eleven years for the same company—four of my coworkers have been there eighteen years—we still haven't received any benefits. That's why, for the sake of our families, we have joined the union of Cesar

■ *Jesus Lopez Avalos, a Watsonville strawberry worker for more than a dozen years, joined a UFW organizing committee. He said workers badly need medical insurance for themselves and their children.*

Chavez. Thanks to the union, the *patrones* have already stopped treating us so badly. But we're going to keep fighting to get medical insurance."

At forty-six, Lopez is well beyond the average age of most strawberry pickers, whose earnings depend in large part on how fast they can pick. The arduous work requires them to bend low between the plants, nimbly selecting the best of the fruit and clearing away old foliage. As they pick, they nudge along miniature carts containing flats in which they arrange the berries. After they fill a flat, they pick it up, race down the furrow, and deliver it to a *poncheador,* a supervisor who punches a card to keep track of how many baskets workers pick.

"Our work is definitely hard," Lopez said. "You get home stiff at the waist, too tired to even be hungry. You go to sleep thinking about how it will be the same thing the next day. But, no matter," he added, with a wave of his hand. "This is the work we are here to do. All we are asking is that since we are glad to work hard, we receive benefits and a better wage."

When Arturo Rodriguez, the new UFW president and Chavez's son-in-law, talks about his responsibility to represent workers like Jesus Lopez Avalos, he thinks about what Cesar faced a generation before when he blazed through the state, challenging growers as they had never been challenged before. Rodriguez can't help himself. He starts to cry. He knows that despite the barriers he faces today, Chavez and other union elders, including vice president Dolores Huerta, have left him with an arsenal of laws and a legacy of persistence that inspire him to go on. "You know," he says, "you work around Cesar for years and years and years, and you're part of the movement, but you never really appreciate the full impact of what he did, and the sacrifice he went through, and the work he did all these years, until you get in the position I'm in now.

"As difficult as our situation is now, and as hard as we have to fight now, we are blessed that Cesar and Dolores spent all of these decades building this foundation for us. That can't be replaced," he says, shaking his head. "People don't realize it. The fact that we have a pension plan now, that we have contracts, that we have an Agricultural Labor Relations Act, that we have unemployment insurance and workmen's compensation for farmworkers. All these changes . . . ," he adds, his voice trailing off, as he tries to imagine the lives of farmworkers before Cesar Chavez.

■ FOLLOWING PAGE: *The Great Depression displaced tens of thousands of people, who poured into California seeking work picking crops. Tents, jalopies, and shacks were the only shelter many could find.*

> *We had never worked for anybody else. We never lived away from our home. Here we come to California, and we were lucky we got a tent. Most of the time we were living under a tree, with just a canvas on top of us.*
>
> —RITA MEDINA,
> *Cesar Chavez's sister*

CHAPTER ONE

THE LAST FAMILY FARM

IN THE SUMMER of 1973, dozens of weary farmworkers gathered around Cesar Chavez in a small park in California's Coachella Valley, a small desert community nestled between craggy mountains in the southern corner of the state. Chavez had brought the workers together in a heat so intense you could see it pulsate; it danced off the bone-dry valley floor and made everyone it touched sweat. This was grape-growing country, and a strike was in full bloom. The workers and Chavez were taking a rare moment of rest between picket lines and meetings to talk about their union.

The forty-six-year-old Cesar was in his element; he'd grown up under this brutal sun, just a couple of mountain ridges to the south and east. Resting under a cluster of trees, the strikers listened intently as he talked about his life, trying to fuel their determination. As he spoke, the afternoon shadows lengthened over the embattled grape fields nearby, where the rows of vines rose from the desert like a vast green mirage. Chavez talked of endless days spent driving California's lonely roads, chasing impatient harvests, and thirsting in dusty fields without water to drink. He'd stayed in the same kinds of run-down labor camps with filthy outhouses as these workers, and paid too much for the privilege. Above all, he knew the lifelong hunger to stand up and say, *enough*. "My grandfather came to this country from Mexico when the frontier was more of an idea than an actual border," Chavez told the crowd in Spanish. "There was no barbed wire and no immigration department. They settled near this valley because the land was very rich, but we lost our lands. . . . Then, in 1962, I came to Delano with the idea of organizing a union. Of every hundred workers I talked to, one would say, 'It's time.'

Everybody said no one could organize farmworkers, that it couldn't be done. But we got a group of forty or fifty, and one by one, that's how we started. . . ."

CESAR ESTRADA CHAVEZ was born the second of five children on March 31, 1927, in Arizona's North Gila Valley, close to the borders of California and Mexico, and on the edge of Quechan Indian country near the Colorado River. At that time, before the big farmers in California's Imperial Valley captured the wild river and diverted it west hundreds of miles, the Colorado flowed deep into Mexico. It also passed near the whitewashed adobe farmhouse where young Cesar Chavez was raised, just outside the parched town of Yuma, Arizona. A shy but strong-willed boy, he was born into a Chicano version of the American Dream, growing up on a rambling eighty-acre farm. His grandfather Cesario, whom Cesar and his siblings would call "Papa Chayo," bought his land from a homesteader at the turn of the century, then carved his farm out of the desert.

To Cesario Chavez, owning his own land was the realization of a lifetime: born a peon, or indentured servant, in Chihuahua, Mexico, he had been trapped on the lowest rung of the country's feudal hacienda system. The hacendados, or lords of the ranches, meticulously recorded a peon's debts from the time he was born, exacting work from him as payment as soon as the child could pick up a shovel. Cesario fled his hacienda in the late 1880s, after an argument with the hacendado's son, an unforgivable offense sometimes punished by flogging or forced conscription in the Mexican army. After crossing the Rio Grande into Texas, Cesario found work around the mines of Arizona, hauling supplies as a mule skinner. Eventually he was able to buy the parcel of land that would become his home, and with the promise of irrigation from a new dam, he was sure he could make the desert blossom.

■ *Librado and Juana Chavez, Cesar's parents, married late in life and began raising their children on the family homestead in Yuma, Arizona.*

Cesario was illiterate, but his wife, Dorotea, "Mama Tella," had learned to write both Spanish and Latin as an orphan in a Mexican convent. She passed her religious instruction on to her children and, later, her grandchildren. Although known as *travieso,* or "a prankster," Cesar would join the other grandchildren in the evenings around Mama Tella's bed, where, thumping her cane to demand attention, she would tell the giggling siblings stories about the saints and the commandments. She prepared them so well, in fact, that an amazed priest in Yuma, struck by the children's knowledge, allowed them to take their first Communion and forgo the requisite catechism classes.

But grandchildren were still far in the future when Cesario, Dorotea, and their fifteen children worked the farm, building a spacious adobe farmhouse framed by dense walls. The structure was overbuilt by design: Cesario wanted it to last, and the eighteen-inch-thick walls would keep the family cool in blazing summers and warm when the desert temperatures plunged on bone-chilling winter nights.

Many of the children moved on when they grew up, but one of Cesario and Dorotea's sons, Librado, stayed on to cultivate the family farm. In 1924, at thirty-eight, he married Juana Estrada, whose

■ *Baby Cesar was baptized into the Catholic Church at nine months.*

family was also from Chihuahua. Librado bought a piece of land a mile away from his parents, and on top of their farm chores, the couple ran three small businesses—a grocery store, a garage, and a pool hall, whose clients included scores of Chavez relatives who lived near the homestead. Juana, who was already thirty-two when she married, had long black hair, a ready smile, and an arsenal of herbs and home remedies to help fight sickness in the family.

As religious as her devout mother-in-law, Juana was also the original inspiration of her son Cesar's interest in nonviolence. She instructed with hundreds of proverbs, *dichos*, which Cesar would employ throughout his life, including "He who holds the cow being killed sins as much as he who kills her." Although the Mexican ideal of machismo, or manliness, required that a man stand up for his honor, she didn't want her male children to strike back physically at those who taunted them. "It's best to turn the other cheek," she would say. "God gave you senses, like your eyes and mind and tongue, so that you can get out of anything."

Although Cesar sometimes scuffled with cousins, he more often lived by his mother's advice, walking away from schoolyard toughs rather than returning the abuse. Chavez did admit, however, to once brandishing a shotgun at a bullying older cousin to stop him from swinging Cesar's pet cat by the tail and throwing it in the canal. As he mused years later, his fleeing cousin "didn't know the gun was unloaded."

■ *Cesar with an unidentified friend on the family ranch near Yuma.*

Librado, who was nearly six feet tall with powerful arms, was a much bigger man physically than Cesar would later become. Cesar recalled he was strict and expected his children to obey him, but he was also given to tenderness, rising in the middle of the night if the children cried out to fetch water for them or to carry them to the outhouse. He spent long hours patiently teaching Cesar and Richard, two years younger, how to irrigate, plant, and hoe the farm properly. Cesar later said he thought his parents were unusually attentive and patient with their children because they married and began a family later in life than most of their peers. His mother always made time to play games with the children, and didn't spare hugs or kisses. Secure in their parents' world, the Chavez children looked back on their years on the farm as idyllic. They were especially fond of the barbecues their family hosted for relatives, who roasted fresh ears of corn over bonfires and spun ghost stories and tales about the Mexican Revolution and rich hacienda owners.

By age eight, Cesar's elder sister, Rita, was taking care of her two younger brothers—making tortillas, washing clothes, combing their hair—while her mother was busy helping her father on the farm. She and Cesar were always very close, calling each other *tita* and *tito* (short for "little sister," "little brother") and going everywhere hand in hand. Rita sometimes wearied of being a second mother, but Cesar helped. He enjoyed his chores, Rita recalls years later, and he was good at planning and setting goals—"Cesar would tell us, you feed those two horses and I'll feed this cow; he always had everybody assigned to something. . . ."

Even as a small boy, Cesar had a stubborn streak that would be put to good use in later years. Sitting in her San Jose home sixty years later, Rita Chavez Medina smiles at the memory of taking Cesar to school for the first time. The six-year-old flatly refused to sit anywhere except with his beloved sister. He warned the teacher that if he couldn't sit with his sister, he was going home. Sure enough, once Rita managed to extricate herself from the arms of a tearful, clinging Cesar, he fled out the door and raced down the road. After Rita caught up to him and coaxed him back, the teacher asked two second graders to pull up an extra desk near his sister. "So Cesar sat by me,"

Rita laughs. "After two or three days—I can't remember how many—he finally agreed to go with the first graders, but he won that first battle right there and then."

Cesar and Richard were also inseparable, hiking all over the ranch, swimming in the irrigation canals, and building elaborate forts. Down at the family garage and general store, the brothers pumped gasoline and waited on customers, but what they loved best was the poolroom. Transfixed by the billiard games, the boys grew adept with the cue stick, often rushing back from chores to play until eleven at night. Three decades later, in the Mexican barrio of Delano, California, Cesar Chavez would choose a pool hall—one also located, curiously enough, next to a store and near a repair shop—to play billiards and brainstorm with new union recruits.

But the good times were about to end. In 1929, thousands of miles away from the North Gila Valley, in New York City, the stock market collapsed, touching off a series of economic tremors that led to the Great Depression. By 1932, the indiscriminate shock waves had touched the Chavez family. Juana and Librado were generous with their many relatives and other customers who bought on credit at their store. But because many had become destitute, bills went unpaid and the Chavezes finally were forced to sell their three businesses. Taking the pool table with them, they retreated to the farm, and the family adobe, to live with Mama Tella, now a widow. On the farm, Cesar said, the family had vegetables, chickens, and fresh eggs and milk: "We didn't suffer the way people in the city or migrants suffered."

■ *Cesar and his sister, Rita, received their first Communion together after their grandmother's religious instruction.*

Despite this diminished state, Juana Chavez held to the teachings of her patron Saint Eduvigis, who gave everything to the poor. She routinely sent the children out to look for *trampitas,* "little tramps," who could use a plate of hot food. For Juana Chavez, the simple meal of beans, rice, and steaming tortillas she handed out to hobos was, in part, a religious gesture; she had long vowed that no one would go hungry around her. One blazing hot day, Cesar and Richard went on their ritual search for hobos. He and Richard were so anxious to be done with it that they startled a tramp sitting under a nearby retaining wall.

"We wanted this one bad, so that we could quit looking and go play. But when we told him about all the free food just waiting for him around the corner, that tramp just couldn't believe it. 'What for?' he said. 'What are you doing it for?' 'For nothing,' we said. 'You just come with us.' We hustled him around the corner and he ate the food, but he still didn't believe it." Chavez said word of his mother spread among the legions of ragged hobos, mostly white, that passed through Yuma on their way to California during the depression years. Because she refused to take anything for her food, "they were very kind about coming at the right hours, and even the ones that talked rough outside took their hats off when they came in, and were very respectful."

Life on the farm was modest but good—with even a little left over to spare. The children never suspected that they would soon be homeless themselves, joining a flood of Okies and other uprooted migrants streaming west.

I T   W A S   I N   1933 that the Chavez family suffered another blow, a natural calamity that led to their greatest loss. Drought hit for the first time in more than half a century, shriveling the Colorado River and the Chavezes' irrigation canals. Crops perished, and Juana and Librado fell perilously behind in property and water taxes, which they were already struggling to pay. Their debt of $3,600 was "just nothing," Chavez would say later, but it was enough to attract the eye of a neighboring farmer who coveted the Chavez property and was powerful enough to try to grab it. Librado "qualified for a loan, a small loan under the New Deal," Cesar remembered, "but the guy next to us who wanted to get the land was the president of the bank, and also of the soil conservation office, so the loan was blocked."

Still confident he could pay off the debt, Chavez's father in 1938 followed other men who went west to California, where he'd heard there were jobs to be had. When Librado found work, he sent word for his family to leave the farm temporarily and join him in California. The Chavezes had five children by then, including Vicky, five, and the youngest, four-year-old Librado, "Lenny." Juana and the children climbed into the overstuffed roadster and headed for Oxnard, a coastal town north of Los Angeles where lettuce, melon, and strawberry fields stretch for miles in tediously precise rows. With two adult cousins along to share the driving, they pulled out of Arizona late at night to avoid the relentless desert sun. Cesar remembered the first day of the trip seemed like an adventure: The children had never been far from their home before. They drifted into sleep as the night wore on, but awoke with a start when the night was shattered by a burst of floodlights. "Suddenly two cars bore down on us," Cesar said. "Uniformed men piled out of the cars and surrounded ours. We were half-asleep, all scared, and crying. It was the border patrol, our first experience with any kind of law. Roughly they asked for identification, our birth certificates, proof of American citizenship. . . . My mother must have died a hundred times that night."

The border patrol agents held the family up for hours, questioning them repeatedly. Apparently, Cesar would say later, a Mexican-American family was automatically suspect and fair game for abuse by agents. "As far as they are concerned we can't be a citizen even though we were born here," he said, recalling the incident. "In their minds, 'If he's Mexican, don't trust him.' "

The family continued on to Oxnard, and considered themselves lucky that Librado had at least found work there, although it was poorly paid. Desperate migrants, mostly "Okie" farmers from the hard-hit dust-bowl states of Oklahoma, Texas, and Arkansas, were pouring into California by the tens of thousands. Buying new land on the coast was out of their reach: Laborers served at the pleasure of growers who could easily find replacements. Wages plummeted, and the Chavez family found a rickety shack to live in. The Chavezes were in such need that they had to forage for wild mustard greens to fend off starvation. When relatives in Arizona sent a few dollars, the family bought enough gas for their old Studebaker to limp to Los Angeles. There, Juana Chavez sold crocheting in the street to raise enough money to follow the crops east to Brawley, California, where the family worked feverishly to keep up with the summer cotton harvest and earn enough for the trip to the next crop. However, they made so little that they were forced to return to Yuma penniless.

In a last-ditch effort to secure a loan for his farm, Librado traveled to Phoenix and managed to win an audience with the governor. It was to no avail. Cesar was eleven years old when the deputy sheriff served the family with a final eviction notice: "He had the papers that told us we had to leave or go to jail. My mother came out of the house crying, we children knew there was trouble, but we were confused, worried. For two or three days, the deputy came back, every day. . . . And we had to leave." Like the Okies and thousands of other Americans uprooted by the Great Depression, the Chavezes were driven off their land and onto the road as migrant workers.

Their last memory of the farm was of its destruction. The day before they left their home, tractors bulldozed the hand-hewn corral where Cesar and Richard had spent hours chasing calves and riding horses bareback. The land was leveled and the irrigation canal filled with dirt while the Chavezes watched helplessly. "Just a monstrous thing, knocking down trees," Chavez said. "We left everything behind. Left chickens and cows and horses and all the implements. Things belonging to my father's family and my mother's as well. Everything." His mother, Juana, wept as the family stuffed a 1927 Studebaker with clothing and bedding, leaving behind what wouldn't fit.

They had just forty dollars left from selling the cows and chickens, and that wouldn't last long. In the hard days that followed, it began to dawn on Cesar and his siblings why their mother had cried when they left the ranch. Neither the Chavez parents nor the children accepted their exile as permanent, and for a long time, Juana and Librado spent hours planning how they would one day save enough money to return to Arizona and redeem the family homestead. As the years dragged on, they gradually stopped talking about it.

The loss of the family farm left an indelible mark on Cesar Chavez, who never forgot the memory of his family's dispossession. "Maybe that is when the rebellion started," he said. The Chavez property was first taken by the state and sold at a public auction, and it eventually fell into the hands of the company that would later become part of Bruce Church Incorporated, a giant vegetable company. This irony did not escape him: "If we'd stayed there, possibly I would have been a grower," he said. "God writes in exceedingly crooked lines." The experience of landownership, Chavez thought, made it easier for his family to stand up to some of the abuses they encountered as migrant workers. "Some had been born into the migrant stream. But we had been on the land, and I knew a different way of life. We were poor, but we had liberty. The migrant is poor, but he has no freedom."

On his family's ranch, it seemed like the whole world belonged to them, Cesar recalled. That first time off the farm, living in a rented shack in an Oxnard barrio, he was "like a wild duck with its wings clipped," he said, tormented by the noise, the crowding, and the fights. "I bitterly missed the ranch," he recalled. "I couldn't get used to the fences; we couldn't play like we used to. On the farm we had a little place where we played, and a tree and we played there. We built bridges and we left everything there and when we came back the next day it was still there. You see, we never knew what stealing was, or to be stolen from. Then we went to the city and we left a ball outside for just a second and boom!—it was gone."

If Cesar felt unhappy away from the family farm, he suspected that his father was suffering more. Librado Chavez missed plowing his own fields, and Cesar watched him admiring the land his family now had to harvest for others as the family traveled

■ *Dozens of Mexican workers gathered for a panoramic photograph outside a labor contractor's office in San Antonio, Texas, in 1924. U.S. contractors recruited Mexican laborers by the thousands to work in California agriculture.*

from the Imperial and San Joaquin Valleys to the Santa Clara Valley near the San Francisco Bay. "He would notice how fertile it was," recalled Chavez. "He would get down and look at the ground, taking some dirt in his huge hands. 'You could really raise things here!' he would say, or 'Look at that plow!' . . . Often he would not agree with the methods being used. . . . He was always noticing mistakes, looking for ways of improving things."

Like Cesar, Rita despised having to compete with other hungry migrants for a chance to pick peas or walnuts in return for pennies. As in their first abortive journey west, the Mexican-American Chavez family found themselves thrown in with tens of thousands of white migrants. Caravans of sputtering cars and trucks were streaming westward from the great Midwestern dust bowl, all headed to California, the land where, as Woody Guthrie sang, migrants hoped "the water tastes like wine."

"We had never worked for anybody else. We never lived away from our home," Rita remembers. "Here we come to California, and we were lucky we got a tent. Most of the time we were living under a tree, with just a canvas on top of us, and sometimes in the car."

Worse, the Chavez family was "green," generous and unused to the fierce competition of depression-era fields, where "there were so many unemployed people looking for jobs, they were like flocks of starlings," Cesar would recall. Arriving to work before dawn at fields of peas and string beans, the bewildered family would find that other migrants had risen even earlier and taken all the baskets, leaving them unable to work. Angry and frustrated, Cesar got up at four in the morning and headed for the fields. "He came back with not one basket, but two," Rita remembers with satisfaction. "Finally we could work for the whole day."

OVER THE NEXT ten years, Cesar and his family would tend and harvest nearly every crop California had to offer, from tomatoes, plums, melons, and berries to

■ *During the depression, people lined up to have baskets of peas they harvested weighed. Picking peas was one of Cesar's first jobs as a migrant.*

grapes, cotton, sugar beets, and lettuce. It wasn't long before he and the other children realized the special drudgery of working on other people's land. This was nothing like working on their father's small farm back in Arizona. Cesar remembered that lettuce, a crop so lucrative it was called California's "green gold," had to be thinned with short-handled hoes. The task required workers to bend low, clawing at row after row of the crop with the hoe. "It's just like being nailed to a cross," Chavez said. "You have to walk twisted, perpendicular to the ground. You are always trying to find the best position."

Despite such problems, the Chavezes had to gain the speed, dexterity, and knowledge that could set them apart from others seeking jobs in the fields. They became experts in wrapping and pruning grapevines and selecting fruit that was neither too green nor too ripe. But some crops were especially hard for the children. The broccoli harvest was in winter, on ground so wet and slippery, Cesar later said, that their shoes would sink into the mud and their hands nearly freeze as they sliced the vegetable's tough stalks.

Everyone was needed to put food on the family's table, so the children had to work, catching bits of school in between their travels. But Librado and Juana didn't push their children as hard as some migrants, Cesar recalled. "Anything you can do is fine with me," Librado would say. The youngest of the boys, Lenny was the only one who could be spared to walk as far as a mile away to refill the family's drinking water jug, usually an old juice or vinegar bottle wrapped in wet burlap to keep the water cold. Despite his small size, Lenny was "a little eager beaver" determined to keep up with his siblings, Vicky Chavez remembers.

Exhausting as field work could be, it was worse for the Chavezes to pull into an area and discover they had arrived too late for a harvest. In between full days of picking, there were weeks of hunger and futile searches for crops. The family was also cheated by unscrupulous labor contractors, including one who disappeared with half their earnings.

In the winter of 1939, Cesar and his family were again among the poorest of the poor in California. Stranded in Oxnard again, they lived in a soggy tent. Juana and Librado rose before dawn in the winter chill to pick late-season peas. Contractors drove them to the fields in buses, charging as much as seventy cents each way—cutting deeply into the day's pay. To help out, Cesar and Richard scoured the roads for tinfoil from cigarette wrappers. They amassed an impressive, eighteen-pound ball of metal foil, which they proudly sold to a Mexican junk dealer for enough to buy two sweatshirts and a pair of tennis shoes. They also got jobs sweeping the local movie theater, for which they received a nickel and a daily pass to their favorite series, *The Lone Ranger.*

Richard Chavez remembers that the harshest work the family encountered during this time was not on small farms, where they often worked in cooperation with owners, but in the vast orchards and vineyards of corporate growers. The depression increasingly knitted farms into large landholdings as family farmers lost their land. These factory farms were run by absentee owners; their anonymous boards of trustees in San Francisco or Los Angeles could easily distance themselves from the starvation wages arbitrarily set by obedient crew bosses in the fields. A young boy interviewed by one reporter in a squatter camp understood—beyond his years—that he was a disposable tool. "When they need us, they call us migrants," he said, "and when we've picked their crop, we're bums and we got to get out."

For the Chavezes, it was a painful shock. "When we joined the big league—I guess you might call it that—where there were hundreds or thousands of people working, that's when we started seeing there was lots of discrimination, a lot of bad treatment of the people who were working," recalls Richard. He is still proud that his father, although a reserved man, refused to tolerate such behavior. "My father never did put up with it—he'd quit. He'd strike or go home or something. He never took the abuse. He'd just say, 'Let's go. Let's not work here.' And we'd go look for something else." His father joined several unions, including the short-lived National Farm Labor Union led by Chicano writer and activist Ernesto Galarza.

Cesar, too, recalled with pleasure that he belonged to "the strikingest family in California." Once, the Chavez family dropped their tools and walked out of a cotton field at the height of the harvest in the southern San Joaquin Valley town of Wasco because they suspected foremen were underweighing the produce they had harvested. During a cherry harvest in Santa Clara, the Chavezes saw their piece-rate pay of two

"We're Human Beings, Not Dogs!"

by Victor Villaseñor

During the Great Depression, and for decades afterward, drinking water was a scarce commodity in the fields. That's what the parents of Mexican-American writer Victor Villaseñor found out in the 1930s, when they worked as young migrants in California. In this excerpt from his epic work Rain of Gold, Villaseñor writes about his parents, Salvador and Lupe, who meet during a showdown over drinking water with an abusive foreman.

The fields of cutting-flowers were in bloom as far as the eye could see: rows of pink, red, yellow and blue. Coming down between the rows, Lupe saw that her father was sweating profusely. It was only eleven in the morning, but already the sun had drained Don Victor and he needed water.

Quickly, Lupe took his arm and started to the truck. But, approaching the truck at the end of the field, Lupe saw the foreman sitting inside the cab. She stopped; they weren't supposed to come for water until noon, but Don Victor was coughing so badly that Lupe didn't care what the foreman might say.

Her father was ice-cold by the time Lupe got him to the vehicle. On the back of the truck was a barrel of water and hanging on hooks was a row of tin cans with wire handles twisted around them. Lupe sat her father down in the shade of the truck and reached for one of the cans.

"Hey, you!" said the big, heavy-set foreman, getting out of the cab with the comic book he'd been reading. "It ain't noon. You get your asses back out there."

"But my father," said Lupe, "he needs water."

"Water, hell!" said the Anglo. He was a huge, fat man, six-feet-four and well over two hundred and fifty pounds. "He looks more like an old wino to me," he said.

Lupe turned red with anger, but she refused to be intimidated. She took one of the tin cans, holding her head high with dignity.

"Eh, girlie, I thought I told you no water 'til noon," he said.

But Lupe ignored him, filling the can with water and handing it to her father, who was now panting dangerously fast, like a tongue-swollen dog.

"Hey, stop that!" yelled the huge man, rushing up and knocking the can out of Don Victor's hand.

"You're fired!" he yelled at the old man. "And you," he said to Lupe, "get back to work or I'll fire you, too!"

But Lupe didn't move. Her father was gasping. He could die if she didn't get him cooled down. "We're not dogs," she said, holding back her tears. "We've been working hard since before five! You have no right to abuse us like this!"

"No right?" yelled the big Anglo. "Well, you got another thing coming, moo-cha-cha-girl!"

And just then, as the big, red-faced Anglo began shouting insults at Lupe, he was grabbed by a blur of motion, spun about and hit in the stomach with such power that his feet came off the ground.

"No!" yelled Lupe.

But it was too late. It was Juan Salvador, dressed in dirty work clothes, who hit the foreman two more times in the face with his huge, iron-driving fists. The big, soft-bellied Anglo went crashing into the side of the truck.

Still moving, still feeling his whole heart pounding with rage, Juan reached down and got the can that the Anglo knocked away from Lupe's father, rinsed it off, filled it with water and handed it to Lupe.

"Here," he said, smiling, "for your father."

"Thank you," she said, "but you didn't have to hit him so hard."

"What?" said Salvador.

"So hard," said Lupe. Her heart was pounding. Oh, how she hated violence. She turned back around to help her father drink the water down.

Salvador stood there, adrenaline pumping wildly, feeling confused, not understanding why Lupe hadn't enjoyed how he'd hit the foreman, especially after how he'd treated them.

He watched Lupe help her father drink. Other people came off the field to drink water, too. They congratulated Salvador, telling him that this big Anglo was one of the most abusive foremen that they'd ever had. Several young women started flirting with Salvador. But then they heard the roar of the boss's truck come rushing up the field and the workers tossed their cans and started back into the fields.

"Hold your ground!" yelled Salvador. "You've done nothing wrong! Drinking water is your right! Don't move! We're human beings! Not dogs! Damn it . . . !"

cents a pound suddenly and arbitrarily dropped to one and a half cents—just because there were so many people willing to work for less. They protested, but to no avail. Another time the family was picking cotton near Corcoran, another sunbaked valley town, when "about a hundred cars came by ringing their horns, you know, like we do now, and the people stopped working and raised their ears like rabbits," Cesar recalled. "And my dad said, 'Let's go.' We didn't say why; we knew what it was. We just went away. . . . We were the first ones to leave the fields if anybody shouted *Huelga* [strike]. . . . It didn't come to us because we knew anything about labor. It came to us because it was the right thing to do."

THE DEPRESSION EXPERIENCE that affected Cesar Chavez and his generation was a profound American tragedy. It was not uncommon for migrant children to die of malnutrition while their parents roamed about desperately searching for work that would pay enough to buy food. Reporter and novelist John Steinbeck, who grew up watching the development of the Salinas Valley's vast vegetable industry, was reviled in his hometown because he bore witness to the exploitation of migrants to millions of readers in 1936. On assignment from the *San Francisco News* to write a series of articles on Okies, Steinbeck visited a squatters' camp and discovered a family that each night rolled up in a carpet to sleep inside a shelter of willow branches and pieces of fabric. As he talked with the parents, the family's malnourished three-year-old weakly brushed flies from his closed eyes and nose. "He will die in a very short time," Steinbeck wrote. "Four nights ago the mother had a baby in the tent, on the dirty carpet. It was born dead, which was just as well because she could not have fed it at the breast; her own diet will not produce milk."

■ *Cesar picking tomatoes alongside an unidentified girl on the migrant trail.*

Corporate farm owners were often too far away—or too indifferent—to notice the misery in their fields, where the surplus in workers made it easy for supervisors to cut wages to pittances. Still, the migrants kept coming: In a single day in 1937, more than three thousand migrants poured across the border from Arizona into California's Imperial Valley. There were dozens of violent strikes in California during the depression,

■ *This striker was felled by a bullet during the Pixley strike. Many strikers were injured and some killed by vigilantes during the mid-thirties.*

as farmworkers rebelled against their impossible circumstances. Many were led by the Communist Party–sponsored Cannery and Agricultural Workers Industrial Union, one of the few labor organizations willing to represent the destitute field and packinghouse workers. Strikes were put down swiftly by local deputies and vigilantes directed by the American Legion and the Associated Farmers, who terrorized and beat organizers, ran them out of town, and even killed some of them. The Associated Farmers was a voluntary association formed in 1933 by ranchers in the Imperial Valley, whose spokesmen included a state senator who was open in his admiration of the Nazi Party in the pre–World War II years. "The wave of violence launched by the Associated Farmers in 1934 swept on into 1935 with organized vigilante groups crushing one strike after another," wrote farm-labor expert and historian Carey McWilliams. "[No] one who has visited a rural county in California under these circumstances will deny the reality of the terror that exists. It is no exaggeration to describe this state of affairs as fascism in practice. Judges blandly deny constitutional rights to defendants and hand out vagrancy sentences which approximate the harvest season. It is useless to appeal, for by the time the appeal is heard, the crop will be harvested." Two U.S. congressional investigations of such abuses echoed McWilliams' indictment, reporting the use of guns, tear gas, and floggings by employers.

Typical of such reprisals was an incident that occurred during the Corcoran cotton strike, which took place only a few years before the Chavezes migrated to California. In 1933, a construction worker named Pat Chambers, a member of the Communist

Party, rose from sweltering workers' camps to organize Mexican and Anglo cotton workers in a massive strike against cotton growers in the San Joaquin Valley. A farmers' vigilante group ambushed a group of the strikers, who were outside a union hall in Pixley, killing two Mexicans, Delores Hernandez and Delfino Davila, and wounding several others. Eleven ranchers were arrested but were never convicted.

Chambers, however, was imprisoned for breaking "criminal syndicalism" laws—vague anti-conspiracy rules designed to squelch organizing by Communists and other groups. (Almost forty years later, Chambers would show up unannounced one day at Chavez's office in Delano after the union's first grape contracts were signed in 1970. Thrilled to meet the veteran organizer he'd heard so much about from aging farm laborers, Chavez asked Chambers why he hadn't come sooner. The old man responded that he didn't want the UFW "red-baited" because of his presence: "I didn't want to get you in trouble.")

In 1936, more close-quarter fighting broke out between striking lettuce workers and deputized citizens in the streets of Salinas, Steinbeck's hometown. Tear gas and bullets left dozens of farmworkers injured. "It would seem that having built the repressive attitude toward labor they need to survive, the [farm] directors are terrified of the things they have created," wrote Steinbeck. "The attitude of the employer on the large ranch is one of hatred and suspicion, his method is the threat of the deputies' guns. The workers are herded about like animals. Every possible method is used to make them feel inferior and insecure. At the slightest suspicion that the men are organizing they are run from the ranch at the points of guns. The large ranch owners know that if organization is ever effected there will be the expense of toilets, showers, decent living conditions, and a raise in wages."

Steinbeck branded the California elite as heartless. "No one complains at the necessity of feeding a horse while he is not working. But we complain about feeding the men and women who work our lands. Is it possible that this state is so stupid, so vicious, and so greedy that it cannot clothe and feed the men and women who help to make it the richest area in the world? Must the hunger become anger and anger fury before anything will be done?"

Populists like Steinbeck were defying a business powerhouse long in the making. The state's big growers had always enjoyed special economic status, but the industry truly began flexing its political muscle during the depression years. In 1934, the corporate farm lobby helped bankroll opposition to progressive candidate Upton Sinclair's promising run for governor on an "End Poverty in California" platform. The campaign, aided by the press, was the first to use attack ads and newsreels produced by Hollywood moguls that branded Sinclair a Communist who would usher into California a dangerous brand of "Russianism."

Mexicans in California during the depression faced an even worse backlash than Okies and would-be reformers. During the depression, thousands of Mexican immigrants were uprooted, either returning voluntarily to their homeland or being forcibly deported. They suffered a devastating loss of property and livelihood. Some of the Mexicans had been brought north by growers seeking cheap labor during previous decades, and they had become residents of California. Others deported were children who had been born in the United States. Mexican workers who managed to stay in the country and dared protest conditions invited the wrath of law enforcement, as promised by this assistant Kern County sheriff during one labor conflict in the thirties: "We protect our farmers here in Kern County. They are our best people. They are always with us. They keep the country going. They put us in here and they can put us out again, so we serve them. But the Mexicans are trash. They have no standard of living. We herd them like pigs."

This was the volatile environment that the Chavez family encountered when they arrived in the Santa Clara Valley in Northern California to pick and dry cherries. Labor organizers from the fledgling Congress of Industrial Organizations, founded in 1936, showed up to help disgruntled drying-shed workers plead for better pay and rest breaks. Cesar's father was among those workers. "They were talking either about an election or strike, and my mother was afraid," Cesar recalled. "I remember Mexican and Anglo organizers came to my house and there was my dad, my uncle, and a couple of my cousins, and I remember my mother was very nervous because she probably associated union with violence." Cesar was sent out of the room because the men were going to talk serious business, and he never found out if the dispute led to an actual strike. He did overhear his father later on talking about how an agreement was reached, and how fantastic it was now that he could enjoy two 10-minute breaks on the job every day.

LIKE POVERTY, discrimination stalked the Chavez family in the twenties and thirties. Cesar's parents, raised in a region that was Mexican soil until just a couple of generations before their birth, were more comfortable speaking Spanish than English. The children, dark-skinned with Indian features, grew up with both languages, but were rapped hard on the knuckles and scolded for speaking Spanish in class in Arizona. Even speaking Spanish on the playground brought swift punishment: Cesar remembered the principal would grab the Mexican children, girls as well as boys, and beat them with a wooden paddle made from a two-by-four. He also remembers a teacher hanging a sign around his neck that read "I am a clown; I speak Spanish." Other teachers humiliated him for making mistakes in English: "Some teachers were really cruel, while a few—a very few—were understanding," he said. The Mexican

■ *Mexican-American and Okie workers honked car horns and blew trumpets to urge destitute cotton pickers to join a strike in Pixley in 1933.*

children deeply resented this degrading treatment, and he and the other Mexican children reacted by speaking Spanish as often as they could, calling English "the dog language" to strike back.

Despite the school's punishment of Spanish-speaking students, the Chavez children felt that they were part of the community. Blacks were segregated, but Mexican Americans and whites played and learned together. It was not until the Imperial Dam project on the Colorado River attracted white Southerners to Yuma that racial tension and name-calling descended on the school, Chavez recalled. "I was returning from the latrine, when a couple of girls said something about 'dirty Mexicans,'" said Chavez. "I never forgot it. Words can be as painful as a switch, and many times those who say them are unaware of how painful their words can be."

The humiliation Chavez experienced in the Arizona school was only the beginning. In California he and his siblings encountered more prejudice based on race and ethnicity, as they were thrust into a bizarre system of segregation that varied from town to town. The state had its own brand of racial bigotry, stemming from early settlers, who were unsure of how to classify the Hispanic Californios and based their assessment on skin color and money: If the Californios, descendants of Spanish or Mexican colonists, were of light complexion and were landed rancheros, they were considered white. At the 1849 California State Constitutional Convention, delegates denied the vote to black and Indian residents of the state. After lengthy debate, it was decided that all male Mexicans or Californios could vote. Dark-skinned Californios often found acceptance in one town, only to be classified as Indian in another and barred from voting or testifying in court. By the time the Chavez family arrived in California, segregation of Mexicans, though not a state law, was certainly an accepted practice, with segregated theaters, schools, and even stores and restaurants that were off limits to Mexican Americans.

Cesar was enrolled in at least thirty-six schools as his family crisscrossed the state in search of crops to harvest and cotton fields to hoe. In Oxnard he went to a Mexican-American school known as Our Lady of Guadalupe, while Anglos attended a Catholic school on the other side of town. In some towns, white and Japanese children attended the same schools, but blacks and Mexican Americans were segregated in another location, where Chavez remembers teachers routinely doubting his intelligence because of the color of his skin. Dozens of masters' theses produced during this era had promoted the view that Mexican students were inferior.

■ *Mexican-American children were ordered to "speak English" at the Southgate School in Corpus Christi, Texas, in 1939 or 1940.*

"In integrated schools, where we were the only Mexicans, we were like monkeys in a cage," Chavez recalled later. "There were lots of racist remarks that still hurt my ears when I think of them. And we couldn't do anything except sit there and take it." The worst of all the schools, he remembered, was in Fresno, where he and Richard had waited all morning to register as new students. As they sat in front of the principal, they were horrified to hear him saying on the phone: "Well, we've got these kids and I don't know what to do with them. . . . Yeah, they're in the fourth grade, but they couldn't do that work. . . . What will we do with them? You wouldn't want them in your class."

After struggling to complete the eighth grade in Brawley, California, Chavez quit school. His father had suffered a serious auto accident, and Cesar told Rita that he

■ *A Mexican migrant worker made a home for his family in this crude shack on the edge of a frozen pea field in California.*

wanted to devote himself completely to work so his mother wouldn't have to labor in the fields anymore. Chavez's mother, who was illiterate and determined that her children would receive an education, was unhappy with his decision. But Cesar refused to change his mind because he felt it was time to become the family's breadwinner. He could read and write English well enough by then, and an uncle had taught him how to read Spanish using Mexican-American newspapers. He always regretted his lost schooling, however, as well as that of his sister Rita, who loved school but dropped out at age twelve because she was too ashamed to go to classes without shoes. Later he would tell his children that his education was a patchwork, apologizing that he couldn't help them with some lessons because he had missed so much school himself. After he started the UFW, Chavez single-handedly steered some young strikers into higher education, telling them they could help farmworkers more if they came back with a degree.

In Brawley, Cesar also encountered his first segregated eating place; just as in the Deep South, many rural towns in the Golden State hung out signs declaring White Trade Only. At age twelve, tired and hungry after a long day of shining shoes, Cesar and his brother Richard walked into a hamburger joint, ignoring a Whites Only sign. They were from Arizona, "from a community that was mostly Mexican or whites too poor to bother about us," Cesar said. A pretty girl behind the counter and her boyfriend were staring at them as, mustering up his nerve, Cesar asked politely for two hamburgers. "The girl said, 'What's the matter, you can't read? Goddamn dumb Mex!' She and her boyfriend laughed, and we ran out. Richard was cursing them, but I was the one who had spoken to them, and I was crying. That laugh rang in my ears for twenty years—it seemed to cut us out of the human race."

Three years later, Librado Chavez and his family were thrown out of a diner after being seated. The boss ordered the waitress to make them leave, then threatened to fire her when she protested. When she came back in tears, the fifteen-year-old Cesar spoke up for the first time, angrily asking the boss why he had to treat people like that. "A man who behaves like you do is not even a human being," he concluded, to which the man replied, "Aw, don't give me that shit. G'wan, get out of here!" In another incident, a café owner cursed and threw out Cesar's father, who had stepped in to buy a cup of coffee. What he never forgot about that experience, Cesar said, was the expression on his father's face.

THE CHAVEZES' STORY was hardly unique. Most migrant families were similarly humbled by the depression, and others who witnessed the depths of the poverty in California were to meet Chavez and become strong allies decades later.

While the Chavez family was still struggling to hold on to its farm in Yuma, a thirteen-year-old girl who, three decades later, would play an important role in the UFW, left Los Angeles to work the fields with her family. Like Cesar, Jessie De La Cruz remembers surviving on mustard greens and scraps to get through the depression. Her family's stove and heater consisted solely of an oil drum; her brothers had torn a round hole in its back and inserted a stovepipe that extended through the flap of the tent. "My grandmother would get us together, and she'd cry with us, because there was nothing to eat. It's a very sad childhood to look back to," says De La Cruz, who in her seventies still grimaces at the memory of a diet of beans and potatoes, spiced only by the rare dash of salt.

De La Cruz was a survivor of a time captured in vivid photographs by Dorothea Lange and other legendary chroniclers in a program sponsored by the Farm Security Administration. The FSA offered rudimentary shelter and food to some migrants, but couldn't keep up with the overwhelming need. At age thirteen, De La Cruz was hunched over a burlap sack, cleaning cotton for the miserable pay of ten cents an hour on the edge of a field near Bakersfield, when a wildcat strike erupted. White and Mexican workers walked out together, and as impressed as she was by that strike, De La Cruz says it was more common for farmworkers of different ethnicities to stay clear of each other. A group of Okies who lived nearby, De La Cruz remembers, "didn't like us. They wouldn't even talk to us. We were all the same age, but they wouldn't talk to us. We were out in the fields picking, doing the same work, and living under the same conditions, but they thought they were better than we were." (Discussing the Okies, veteran labor organizer Dorothy Healy recalls: "I can remember the biggest impression I had of those days was watching those white people coming in from Oklahoma and Arkansas and Texas, coming in with their ingrown prejudices and hatred, and learning in the course of the strike that they had more in common with that worker with the brown skin and black skin than they had with the vigilantes with the white skin who were beating everybody up.")

Another firsthand witness to the depression in California was Fred Ross of San Francisco, an unemployed teacher who in the 1930s found himself working in an FSA relief station in the Coachella Valley. There, he handed out sacks of flour and other staples to hungry migrants who subsisted for months on nothing but fried dough and mush. Ross moved up to the cotton fields in the San Joaquin Valley to supervise the Sunset Labor Camp, a New Deal migrant camp near Arvin that was immortalized in Steinbeck's *The Grapes of Wrath* as the fictional Wheatpatch camp, a place of refuge from the depression. Two decades later, in San Jose, Ross would meet Chavez and pass on to him the community organizing skills Cesar would later use to forge the UFW.

IN PURSUIT OF CROPS, the Chavezes often ended up in Delano, a little town in the heart of the vast San Joaquin Valley. There Chavez became enamored of the Mexican-American *pachuco* look—baggy zoot suits and wide-brimmed hats, a style cut by the coolest of youths in East Los Angeles in the early 1940s. "When I became a teenager, I began to rebel about certain things—the home remedies and herbs my mother used, Mexican music, religious customs," Chavez said later. "I also got into that trap of thinking [older people] were dull and uninteresting." Although Chavez would later celebrate his Mexican heritage, the teenage Cesar passed up mariachi music for the big-band sound of Duke Ellington and Billy Eckstine. Sporting a zoot suit, he smoked cigarettes and wore his hair long and cut in a ducktail, noting that "little old ladies were afraid of us"; an elderly Mexican woman once asked him why he was wearing a "monkey suit" since he was so well-mannered. Although Cesar's parents were a bit worried, and the police harassed him and his friends for their *pachuco* look, "I didn't want to be a square," he later said, fondly remembering his rebellious years. "We had a lot more fun than they did."

In 1943, Chavez met his future wife, Helen Fabela, at a Delano malt shop called La Baratita. Just fifteen, Helen was still in high school, but she, too, would soon drop out because her family had been under financial strain since the death of her father a few years before. Like the sixteen-year-old Cesar, Helen was keenly familiar with the backbreaking routine of farmwork. At age seven, she was already out with thousands of other small children helping pick grapes, berries, tree fruit, and walnuts. Her family

■ *Cesar Chavez and Helen Fabela were married after World War II and honeymooned on California's coast.*

drove the length of the San Joaquin Valley looking for work, pitching a tent by the side of a field as long as there were plums to pick or cotton to chop.

The social world of the Fabelas and the Chavezes was mostly limited to the Mexican side of the railroad tracks cutting through Delano. Segregation affected everyone: Delano had a large population of Filipino farmworkers, mostly male, who had been brought there in previous decades to tend vineyards and lettuce fields. Theirs was a lonely life: Relatively few Filipina women had been permitted to emigrate, and California had imposed strict miscegenation laws that prohibited whites from marrying blacks, Chinese, or people of the Malay race. Filipino men were known to suffer beatings for mixing openly with Anglo women. Some of them married Mexican women or moved to other states that allowed unions with whites, but many died, not by choice, as bachelors.

Mexicans, too, were often barred from white-owned stores in Delano and were forced to sit in the "colored" section of the downtown movie theater. Young Mexicans, regardless, considered American popular culture their own, and Helen was no exception. Richard, Cesar's brother, remembers that Helen became part of their *pachuco* circle. "She loved to dance, and it was the time of the jitterbug and the boogie-woogie. She was very good at it." Chavez was separated from Helen when his family traveled to Sacramento that summer to harvest tomatoes, but they returned to Delano to work in cotton fields in the fall. They couldn't find any housing in town, so they lodged in a government-run tent city nearby, and drove into Delano for supplies. Cesar saw Helen again, working at the malt shop and People's Market, where he soon became a very good customer. The young Helen, Cesar would recall later, "had flowers in her hair all the time. I remember that very well. It was during the war, and I used to kid her that the only reason I went with her was that she saved cigarettes for me."

Today at People's Market, still a hub of activity in Delano's west-side barrio, the elderly Chicano owners and employees remember Cesar and Helen. Johnny Serda, a Delano resident who, in his seventies, is still strong enough to stock boxes at People's, said he used to cruise the town with Cesar and other *pachuco* friends in his '38 Chevy. "He was just like an ordinary guy. But he didn't take no bull," says Serda, a wry smile hinting at memories of Friday and Saturday night adventures.

Lucio Gonzalez, who owns People's Market with his wife, Butter Torres Gonzalez, remembers the prejudice that divided the races. A cousin, Eddie Rodriguez, who had volunteered in World War II, was once visiting back home in the town of Blythe, near the California-Nevada border. "The sign in the store window said 'No Mexicans or Colored Trade Wanted,'" Gonzalez says. Rodriguez ignored the warning and entered the store clad in his U.S. army uniform. He refused to leave, and was ejected, bloodied by a punch to the mouth that left him with a permanent scar.

During the final year of World War II, Cesar joined the other young men who'd marched off to war. He enlisted in the navy in 1944, at just seventeen, a decision in which farmwork played no small part. "I was doing sugar-beet thinning, the worst backbreaking job, and I remember telling my father, Dad, I've had it! Neither my mother or father wanted me to go, but I joined up anyway." After training camp in San Diego, Chavez was ordered to the Mariana Islands, in the Pacific, and to Guam, where he worked as a navy painter.

Although the navy exposed him to a new world—and afforded him his first visit with a real doctor—Chavez regarded his experience in the military as the worst two years of his life. He observed that black men and Filipinos in the navy could advance no further than the kitchen or jobs as painters. The average Mexican-American sailor, he said, seemed destined to remain a

■ *Cesar Chavez (second to left) joined the navy near the end of World War II. Here he relaxes with buddies under the shade of a palm tree.*

low-ranking deckhand. But he discovered whites could also discriminate among themselves: "I saw this white kid fighting, because someone had called him a Polack and I found out he was Polish and hated that word Polack. He fought every time he heard it. I began to learn something; [I saw] that others suffered, too." Following an honorable discharge, he returned to Delano and he and his sweetheart "got serious about each other," Richard Chavez recalls. Within a year he and Helen were married.

■ *As late as the 1950s, towns throughout the West made their own rules for segregating the races. This placard was in a public park in Texas.*

After a brief honeymoon, the newlyweds came back to Delano, where Cesar worked in grapes and cotton. Their first home was a one-room shack in a labor camp, where the kerosene stove was no match for the bitter cold. Cesar and Helen were miserable. "When we stepped out of the shack, we stepped right into mud—thick, black clay." The following year, the family finally borrowed enough money to escape with Cesar's parents to San Jose, where Richard said he'd try to hustle up some work for Cesar—now the father of baby Fernando—on the apricot ranch where Richard had found a steady job.

To Cesar's dismay, the ranch could only fit him in two or three days a week. Lacking even money for a streetcar, he spent the rest of the time tramping around town looking for work. When he and his parents were offered jobs as sharecroppers—growing strawberries for a company that provided seeds, fertilizer, two small homes for each family, and twenty-five dollars a week for food and gas—they jumped at it. It turned out to be exhausting work, with all the profit going to the company. After working every day, even Christmas, for more than two years, and adding another child to his family, Cesar had had enough. He convinced his reluctant father that it wouldn't be a dishonor to break their agreement.

Back at the employment office, Cesar learned that a lumber company was recruiting workers for jobs in Crescent City, near the Oregon border. Cesar and Richard packed up their families and headed up to the majestic redwood and pine forests of Northern California. Cesar loved the woods, but Helen Chavez, who had two children and was pregnant again, grew tired of the relentless rain and wind. "We had this woodstove and you had to chop wood all the time, to cook, to keep warm, everything," she remembers. After a year and a half, the two families headed back to familiar territory. With Fernando, daughter Sylvia, and baby Linda in tow, Cesar and Helen moved to the barrio Sal Si Puedes in southeast San Jose. There, the Chavez brothers got jobs working in a nearby lumber mill.

The young Cesar was used to both poverty and hard work, but discrimination was something he could never grow accustomed to. Even before he was married, he had taken his own first step into what would be a long battle for civil rights.

It happened in a movie theater in Delano when Cesar, who was on a seventy-two-hour leave from the navy, declined to sit in the "colored" section. Segregated movie theaters were common in California farm towns at the time, and in Delano, Mexicans, blacks, and Filipinos were confined to a quarter section of the seats on the right near the entrance of the movie house. "It had been like that since the theater was built, I guess," Chavez later said. "I thought an awful lot about it, but I hadn't planned anything. . . . It wasn't a question of sitting there because it was more comfortable or anything. . . . I wanted a free choice of where I wanted to be. Or at least this is how I rationalize it now. It think it was a coincidence that I decided to challenge it. But I was very frightened."

Chavez wasn't in uniform when he deliberately chose a seat in the white-only section. For many years, Cesar and his parents had stood up to farm supervisors who had tried to cheat them or push them around, but the choice he made in the movie theater was his most brazen act of civil disobedience. Many Mexicans at the time accepted the segregated seating because they had known nothing else, Chavez said. Perhaps those who were lighter skinned had not felt the sting of discrimination quite as much. For his offense that night at the Delano Theater, Chavez was taken into custody and driven to the town jail. "The desk sergeant didn't know what to charge me with. They made a couple of calls. I think the first call was to the chief," Chavez said, recounting the sergeant's end of the phone conversation: "What about drunk? Well, he's not drunk . . . Well, he wasn't really disturbing the peace . . ."

After an hour, the sergeant gave Chavez a lecture and released him. "Oh, he tried to scare me about putting me in jail for life, you know how they do these things." Almost twenty years later, Chavez would return to Delano and return to jail, many times over. By that time, however, the question of what to do with a new Mexican-American civil rights movement would not be so easily and quickly resolved.

■ FOLLOWING PAGE: *The Community Service Organization, for which Cesar Chavez worked for many years, registered tens of thousands of Chicanos to vote during the 1940s and 1950s.*

> If you talked about civil rights you were a Communist, no doubt. Organizing, you were a Communist. Police brutality, you were a Communist. It was just fantastic. . . . Imagine letting us do some real blatant red-blooded American work like registering <u>people</u> to vote.
>
> —CESAR CHAVEZ

CHAPTER TWO

SAL SI PUEDES

CESAR CHAVEZ NEVER intended to let the bespectacled Anglo lecture him for so long that evening in June 1952. Why was the stranger so anxious to hold a meeting at his home, anyway? He was sure this *güero,* this white guy, was just another meddling graduate student in anthropology down from Stanford, Berkeley, or San Jose State. They came to study Cesar's barrio, probing its inhabitants like so many specimens in a zoo: Why do Mexicans have so many babies? Why do they eat beans and chiles? We've had enough, Cesar decided, not realizing that this was to be one of the most fateful nights of his life.

The self-invited guest was Fred Ross, a thin man in wrinkled clothes who identified himself as an "organizer" with a group called the Community Service Organization in Los Angeles. He had been trying vainly for days to meet the reluctant Chavez. It was a balmy night when he pulled up in front of Cesar's house in the southeast San Jose barrio of Sal Si Puedes, literally, "get out if you can." The name of this West Coast ghetto was both a judgment and a challenge: Paychecks were slim, and opportunities for advancement few. The roads were unpaved, and every winter when the rains came, the nearby creek spilled over its banks and backed up against the neighborhood's dead-end roads, melting them into thick mud that trapped any car parked there. The locals knew to leave their cars parked on the nearest paved road and slog through the muck to get home. There were no sidewalks in this part of town, and no politicians to answer why.

Cesar and Fred Ross might never have met were it not for Ross's persistence—and Helen Chavez's favorable impression of the tall, gangly white man. Soon after he had arrived in town from Los Angeles, Ross had started going door-to-door with near-missionary zeal, looking for recruits who could lead an ambitious voting-rights effort among Chicanos, who were all but absent as a political force in San Jose. Ross had been talking to shopkeepers, veterans, mothers, *pachucos* as well as the underpaid laborers who harvested the plum and apricot orchards that enriched California's Santa Clara Valley.

■ *Sal Si Puedes, the neglected San Jose neighborhood where Cesar and Helen Chavez settled, was plagued by poor services and pollution.*

Three weeks later, Ross had almost exhausted his list of people to contact without finding the person who could direct the registration campaign. Then he heard about Cesar and Helen Chavez from a San Jose public health nurse and from Father Donald McDonnell, a neighborhood Catholic priest. Chavez had befriended McDonnell after settling in San Jose, and he had started accompanying him to nearby labor camps to talk to Mexican braceros about their problems. Cesar enjoyed helping the priest. But when Helen told her husband that a studious-looking white man had come by the house looking for him, he assumed it was another university researcher. He instructed Helen to tell Ross that he wasn't home. He said he would hide out at his brother Richard's across the street until the interloper got back in his car and left.

Helen reluctantly agreed to do as Cesar asked. Always polite, Ross told her he'd come back the next night after Cesar got off his job at a lumberyard. When he did, an irritated Cesar asked Helen to cover for him again, but when Ross said he'd swing by yet again the following evening, Helen told Cesar enough was enough. The gringo seemed friendly enough, she thought, and *quién sabe*, he might know about a good-paying job. She persuaded Cesar to meet with Ross, and he agreed only because he was already planning a scheme to run the man out of Sal Si Puedes for good. But after that brief initial meeting, Cesar felt uneasy about going forward with his plans to humiliate Ross. "Somehow I knew that that gringo had impressed me, and that I was being dishonest," he recalled later.

Cesar's plan was to invite a few of the neighborhood toughs over to the house on the night the Anglo was to come by. As Librado "Lenny" Chavez Jr. remembers it, they'd drink a few beers and let him ask a few insulting questions about Mexicans. Then Cesar would give the signal, passing a cigarette from one hand to the other, to tell the chosen enforcer it was time to demand "real tough-like" that the gringo get out of town—and fast.

For Fred Ross, that night was as memorable "as if it were caught and chiseled forever somewhere deep in my retina: the tiny, narrow house on Sharf Street with the proud little gate, Cesar's wife, Helen, opening the door." The house was jammed with neighbors and cribs, and old couches sagged audibly under the weight of too many people. Out in the kitchen, more of Cesar's friends strained for a better view of the soft-spoken Ross, while children sneaked peeks from between legs. The atmosphere in the small house was chilly with doubt and mistrust. "When I put my hand toward Cesar, instead of getting up, he just sort of rocked forward on his behind, raising it an inch or two off the couch," Ross recalled. "When he shook my hand, it felt like a small piece of pig's liver." Ross offered cigarettes as an icebreaker, but no one moved. Minutes later, he noticed everyone lighting up their own.

Once Ross began speaking, though, Cesar was surprised to find his hostility slowly evaporating. The gringo was trying his best to pronounce the little Spanish he knew correctly, and his accent was actually quite good. More impressive, it was clear he'd taken time to learn about the real problems that plagued Sal Si Puedes. Ross knew about the creek that overflowed, about how it gave kids sores on their feet and legs because it was contaminated by waste that regularly leeched from a nearby fruit-packing shed. "He took on the politicians for not doing something more about it," Cesar said later. This was a surprise, since Sal Si Puedes "never heard anything from whites unless it was the police."

"I told Cesar and his buddies I had worked all over Southern California, and wherever I went the conditions among the Mexican Americans were as bad as in Sal Si Puedes," Ross recalled. "The same polluted creeks . . . for kids to play in. The same kind of cops beating up young guys and 'breaking and entering' without warrants. The same mean streets and walkways and lack of streetlights and traffic signals. The same poor drainage, overflowing cesspools, and amoebic dysentery. Cesar was impressed: I knew his problems as well as anyone."

Remembering Fred Ross

by Susan Ferriss and Ricardo Sandoval

"There was once a reporter who likened my father to Gary Cooper coming into town and taking on the biggest, baddest bully," says Fred Ross Jr., discussing his father's legacy. "But he was not a solo operator like a Gary Cooper. What he did was go out and teach people [so] they could do it for themselves—because he understood that's a lot more respectful of people. And even more important, it's a lot more enduring because when you're gone, you want to leave something in your place."

America's Great Depression gave rise to Fred Ross's eventful career in community organizing. He

---

■ *Fred Ross, right, worked with Okie migrants during the depression and befriended singer and songwriter Woody Guthrie, left.*

had originally hoped to become a teacher in his native San Francisco, but after college he found himself managing migrant-labor camps for the New Deal's Farm Security Administration. There, he befriended singer Woody Guthrie and persuaded First Lady Eleanor Roosevelt to visit the Okies' squatter camps, so she could see firsthand the migrants' misery.

By the end of World War II, Ross was working for the government, helping resettle thousands of interred Japanese Americans in new homes. He then began working with veteran organizer Saul Alinsky of the Industrial Areas Foundation, who sent him to organize Latino communities—even though he couldn't speak Spanish. Ross was well aware that he had that strike against him. "So he had to prove his commitment by his actions," says Fred Ross Jr. "He had a lot of anger about injustice, and he couldn't live with himself if he wasn't doing something about it. And there was plenty around."

At the time, Latinos were plagued by poor housing, discrimination, and police brutality. To combat these conditions, Ross helped start a Latino civil rights group known as the Community Service Organization. Borrowing from Alinsky's lessons about organizing "one person at a time," Ross developed his own brand of outreach—one house at a time. (He also taught himself Spanish.) "My strongest memories of growing up are of going to CSO meetings as a child," says Jessica Govea, a former UFW organizer whose parents had joined Ross to create a CSO chapter in the tiny farmworker enclave of Bakersfield. She credits Ross with encouraging women to shake tradition and jump into community organizing. "Fred in particular, and then later Cesar, when he became an organizer with CSO, emphasized the importance of this not being an organization of men. Our community was very traditionally father-dominated, but [the CSO] became an organization of men and of women and of children."

After Ross met Chavez in San Jose in 1952, and helped Cesar recognize his own organizing and leadership skills, the pair expanded the CSO by twenty-two chapters, turning whole neighborhoods into potent lobbying groups, and shy homemakers into proud voters. "Fred Ross was a real gentleman," says Beatriz Bedoya, who became his assistant in 1952 when she was a young Chicana newlywed in the San Francisco Bay community of Decoto, now Union City. Bedoya, today in her seventies, lives in the same neighborhood where she walked precincts with Ross. "He had this way of making you feel like the organizing work was *the* most important thing you could be doing. I liked that and became really involved. I thank my late husband for being so understanding. He let me go out with Mr. Ross almost every night to go talk to people about registering to vote."

Not long after Cesar left CSO to begin organizing farmworkers in Delano, Ross moved on to work with the National Presbyterian Church, setting up self-help groups among destitute Yaqui Indians and Latinos in southern Arizona. In 1965, Ross was teaching courses in community organizing at Syracuse University in New York when he got a call from Delano, where Cesar needed help with his union's ambitious plans for a strike against San Joaquin Valley grape growers.

During his many years of friendship with Chavez, Ross trained a new generation of UFW leaders and proved that he, too, could learn from his protégé. Ross, like Cesar, became a health food aficionado and adopted a tough regimen of exercise and yoga. Staff members recall that Ross would also carve out time from even the busiest of days to meditate.

Ross retired from the UFW in the early 1980s, but he never stopped organizing. He helped teach young activists how to take on the government over the

■ *Fred Ross and the first class of negotiators, La Paz, 1978.*

nuclear arms race and, with his son, Fred Jr., founded Neighbor to Neighbor, a group dedicated to ending U.S. intervention in Central America. Ross died of cancer at age eighty-two in 1992.

■ *Hiram Samaniego, Cesar Chavez, and Fred Ross working together on Community Service Organization strategy.*

As Ross remembers it, he didn't "rabble-rouse" that night. He just talked plainly about what the Community Service Organization had done for Chicanos in Southern California: They'd gotten rid of segregation in schools, theaters, and school buses. The CSO was formed five years earlier in East Los Angeles, emerging out of frustration with police abuse of Mexicans and the defeat of Edward Roybal, a popular and capable Chicano who had run unsuccessfully for city council in 1947. Latino activists and Fred Ross, an expert at voter-registration drives, formed CSO to boost the neighborhoods' political clout and try again to get Roybal elected. Saul Alinsky, a teacher of community-organizing techniques who ran the Chicago-based Industrial Areas Foundation, offered to pay Ross a small salary while he helped establish CSO. By 1949, CSO had registered fifteen thousand new voters who helped catapult Roybal to victory. He became the first Mexican American to sit on L.A.'s city council since the 1880s.

Ross also told Chavez and his friends that CSO had pushed for the punishment of drunken police officers who'd nearly killed seven young Chicanos in a 1951 incident known as Bloody Christmas. "Never before in the whole history of Los Angeles had any cop ever gotten 'canned' for beating up a Mexican American," Ross said. But the CSO, with its newly developed strength, had made sure the police were held accountable. "If the people of Los Angeles could do it," Ross said, "there was no reason why we couldn't do the same sort of thing in Sal Si Puedes, if we wanted to badly enough."

That part of Ross's story perhaps impressed Chavez most, since news of the Los Angeles incident had already spread to every corner of Chicano California. Mexican Americans had had their fill of police brutality, being only too familiar with abuses like the Zoot Suit riots of 1942, in which police had joined drunken sailors in savagely beating dozens of young Chicanos in East L.A.

Mesmerized by Ross's tale, Cesar forgot about the cigarette signal. Cesar's bouncer waited impatiently in back of the living room to make his move, growling under his breath, "Come on, man, the signal!" Finally this beer-stoked rowdy, tired of all this talk, suddenly shouted out that Ross should mind his own business. At that, Chavez had the man ushered out of the house. Winding up his talk, Ross told Cesar that he was looking for volunteers to get involved in a local chapter of the Community Service Organization.

"I didn't know what CSO was, or who this guy Fred Ross was, but I knew about the Bloody Christmas case, and so did everybody in that room," Cesar recalled later. "Five cops actually had been jailed for brutality. And that miracle was the result of CSO efforts. Fred did such a good job of explaining how poor people could build power that I could taste it. I could really *feel* it. I thought, Gee, it's like digging a hole; there was nothing complicated about it. . . .

"I'd never been in a group before, and I didn't know a thing. We were just a bunch of *pachucos*—you know, long hair and pegged pants. But Fred wanted to get the *pachucos* involved—no one had really done this—and he knew how to handle the difficulties that came up. He didn't take for granted a lot of little things that other people take for granted when they're working with the poor. He had learned, you know."

Ross immediately invited the baby-faced twenty-five-year-old to a meeting that same night in East San Jose, and by the end of the evening, Cesar found himself volunteering to work the next day as a "bird dog," someone who knocks on doors and asks people to register to vote. The drive's aim was to increase Chicano turnout in the November 1952 election and lay the foundation for the new San Jose CSO chapter, the first one outside of the Los Angeles area. Ross had already found election deputies in the barrio, some with college educations, and had had them sworn in by the Santa

---

■ *Third from left, young Cesar Chavez favored the baggy zoot-suit look popular among Chicano* pachucos, *"tough guys."*

The Zoot Suit Riots

by Carey McWilliams

The Zoot Suit riots in Los Angeles began with two minor clashes between Mexicans and white youths on June 3, 1943. After inflammatory newspaper stories and a well-publicized police raid on Mexican neighborhoods whipped up racism and public anger against "zoot-suiters"—Chicano youths wearing the long jackets and porkpie hats in vogue at the time—rioting against Mexican Americans erupted across the city. Historian Carey McWilliams, author of the groundbreaking books Factories in the Field and North from Mexico, gave this account.

On Monday evening [June 7, 1943], thousands of Angelenos, in response to twelve hours' advance notice in the press, turned out for a mass lynching. Marching through the streets of downtown Los Angeles, a mob

■ *Navy men, stationed in Los Angeles during World War II, caught by a photographer in the midst of the Zoot Suit race riots in June of 1943. During these riots, hundreds of unarmed Chicano "zoot suiters" were beaten by sailors carrying sticks, clubs,*

of several thousand soldiers, sailors, and civilians proceeded to beat up every zoot suiter they could find. Pushing its way into the important motion picture theaters, the mob ordered the management to turn on the house lights and then ranged up and down the aisles, dragging Mexicans out of their seats. Streetcars were halted while Mexicans, and some Filipinos and Negroes, were jerked out of their seats, pushed into the street, and beaten with sadistic frenzy. If the victims wore zoot-suits, they were stripped of their clothing and left naked or half-naked on the streets, bleeding and bruised.

Here is one of numerous eye-witness accounts written by Al Waxman, editor of the *Eastside Journal:*

"At Twelfth and Central I came upon a scene that will long live in my memory. . . . Four boys came out of a pool hall. They were wearing the zoot-suits that have become the symbol of a fighting flag. Police ordered them into arrest cars. One refused. He asked: 'Why am I being arrested?' The police officer answered with three swift blows of the night-stick across the boy's head and he went down. As he sprawled, he was kicked in the face. Police had difficulty loading his body into the vehicle because he was one-legged and wore a wooden limb. . . .

"At the next corner a Mexican mother cried out, 'Don't take my boy, he did nothing. He's only fifteen years old. Don't take him.' She was struck across the jaw with a night-stick and almost dropped that two and a half year old baby that was clinging in her arms. . . ."

Throughout the night the Mexican communities were in the wildest possible turmoil. Scores of Mexican mothers were trying to locate their youngsters and several hundred Mexicans milled around each of the police substations and the Central jail trying to get word of the missing members of their families. Boys came into the police stations saying: "Charge me with vagrancy or anything, but don't send me out there!" pointing to the streets where other boys, as young as twelve or thirteen, were being beaten and stripped of their clothes. From affidavits which I helped prepare at the time, I should say that not more than half of the victims were actually wearing zoot-suits.

Unfortunately, the [rioting] spread to the suburbs, where it continued for two more days. When it stopped, the *Eagle Rock Advertiser* mournfully editorialized: "It is too bad the servicemen were called off before they were able to complete the job."

Clara County registrar's office. Cesar probably needed to learn more—and sharpen his writing skills—if he wanted to be a registrar and chief organizer, Ross thought. But soon the unassuming blue-collar laborer became Ross's greatest asset: Cesar threw himself into the San Jose registration drive with gusto. After a day of heavy lifting at the lumberyard, he'd rush home, wolf down a meal, and wait impatiently for Ross to pick him up. After weeks of Cesar's tireless bird-dogging, Ross finally had to tell him to take a night off so he could spend time with Helen and the kids. "For whatever reasons," Ross later explained, "all of his actions were invested with a tremendous amount of urgency."

The first few times Cesar knocked on a prospective registrant's door, he was so nervous that he was left speechless for a few seconds, and had trouble explaining why he was there. He often stumbled through an initial meeting—perhaps a little self-conscious about having only an eighth-grade education. Herman Gallegos, another early CSO recruit who met Chavez in San Jose, noticed that he had a gold tooth that flashed when he smiled, "and a rather anemic handshake." He advised Chavez to beef up his grip: "Look 'em in the eye, and shake them." This was a politician's technique that Cesar never really cared to master. Gallegos, who would go on to lead the Mexican-American Political Association, a prominent civil rights group, was one of the few people in the CSO who had a college degree, and he hid that as much as he could. "You couldn't have middle-class social workers going into the barrio to assume leadership roles in place of the people affected," he says. Like some others who met the young Cesar back then, he remembers Chavez had "a very deep-rooted suspicion of the middle class."

But Ross sensed in Chavez a leader who would bloom, and he soon asked Cesar to chair the registration campaign. "So here I am in charge," Cesar remembered, "and where do I start? I can't go to the middle class, or even the aspiring middle class, for my deputy registrars; I have to go to my friends in Sal Si Puedes. So I round up about sixteen guys, and not one of them can qualify as a deputy registrar, not one. They can't even vote! Every damn one of these guys had a felony!" Nonetheless they could, by law, knock on doors and encourage people to register. "They were my friends. I grew up with them and I knew what they were up against, and I always thought they were in the right except when they got sent up someplace to do their time."

Fred Ross would become Cesar's mentor, his hero, and one of his closest lifelong friends. A San Francisco–born community organizer twice Cesar's age, Ross could only speak a few words of Spanish when he knocked on the Chavezes' door in 1952. But it was he who would first pull Chavez into the political arena by getting him involved with CSO. For Cesar the experience would prove intoxicating; CSO was a place where he would cut his organizing teeth and go on to challenge the Goliaths of California politics and industry. As the two walked the barrio together, Ross got to know more about Cesar and what it was he most cared about. Chavez asked Ross, "What about the farmworkers?" and Ross told him he thought CSO could serve as a base for organizing field laborers, if that's what he wanted. In a diary that Ross kept to chronicle those days, he wrote: "I think I've found the guy I'm looking for."

EVEN BEFORE MEETING Ross in 1952, Cesar had shown flashes of the activist he would become. In the late 1940s, after he and his family had walked with pickets during a particularly acrimonious strike, Cesar went out of his way to sit in on congressional hearings in Bakersfield that had been called to investigate a different labor walkout against the big DiGiorgio farming company. A young representative named Richard Nixon had set the stage at the hearings by insinuating that the strike was nothing but a Communist plot.

Later Cesar began informally studying the art of organizing with the other pivotal influence in his early life. One of a group of socially aware and active priests called the "mission band," Father Donald McDonnell had convinced San Francisco's Archdiocese to assign him to Sal Si Puedes. Lenny Chavez remembers when the family first met the earnest McDonnell as he walked the barrio in his starched priest's collar, surveying Catholics and asking them if they would support opening a new church in the neighborhood. Cesar was most enthusiastic at the prospect. He told the priest his family had to go across town to attend mass at a Portuguese church, where he could sense that his dark-skinned Mexican family was not welcomed by the other worshipers.

McDonnell was part of a new wave in Catholicism that encouraged clerics to live among the Spanish-speaking Mexican and Chicano farm laborers, for whom wages and conditions had deteriorated since the 1940s. The U.S. Department of Labor had brought in Mexican braceros—literally, "a pair of arms"—during World War II to fill jobs left suddenly vacant when the Okies and others rushed off to join the military or to work in urban factories; in the postwar years, the program was riddled with abuse. Father McDonnell demanded that he be allowed to enter the bracero camps to offer mass on portable altars, and Chavez asked to accompany him. Although Chavez enjoyed hanging around with his pals or shooting pool over a beer or two, he was a family man now and, to his core, devoutly Catholic. As a young boy in Brawley, he had been a *crucero,* assisting the priest with mass, so working with the church came naturally to him. McDonnell introduced Cesar to a brand of Catholicism he would later embrace even more fervently. "He told me about social justice, and the Church's stand on farm labor and reading from the encyclicals of Pope Leo XIII, in which he upheld labor unions. I would do anything to get the Father to tell more about labor history. I began going to the bracero camps with him to help with the mass, to the city jail with him to talk to the prisoners, anything to be with him. . . ."

■ *Second from right, Cesar Chavez wore suits and grew a mustache to look more mature after becoming an organizer for the Community Service Organization. Also pictured: Fred Ross, center of the front row; Saul Alinsky, far left; and Helen Chavez, third from the right in the back.*

Cesar was a slow but disciplined reader, and McDonnell introduced him to books he couldn't help but absorb. He read about Saint Francis of Assisi and read the teachings of Saint Paul. Later Chavez discovered biographies of American labor giants like John L. Lewis and Eugene Debs, as well as the political philosophies of Machiavelli and de Tocqueville. But the most influential books were biographies of the Indian independence fighter Mahatma Gandhi, and Gandhi's own writings. Gandhi's uncompromising pacifism, his use of fasting as protest, and the imaginative ways he confronted British colonial authorities fascinated Chavez throughout his life. "I knew very little about him except what I read in the papers and saw in newsreels," Cesar would say. "There was one scene I never forgot. Gandhi was going to a meeting with a high British official in India. There were throngs of people as he walked all but naked out of his little hut. Then he was filmed in his loincloth, sandals, and shawl walking up the steps to the palace."

After his experience with McDonnell, Chavez was ready to learn something new from Ross—especially if it yielded political change and not just charity. As his brother Richard, who also joined CSO, recalls, "I guess we were ripe for it." Cesar's sister Rita and Lenny, the youngest, also got involved. Cesar knew he was in training for the future: "All the time I was observing the things Fred did, secretly, because I wanted to learn how to organize, to see how it was done. . . . I was impressed with his patience, and understanding of people. I thought this was a tool, one of the greatest things he had."

A few months after meeting Chavez, Ross received more money from Saul Alinsky so he could hire Chavez at thirty-five dollars a week. The core of CSO's work—coaxing ever greater numbers of Mexican-born residents to become American citizens and vote—was a wholesome enough activity: an exercise in the most basic of democratic rights. However, Cesar's activism emerged right in the middle of the McCarthy era, when it seemed that every local government in the country was waging its own witch-hunt to persecute anyone with a trace of suspected "red" sympathies. Conveniently, charges of un-Americanism also became an easy weapon to intimidate people and beat back challenges to the status quo. Americans who were pro-union, against segregation, and for voting rights all became suspect, and Chavez was no exception. It was shortly after he plunged so enthusiastically into voter registration that he had his first brush with the Federal Bureau of Investigation, which would later shadow Cesar and the UFW through two decades, as their farmworkers' rights movement moved inexorably across the country.

Cesar's initial confrontation with red-baiting took place in 1952, a year that saw a gratifying surge in the numbers of Mexican Americans at San Jose's polls. Fearing a Chicano voting bloc would help the Democratic Party, the local Republicans decided to take matters into their own hands, dispatching representatives to polling places. They walked up to Mexican Americans as they emerged from voting booths, Cesar recalled, and demanded to know if they could read, or if they were indeed U.S. citizens. Ross was alerted to come help local CSO leaders decide how to respond to the intimidation.

Ross suggested wiring an urgent complaint to the U.S. attorney general, demanding an investigation. Chavez became angry when the more middle-class CSO activists were afraid to sign the telegram. He stepped forward to volunteer his signature, and for the first time, controversy pushed him to the front line of a social fight—and onto the pages of local newspapers. Republicans accused CSO of registering illegal immigrants, and the civil rights group fired back, calling the GOP activists bigots and racists.

It didn't take long for the public spotlight to follow Chavez into the lumberyard where he worked. Coworkers had taken to teasing him and calling him "the politician," and an Italian foreman who was fond of Cesar warned him to stay away from politics because it could bring him trouble. Not long after the conflict at the polls erupted, men in suits appeared at the lumberyard and identified themselves as FBI agents—looking for Cesar Chavez. The foreman became nearly hysterical, rushing out to warn his friend, his *compagno*. Cesar was bewildered by the scrutiny. "The agents started asking me a lot of questions about Communism," he recalled. "I said, 'You know damn well I'm not a Communist!' But what they really wanted to talk to me about was the complaint CSO had filed against the Republican Central Committee. So I relaxed, and we talked about a half hour."

Chavez was convincing enough to temporarily transform the FBI agents into arbitrators; they drove Chavez to a negotiating meeting with the Republican Central Committee that same afternoon. It was the first time he exploded under pressure and shouted back at Anglos. "The agent in charge was trying to work the thing out," Cesar recalled later. "Finally he said, 'Well, we have enough of these problems in Mississippi and the South, and we don't want to have any of this nonsense in California.' I felt pretty good by then. It was really reassuring. The Republicans were told that they couldn't intimidate people at the polls, and that the investigation would continue."

It was a victory: The Republicans were put on notice by federal agents not to interfere with CSO's voting rights work. Most important, Chavez had proven, to himself and to the local power structure, that he wasn't afraid to fight back when he felt under unwarranted attack. Perhaps, he later said, he stood up because he was so "damned naive" in the beginning, especially when he first confronted accusations that he was a Communist sympathizer. He had an indignation and an innocence that made him fearless when others, with professional careers and reputations to protect, would wither under pressure.

Chavez, however, also discovered just how tenacious federal agents could be, especially if they believed they were on the trail of another subversive. After the clash with the local Republicans, Cesar remained under FBI scrutiny. "Another FBI agent came to investigate me at home and wanted to know how many organizations I had belonged to, and I was pretty mad," Cesar remembered. "I told him to go to hell."

Coworkers began to shun Chavez because they heard repeated rumors he was a Communist, an accusation that flew in the face of his Catholic conservatism. "The Chicanos wouldn't talk to me. They were afraid," Cesar said. "Everywhere I went to organize they would bluntly ask, 'Are you a Communist?'" The observant Cesar hit upon a countermeasure: he already knew the church wielded great authority and power in the Mexican community, so he went to Father McDonnell and some of the other priests in San Jose and asked them to issue a statement in his defense and offer him their blessing publicly before worshipers.

The church's endorsement helped Chavez grow into an authority figure in his new job as a full-time organizer. Ross believed every CSO chapter should help immigrants to become U.S. citizens, and Cesar wholeheartedly agreed. He spent many long evenings patiently sifting through documents and tattered receipts that old Mexicans brought to him, hoping to establish that they'd lived in the United States long enough to become citizens. Some had arrived before the turn of the century, yet no one had ever bothered to tell them how they could naturalize, let alone vote. Mexicans and Chicanos began coming to Cesar with other problems. They told stories of being swindled by salesmen, or asked for help with an insurance claim or simply needed him to translate for them. Chavez also collected food donations for families poorer than his own; longtime San Jose resident Ernie Abeytia still remembers Cesar would regularly walk into his Sal Si Puedes general store, buy dozens of tamales, and then pass them out in the neighborhood.

Yet Chavez loathed the idea that his work might be perceived as charity or a "gimmick." Like Ross, he expected people who received from the CSO to turn around and give something back—time, food, a little cash, or a signature on a letter of protest. Quietly stubborn, Chavez also began to break with some established CSO policies. He thought the group needed a tangible presence, so he suggested opening a small office in Sal Si Puedes. Others involved with CSO disagreed. Saul Alinsky, for one, thought any organizer who was hiding out in an office wasn't doing his job. When Cesar threatened to quit, Ross and Alinsky relented, and the office was opened. The center proved a worthy experiment, establishing a comfortable home for the group. Cesar would replicate the office model when he started the UFW's farmworker service centers years later.

R OSS'S MISSION was to start more CSO chapters in Northern California, and he knew Chavez would be a valuable assistant. Together, the pair walked neighborhoods and organized house meetings in small cities north of San Jose, where Chavez gained experience working on unfamiliar turf. With the Sal Si Puedes CSO chapter firmly established, Chavez told Ross he wanted to try setting up another chapter on his own. Ross agreed and assigned him to Oakland, a blue-collar seaport across the bay from San Francisco that had grown into a metropolis during World War II. While Cesar believed he was ready to organize on his own, his solo flight into Oakland terrified him. He always got lost whenever he drove into the sprawling, unfamiliar city. When a local priest set up Cesar's first house meeting in West Oakland, Chavez was so nervous he kept circling around the block until he finally mustered the nerve to park and ring the doorbell. Inside, he found a dozen middle-aged women, and felt so awkward he neglected to identify himself until one of the women wondered aloud when the CSO organizer would finally arrive.

" 'Well, I'm the organizer,' " Chavez remembered answering tentatively. "She looked at me and said, 'Umph!' I could tell what she meant, a snotty kid, a kid organizer, you're kidding! That meeting was a disaster, really a disaster. I fumbled all over the place. . . . But toward the end of the meeting they were listening to me, and I got them to promise to hold house meetings, a lot of house meetings, and to commit themselves. Probably they felt sorry for me more than anything else."

After clearing this initial hurdle, and another three months of canvassing and house meetings all over Oakland, Chavez called together his new recruits for their first mass gathering at St. Mary's Church. He was nervous to the point of illness, afraid no one would come. Unable to tear his eyes from the door, he watched a slow trickle eventually swell to a crowd of over four hundred. He was exultant. "After the meeting, the first thing I did was sprint to the phone and report to Fred," Cesar recalled. Now that he was a leader, ready to take on big-time politicians and city bureaucrats, Cesar became convinced that it might serve him well to cultivate a more mature image. He took to wearing suits and, for a brief time, grew the kind of manicured, pencil-thin mustache that was popular among Latino matinee idols in the early 1950s.

In 1953, with the CSO chapter flourishing in Oakland, Fred Ross dispatched Cesar into the heart of the San Joaquin Valley, an assignment that required that he and his family be ready to move every few months if necessary. Cesar approached the job with great eagerness: He would be closer to farm laborers, even though his mission was to work with all the Chicanos in town, and it would mark his return to country he knew well. Helen had been raised in the San Joaquin, and he had spent countless hours as a migrant child pursuing the harvests there. The insular farm towns were fierce, though, and could be hostile to "outsiders" who came in declaring that things needed fixing, especially when it came to field laborers or life on the Mexican side of town.

To Cesar's surprise, however, he encountered other prejudices he hadn't expected to find—among priests as well as among his own people. On his first stop, Madera, a small community near Fresno surrounded by thick walls of grapevines, a priest declared publicly that he suspected the young CSO organizer was—yet again—a Communist. Besides, the priest said, CSO seemed to have far too many Chicano enthusiasts who were Protestants. Chavez may have been Catholic, but Protestants had saved his mother and father from going hungry more than once when the family was destitute, and he told the priest the Protestants would definitely stay. That was settled, but the unrelenting current of red-baiting continued to sweep Cesar along with it. A top immigration official in Fresno refused to assist CSO's Spanish-speaking citizenship applicants, and joined in spreading rumors among the young new recruits of Madera's CSO that Chavez was a Communist. "Everybody went running for the woods," Cesar would remember. "It taught me the most important lesson in my life about organizing. When people are fearful, when it's their skin, they don't care about anybody."

The climax of CSO turmoil in Madera came when a number of the young recruits, many of them aspiring white-collar workers, turned against Chavez, refusing to talk to him and plotting how to run him out of town. In this case, Chavez found many loyal allies among the maligned Protestants, the immigrants for whom he had started citizenship classes, and the farmworkers. And he had the unswerving support of Fred Ross, who, in one of the many letters the two would exchange over the years, wrote Cesar in 1954 to tell him he was "doing a swell job up there." Ross advised Cesar to stay in Madera long enough to finish a voter-registration effort, and make sure that all those new "registrants don't get flushed down the drain for failure to vote." He also broke the news that Merced, a town farther south in the valley, would probably be Cesar's next assignment, adding, "I hate to keep moving you around because of the family, the kids in school, etc. But I don't see how else we can keep CSO moving forward."

Red-baiting by his own peers wasn't the only unhappy experience Chavez had during his early years with the CSO. He was astonished when Felix Leon, a Chicano contributor who ran what Cesar considered the best Mexican restaurant in San Jose, refused to allow several black CSO supporters into his restaurant. (The local chapter had about fifty black members at the time.) Cesar and his sister Rita, also active in the CSO, at first thought there had been a mistake, then they realized that Leon was serious. They tried to reason with him, but he wouldn't budge; and more shockingly, many CSO members sided with him. Anonymous leaflets were even printed up—in a print shop Cesar later traced to Tijuana—and distributed in churches calling Rita and the CSO "a bunch of nigger lovers." The CSO branch was nearly crushed when hundreds of Chicanos fled the organization over the controversy.

Still, Cesar had great faith in the CSO, and went on to hold house meetings throughout the San Joaquin Valley, working long hours in all the flat, dry farm towns he would later return to as a union organizer—forgotten places like McFarland, Delano, Wasco, Buttonwillow, and Arvin. Writing to Ross from the scruffy Okie town of Oildale on December 22, 1954, Cesar discussed the concerns of people in the house meetings and clubs he had spoken to: "Some of them thought the CSO was a fly-by-night deal and that it would die before it got started. . . . I explained to them that the CSO is an action group and not a talking one."

"The holiday fever is running very high," he admitted in the same letter, almost apologizing for not doing more. "When I speak of the CSO, citizenship, and voting, they are thinking of tamales, *buñuelos* [buns], and tequila. . . ."

Chavez's experience in the San Joaquin Valley would season him for one of his last and most meaningful CSO assignments, which was to be in the city of Oxnard. This would be the only place where Chavez was able to steer CSO not just into civil rights work, but into a prophetic confrontation with California's powerful agricultural establishment.

IN THE SUMMER of 1958, Chavez was summoned to meet with Saul Alinsky and CSO officers to consider taking a challenging new project in Oxnard, a small city on the coast north of Los Angeles, where huge tracts of lemons and vegetables were

■ *The Mexican braceros were guest workers first brought in seasonally during World War II to harvest fields in the West. Their rights were restricted, and they were sent back to Mexico if they tried to protest conditions.*

the area's lifeblood. His mission in Oxnard would be unique: the United Packing-house Workers union, whose president was a friend of Alinsky's, had promised CSO $20,000 to start a chapter in Oxnard. The union had been stumbling in its attempt to organize laborers who worked in sheds, packing lemons for nationwide shipping. The workers were frightened and didn't know their rights, and the union thought CSO might embolden them while also doing the Chicano community some good.

Though Cesar was frustrated that he wouldn't be able to get involved in union business himself, he was excited about returning to Oxnard. He and his family had picked peas and beans there, enduring one cold, wet winter in a tent. Organizing in Oxnard would almost be like settling the score for his family's suffering. But the thought of having his salary paid for by the Packinghouse Workers union weighed heavily on Cesar, who felt a tremendous responsibility not to fail. "See, Fred," he would later tell Ross, "I was deathly afraid I would take that money and not use it right. . . . You sorta see this big mass of people, red and white and black and brown, all reaching into their pockets and pulling out their pennies and nickels and dimes."

The Braceros
by Ernesto Galarza

Among those investigating the braceros' plight was Ernesto Galarza, a writer, academic, and activist who was the first farm-labor organizer with a Ph.D., which was awarded by Columbia University. Born in Mexico, he migrated to California with his family in 1910, later going to college on a scholarship and joining the National Farm Labor Union as an organizer. A tireless farm-labor advocate, Galarza sought to topple the alliance between ranchers and government bureaucrats by exposing the scandalous abuses of the bracero program. His work not only showed how the bracero program had hurt domestic workers, but also examined the danger and indignities routinely faced by Mexican braceros, as in these excerpts from his book Merchants of Labor.

The few complaints registered by braceros directly have given an impression of close harmony between the workers and the [farm] associations. In 1960, there was an average of one grievance for every 300 workers.

These statistics are worthless. Many complaints were scratched by the field men of the associations, or ignored by the managers. The conversations in the camps, the work stoppages in the fields, the desertions, the quantity of mail addressed to their consuls, the pilgrimages of men from their camps to nearby towns in search of advice from anyone who would listen to them—all these were symptoms of a distress which was not officially recorded.

No single aspect of the bracero program was the cause of more irritation than the food services. These were all too often operated by catering companies inured to complaints, or by canny concessionaires who cut corners freely. It was in the latter class that most of the cases of food poisoning occurred, and where the menu more often included sheep heads and chicken necks, and where moldy leftovers were apt to appear in the next day's lunch boxes. Work stoppages occurred as quickly over objectionable food as over low earnings. Braceros were known to have packaged and mailed an ill considered meal to the nearest [Mexican] consul, with an invitation to dinner at the camp.

The quality of housing remained a sore spot for years. . . . In the San Joaquin Valley, temporary camps were set up among mosquito-infested sloughs. Barns and stables were converted to human habitation. Of some 700 camps inspected in 1957, over a third were judged to be beyond salvaging.

Physical abuse was a grievance that could cause uncommon solidarity in a field gang and lead to an immediate stoppage of work. . . . [In one case] more than 200 men walked off the job and 26 arrived at the Delano police station to complain. The danger of transportation accidents was also the subject of many complaints. Vehicles overcrowded with men overturned, rolled into irrigation ditches, stalled at railroad crossings or burned with their occupants locked in. . . . Disabling or fatal accidents on the job also caused uneasiness in the camps. In 1959 [California records] reported the investigation of 143 cases of organophosphate poisoning: the typical victim was of Mexican descent, did not speak or write English, and knew nothing of the hazards to which he was exposed.

He was, in other words, a bracero.

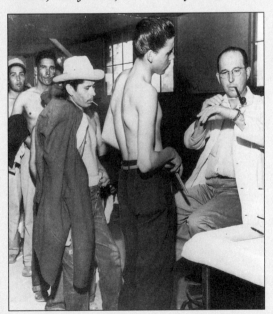

■ *A male nurse examining Mexican braceros at a U.S. Labor Reception Center in El Centro, California, in February 1954.*

That summer, Cesar and Helen took their seven kids to a beach south of Santa Barbara—a quiet seashore both loved—for a long overdue week of vacation before they would pack up and make another move. Characteristically, Cesar felt he had to interrupt his vacations for an initial round of meetings to find out what irked people most in Oxnard's barrio, known as the *colonia*. What gnawed at them most, he found, was that farmers often denied local people jobs in the fields, turning instead to the five thousand braceros that the Ventura Farm Labor Association brought in every year from Mexico. If the local workers were hired at all, the wages were dirt-poor. Cesar wondered how the farmers got away with it, since by law, local people were supposed to get first crack at all farm jobs.

The bracero program had long been controversial. Delighted with what turned out to be a compliant, dutiful, and cheap workforce during World War II, western farmers had successfully lobbied federal agencies to continue the bracero program into peace-time. Braceros were supposed to be protected from servitude, and they were to be imported only when domestic workers were not available. In practice, the system was corrupt: It was routine for labor department officials to recruit more braceros than were needed, and local workers would somehow find themselves unable to land field jobs.

Over a beer and a taco at El Mirador cantina in Oxnard, Cesar met an old Chicano lemon picker who showed him the permanent indentations on his shoulders left by carrying harvest bags for so many years. During the war, the braceros were needed, the old man said, but then the growers discovered they could make *puro dinero*—"big bucks"—by pretending they couldn't live without them. "The grower has little expense of recruitment; the government brings them practically to his gate. He gouges them on wages. He works them twelve or fifteen hours a day, on Saturday and Sunday and holidays. If they complain, or join a union, or do anything but work their ass off, *pfft!* He slams them back to Mexico."

Do the locals want to stop this abuse? Cesar asked. "Who knows?" the old man shrugged. But he promised to introduce Cesar to others who could tell him more once Cesar found a house and settled down. Cesar returned to the coast south of Santa Barbara and later told Ross that he had been so engrossed in the talk that he neglected to look for a home to rent. "Helen was hurt, but she took it pretty well," he told Ross. "'You know, Cesar,' she says, 'there's just something in you about working. You can't seem to stay still.'

"'But Helen,' I tell her, 'If you only knew what's happening to me. There's something tearing at me inside. All this excitement and expectation! I just want to get started right now.' A couple days later, he told Ross, he was so busy he even forgot to come home for dinner. "I just completely forgot! 'You might as well go then,' says Helen. 'Even if you stay here with us, you're not really going to be here.'"

After he finally found a home for his large family—always a difficult task because few would rent to a family with so many children—Cesar stuck to the regular CSO agenda in spite of what he'd heard from the old lemon picker. In his new office, he saw a parade of Mexican Americans who were seeking help, some with problems Cesar felt were more suitable for a lawyer or priest. One heartsick woman wanted to have her husband deported because he had run away with her cousin. "We have a long talk, and she tries to forgive and forget," Cesar remarked to Ross. That fall, he also typed reports for Ross detailing the usual voter-registration meetings over coffee and pastries. "Can't get over how these old-timers who aren't citizens realize the importance of voting much more than the younger citizens," he wrote in November. But the stream of letters Cesar produced showed that the nagging question of braceros wouldn't leave him alone.

In house meetings with angry field workers, he learned that whole crews of locals had been instantly displaced by braceros. Impatient to take action, Cesar knew that taking on the farmers would seem to pit Mexicans against braceros, but he became convinced the bracero program was evil. "There's an old *dicho, no puede dejar Dios por Dios*—you can't exchange one god for another. This was a question of justice, and I've never had any problem making a decision like that.

"The jobs belonged to local workers . . . ," he said. "Braceros didn't make any money, and they were exploited viciously, forced to work under conditions the local people wouldn't tolerate." Cesar and the new CSO recruits formed an employment committee, and gradually he began to learn just how far Ventura's growers would go to keep the braceros—and how corrupt local state officials would rush to help them.

CSO's voter-registration campaign in Oxnard barreled on that fall, tripling the turnout of voters in the barrio by the next election and boosting Cesar's stature as an effective leader. More locals were showing up at his door to try to get some immediate action out in the fields—like Al Rojas, a migrant worker from the San Joaquin Valley, who recalls meeting the thirty-two-year-old Chavez during a wet, cold day after harvesting vegetables.

Rojas had arrived in Oxnard about the same time as Cesar. He was lucky enough to find work, and at eighteen, was putting his strong back to use, trimming endless rows of lettuce plants with the tomahawk stroke of his short-handled hoe. He was good at lettuce, but in Oxnard, Rojas came across things he'd never imagined while growing up in the San Joaquin Valley. In Oxnard he learned to cut broccoli in the winter, even when it rained. "In the valley, we'd just go home. But here, the contractor told me: 'Welcome to the new world, man. Out here you work in the rain.'"

Rojas, who later became an aide to Chavez in the 1960s, would play an important role in attracting consumer support on the East Coast for the UFW's grape boycott.

But his first taste of organizing came when he found himself shin-deep in mud in Oxnard's soggy broccoli fields. The first time it rained, the contractor tossed Rojas a pair of boots that were riddled with holes and a raincoat so cracked he was quickly drenched. "And right next to us were the braceros, wearing brand-new yellow rain gear. I mean bright yellow!" Rojas and his crew protested, but were quickly warned to work with what they were given, or go home. Afterward, on the contractor's bus, the crew chose Rojas to lead their effort to get new gear. "This fat guy, Efren, who looked like Santa Claus, only dark, said 'Hey, I know a guy! He works with this CSO in the *colonia*. I think they call him Cesar, Cesario. Go see him.'"

Rojas went straight away to Chavez's small office, still dripping wet. "I looked at him and I thought, Hey, this guy really looks Indian. He's dark, too. And it was like we all thought at that time: I wonder if this guy is really that bright?" Rojas tried to shake off a common prejudice among Mexicans that light-skinned *paisanos* were smarter or more sophisticated. "I thought to myself, Well, Rojas, you're not all that bright, either, being where you are right now." He told Chavez about the rain gear. "It's very simple, Albert," Rojas remembers Cesar saying in response. "When you guys next get on that bus, you make them understand that they have to be ready to make a sacrifice and they have to make a commitment. Are you the leader? Someone in there must be the leader. . . . The guys must be looking to you, or else why are you here?"

That challenge sent Rojas back to the fields with a plan: The crew boarded the contractor's bus, rode to the crop, but then refused to get off or let the bus return until they got the new rain gear. The foreman, who had been enjoying a morning smoke inside the cabin of his truck, blew up at the audacity.

"What? You're not pulling that shit!" the foreman yelled at Rojas, who was standing in the pouring rain outside the truck. "You're all fired!"

"Bullshit," Rojas remembers answering back, half scaring himself with his boldness. "We are not leaving, the bus is not leaving, and we're not getting off that goddamn bus until you call in our demand."

The standoff continued until the farm boss came out, saw the tattered rain gear, and invited Rojas into his car. "Is this all you want? There's nothing else?" he asked Rojas, who responded with a self-assured, "That's all, just new gear, just like those guys over there, the braceros. They're Mexicans, just like us." The boss ordered the foreman out of his truck to listen to his order: "Call the office and have them call the war surplus and tell them you're going down today to pick up all the rain gear these guys need; a complete set for all of them. Got that?" With that, Rojas turned and delivered the good news to the crew. "That was my first taste of what I would call a little victory. That made me think, Hey, there's something to this."

SMALL VICTORIES like that of Al Rojas were much easier to come by than successes in taking on the whole farm-labor system, which was controlled by a cabal of growers and the state Farm Placement Service. This agency, along with the federal Labor Department, was supposed to supervise how braceros were deployed and make sure requests were based on legitimate shortages of local laborers. In reality, braceros were given preference. It took Cesar months to understand just how local residents were swindled out of jobs. A field worker named Mejia, with callused hands and blackened nails, provided him with a crucial piece to the puzzle.

After one house meeting, Mejia stayed late and told him how a contractor named Zamora would tell locals they had to travel eight miles away to the farm placement office and get referral cards. When they would finally make the trek back, Zamora would tell them they were too late and that he had to send braceros out to the fields—at the growers' request, of course.

At last Cesar saw his opening. When a furious lemon picker rushed into the CSO office not long afterward, complaining he and five others had just been fired and replaced with braceros, Cesar took advantage of the momentum. He convinced the men to accompany him to the farm placement office, where officials lied to Chavez and told him the men quit. A man named Turner then pulled him aside, whispering that they were bums and didn't want to work anyway. When Cesar persisted, Turner forced all the men to fill out long job-application forms, a tactic that kept them busy for hours. By the time the group drove back to Oxnard and to the office of the Ventura Farm Labor Association, the day was half gone—and out at the ranch, the labor contractor told them they were too late for the job. If an enterprising farmworker obtained a card for the next day, the association would refuse to honor it.

Cesar told Ross that the same maddening cycle continued for forty days straight. Each day, the locals would be forced to fill out new job applications, and each day, a bald association official they called *Pelón,* "bald head," would smirk and tell them they were too late. The battle had begun. In grueling fourteen-hour days, Chavez made time to study reports on the braceros and became fluent in the law governing their employment, Public Law 78. He managed to interest state employment officials and federal labor department investigators in the fraud, but was unsure when—if ever—they would make their move; he asked statewide political contacts to press Governor Pat Brown for action.

After Cesar organized pickets outside the bracero camps, local authorities reluctantly forced farmers to hire a handful of his recruits. Soon afterward, the federal labor investigators contacted by Chavez shocked growers by taking action, conducting surprise inspections and ordering farmers to hire local workers. The growers fought back, though, firing the workers after falsely accusing them of incompetence. With a persistence he learned from Ross, Cesar kept bringing men down to the farm placement office. He continued to quietly compile hundreds of copies of spurned—but notarized—referral cards, the evidence he would need to prove the fraud. One employee at the farm placement office even began to sympathize with the stubborn Chavez, and risked his job by secretly advising him to file some complaints.

Cesar stopped at nothing, and his best weapon was, perhaps, his ever-fertile imagination: In April, he persuaded a local barmaid who owed the CSO a favor to flirt with a group of braceros and find out what farm they would be dispatched to the next day and what their government-issued bracero numbers were. Then before dawn, he led a group of unemployed workers to the ranch where the braceros—already hard at work thinning tomato plants—good-naturedly handed their tools over to the locals. "Surely, it is your right," one told Chavez politely. Cesar then confronted the farmer with the names and numbers of the braceros, demanding that locals be hired to replace them.

The farmer called sheriff's deputies. But Chavez asked a protester to run to a phone and alert a sympathetic state employment official, who persuaded deputies not to arrest the group. The farm was briefly forced to hire locals, but promptly fired them when the spotlight faded. The locals were ready to give up: "He can keep his damn tomatoes!" "Yeah, and his ninety cents an hour, too!" "No more of this going back and forth. We're through!" But Cesar had another idea, something dramatic, something sure to attract the voracious appetite of the media.

"Get all the publicity you can," he ordered. "Press, radio, TV, the works . . . Oh yeah, and get ahold of a bunch of the housewives to bring some cars and fix something for us to eat. It's liable to be a long day."

The workers piled into cars with their wives and children, and a ten-car caravan converged on the tomato ranch, with occupants singing a Pancho Villa marching tune and carrying pictures of the Virgin of Guadalupe. The police arrived, and the press, and then alarmed state labor officials showed up, refusing to answer reporters' questions about the disturbance. One of the Packinghouse Workers union organizers tried to lead the unemployed in a charge onto the ranch, but Cesar stopped him. He noticed some of the men warming themselves around a bonfire they'd built on the edge of the ranch. Suddenly the idea for a much better and more symbolic protest dawned on him. He asked each man to pull out the wads of useless referral cards they had collected, and set them on fire over the bonfire. "As soon as it happens, everyone—even the growers—get real quiet," Chavez told Ross later, remembering the flames and the burning paper. "You see, Fred, they're doing what they marched out there to do—they're taking a crack at the enemy. Not only that, but each one, when he burns that card, is sort of committing himself never to go back for another one."

After more than a year of fighting the system and deluging the state with hundreds of written complaints—a feat that left the Ventura Farm Labor Association reeling—Cesar and the local workers won their campaign to be hired before braceros. The state took action, firing the regional director of the Farm Placement Service. Growers abandoned the referral cards and reverted back to hiring unemployed workers in an early morning "shape up" outside the CSO office. Cesar was proud that CSO even managed to drive the prevailing wages up, from sixty-five to ninety cents an hour.

"We could have built a union there, but the CSO wouldn't approve," Cesar later commented. He went along with their wishes, hoping that the Packinghouse Workers union would take advantage of the movement he had so painstakingly developed. But after Chavez was selected to move to Los Angeles in late 1959 to work at CSO headquarters, the braceros returned to the fields again and the farmworkers' fragile victory collapsed. Cesar was saddened and a little bitter, but he knew it wouldn't be the last time he would organize farmworkers. "This has been a wonderful experience in Oxnard for me," he wrote to Ross. "I never dreamed that so much hell could be raised."

In 1959, Chavez was promoted to executive director of the CSO. The organization, by 1962, had blossomed into twenty-two chapters throughout California and Arizona. It had registered tens of thousands of Chicano voters, and offered citizenship training to thousands more. For its time, the CSO was one of the few, if not the only, forums where Chicanos and Mexicans could break through racial barriers and become community leaders, fighting city hall on everything from potholes to police brutality.

It was no accident that in the CSO of the 1950s, Chavez would meet other well-trained Chicano rights organizers who would later help him lay the foundations for his farmworkers union. One of the most important was Dolores Huerta, an extroverted teacher and young mother.

Fred Ross had told Huerta that Cesar was a great organizer, one of just two paid staff people in the CSO at the time, but when Huerta first met Cesar at a CSO fundraising dance in Oakland in 1956, he seemed so extraordinarily shy and unassuming that afterward she couldn't remember what he looked like. "I mean, we saw him for about five seconds." She saw him again at other CSO conventions, where "he always just gave a very impressive report and then sat down and that was the end of Cesar. . . . He was very private." Still, both gravitated to projects involving farmworkers, a passion that would later draw Dolores deeply into the infant UFW.

Also known as "Lola," Dolores Huerta was born in 1930 in New Mexico, but grew up in the city of Stockton, a rough-edged Northern California river port. Huerta was raised there by her mother, a divorcée who was proud of her Southwestern Hispanic heritage. Alicia Chavez Fernandez ran her own businesses, including a boardinghouse for farm laborers. When her boarders were destitute, she would allow them to stay on for free or pay in onions and other field crops. Alicia made sure that Dolores was sensitive to the poor, and urged her to practice *servicio,* a religious ideal of modest generosity. "You were not supposed to talk about what you did because if you did, then that removed the grace for that good deed," recalls Dolores.

Nonetheless, Alicia, whose own parents had been union loyalists and political activists, was determined that Dolores would grow up a bold, confident woman. "I was raised with two brothers and a mother, so there was no sexism," Huerta recalls. "My mother was a strong woman and she didn't favor my brothers. There was no idea that men were superior. At home we all shared equally in the household tasks: I never had to cook for my brothers or do their clothes like in many traditional Mexican families." With a self-confidence rare in many girls of her background, as a teenager Dolores displayed a flair for organizing people for celebration or *servicio.* With women's groups, she helped put on the Mexican Independence Day fiesta and gathered food donations for the down-and-out. She also persuaded local businesses to donate a jukebox and Ping-Pong tables so she and other teenagers could have somewhere to go after school. The police unplugged the jukebox soon enough, however, Dolores says, "because they seemed uncomfortable that Chicano, black, white, and Filipino kids were mixing together and having fun."

Although she was popular in high school, Dolores remembers, she first began to notice the sting of discrimination. "The rich kids always got special treatment in our high school," she explained, years later. "So I think that's when I first started being aware of injustices that happen."

By the time Fred Ross rolled into Stockton in 1956, Huerta had married twice and started a family, and she had a reputation among clerics and others in the Chicano community as an especially dynamic woman who might have a future in politics. Ross quickly looked her up and told her about CSO. At first, Huerta wondered if he was a Communist. In recollection, she marveled that she went straight to the FBI "and had him checked out. I really did that. I used to work at the sheriff's department. See how middle class I am?" Ross's invitation to join the CSO, however, proved irresistible. "This was, of course, something I had been looking for all my life," Dolores recalls. "When Fred showed us pictures of people in Los Angeles that had come together, that had organized, that had fought the police and won, that had built health clinics, that had gotten people elected to office, I just felt like I had found a pot of gold! If organizing could make this happen, then this is definitely something that I want to be part of." Along with her mother and aunts, she became a key CSO organizer in Stockton, where she weathered blistering criticism from those who thought that a young mother shouldn't devote herself to organizing farmworkers—her greatest passion while in the civil rights group.

IN THE SPRING OF 1962, as CSO was preparing to hold its annual convention in the searingly hot desert town of Calexico, Cesar had already made one of the most important decisions of his life. He had talked about it with Helen and a few close friends, including Fred Ross, who backed him up as always. He had made up his mind to devote himself full-time to an improbable dream that was rooted in his migrant childhood and his most rewarding experiences as a political activist—the creation of an independent farmworkers union that would force growers to sit at a bargaining table face-to-face with the people who helped make them rich.

Now living in East Los Angeles, Cesar was still CSO's executive director. Three months before the March convention, he had asked CSO's board members if they would agree to a pilot project, an experiment, to organize farmworkers. CSO would endorse the project, but Cesar's salary would be paid by the workers themselves. Ross and Saul Alinsky thought the timing was perfect, and the CSO board agreed—as long as the full membership at the convention voted to approve the idea. But when the proposal came to the convention floor in Calexico, it was turned down. A majority of CSO members wanted to keep the focus on the growing population of Chicanos who lived in cities—those who had left field work behind them. Cesar rose and in a quiet voice told the assembly: "I have an announcement to make. I resign."

Pandemonium broke out. CSO president Tony Rios was flabbergasted and refused to accept Cesar's resignation. Others stood up and tried to argue with Cesar and talk him out of his decision. Ross and Huerta wept, but they knew Cesar wouldn't budge. He had already decided what he would do next. He and his wife had chosen Delano as their base to build the union. That night, Cesar, Ross, and Huerta crossed the border into Mexico and huddled together in a cantina to commiserate. "It took me six months to get over leaving CSO," Chavez would later admit. "I was heartbroken. CSO was my home. I wasn't angry when I left—if I had left angry, my anger would have sustained me—but I was homesick for a long time."

Helen knew leaving CSO would mean even more sacrifice. The family had about a thousand dollars in savings, but neither Helen nor Cesar had a firm idea—other than laboring in the fields themselves—of how they would support eight children, the oldest of whom was only thirteen. But the couple thought if they could survive anywhere it would be in Delano, Helen's hometown. Linda Chavez Rodriguez, who was twelve when her parents left CSO, says Delano was really her mother's choice. "He asked her what area would she like to go to—because he knew it was going to be hard on her. And she chose Delano because she had family there. She had two sisters and she had two brothers who lived nearby. And she figured that if we needed help she'd have her family nearby." Others say Cesar went to Delano because his brother Richard had moved there, and the two brothers had rarely been separated their whole lives.

Cesar's last day at CSO was on his thirty-fifth birthday. He and Helen took the kids to the coast again, to camp on the beach and reflect on the risks ahead. He knew that in Delano they'd always have shelter and food and family to turn to. He knew he had been well trained under Fred Ross. "What I didn't know," he would say later, "was that we would go through hell because it was all but an impossible task."

> *Anyone who comes in with the idea that farmworkers are free of sin and that the growers are all bastards either has never dealt with the situation or is an idealist of the first order. Things don't work that way.*
> —CESAR CHAVEZ

HUELGA

ERM 808

CHAPTER THREE

DELANO

CESAR CHAVEZ and his family drove into Delano in early April of 1962, just as the sun was beginning to warm the great flat plain of the San Joaquin Valley, ripening its crops and turning its fields of grass into pungent golden pastures. The weeks ahead would be a lonely time for Chavez; he deeply missed working with his friends in the CSO. He was also worried that he had no sure means of supporting his family. "For the first time, I was frightened—I was very frightened," he said later.

The family rented the cheapest house they could find, at 1223 Kensington Street near downtown Delano, for fifty dollars a month. Within days, Cesar was back in the simple work clothes he preferred over suits and ties, and he was out quietly talking to local people about a new project: the National Farm Workers Association, a grassroots group that would build strength slowly, almost one worker at a time. Rather than a traditional union, he envisioned a social movement—a crusade he later called *"el movimiento"*—that would inspire farmworkers, the poorest of America's laborers, to organize themselves and change their lives.

Cesar and Helen, along with Dolores Huerta, who had agreed to help the Chavezes, made a calculated decision to avoid calling their association a union at first. They knew the word provoked one single image in the minds of growers and workers: a strike. And strikes, while sometimes leading to quick pay increases and a fleeting sense of power, often led to firings, blacklistings, and violence. Another group that had recently started organizing farm laborers was the AFL-CIO-sponsored Agricultural Workers Organizing Committee, which had led dozens of strikes in the Imperial Valley and in other areas outside Delano. Sometimes striking workers were given a quick boost in wages, but none of

■ *Cesar and Helen Chavez in the 1960s with six of their eight children. They moved to Delano in 1962.*

the strikes resulted in contracts or better conditions or tipped the balance of power with the growers. AWOC's mostly white leadership had little success organizing Mexican workers, and Chavez believed he knew a better way.

Chavez knew farmworkers were hungry for change—he came from the fields—but he also knew that even some of the toughest were afraid or were discouraged by past failures. One old Chicano farmworker, remembering strikes that ended in spilled blood, warned Chavez: "They come and they go, good organizers and would-be organizers. But one thing they all have in common is that all of them have failed and all will fail."

In 1962, the San Joaquin workforce was a mix of Latinos, aging dust-bowl Okies, Filipino immigrants who had arrived as far back as the 1920s, and American blacks. Mexicans and Chicanos were already beginning to predominate in the fields, however, and growers were still bringing in braceros when authorized to do so. Cesar tried to bridge divisions that kept ethnic groups apart and competing with each other for jobs, but it wasn't easy. Even the Mexicans were a tough sell, although they were more likely to listen.

Chavez thought if his association were to succeed, it would need many years to build a solid membership throughout the valley before it could take on California's growers, whose generous campaign contributions had bought the support of legislators from Delano to Washington, D.C. This new movement would have to be stealthy, even clandestine. It would meet in homes, not in union halls, and be a "marvelous secret," as Fred Ross put it. One of the workers' advantages was that they communicated in a code the growers didn't understand—Spanish.

"Everyone but everyone wants to know if this means 'strikes' and I have been saying—no strikes, unless we know we'll win," Chavez wrote Ross on May 10. The workers' most bitter complaints were about the labor contractors, who worked closely with the growers. In the first house meetings, over beer or *pan dulce*—Mexican "sweet bread"—they freely vented their loathing for the contractors, who bused them to fields for a price, took a slice of their wages, and often charged them for water to drink. Some contractors even demanded bribes or sex from women looking for work. The early meetings gave Chavez a boost because the workers seemed so bold and clearly identified what their problems were. In a postscript to one of his first letters to Ross, Chavez added that he was "a heck of a lot more confident now than a week ago."

There were other basic reasons why farmworkers would respond to overtures for change. The state of California had already documented the inadequacy of farm wages. In a 1960 survey of one hundred field laborers' households in Fresno County, a quarter of the families had no place to refrigerate food. Another 25 percent had to use privies instead of flush toilets, and fewer than half had a water tap in their homes. Many lacked access to the preventative medicine that most Americans took for granted: More than half of the farmworkers' children under eighteen had not been immunized against polio.

To reach people in homes as dismal as these, Chavez drew a map of the valley, and pinpointed eighty-six towns and farm labor camps between Stockton, to the north, and Arvin, to the south. The plan was to distribute tens of thousands of mimeographed registration cards that asked for a name, address, and answers to a few simple questions: How much do you think you should be paid? What do you think about having no guarantee to Social Security, unemployment, or minimum wage? Once the cards rolled in, Cesar and other volunteers would quickly fan out to form committees in as many towns as possible.

Julio Hernandez, a Mexican field worker married to a Chicana from Yuma, remembers doing his best to avoid Cesar. Their encounter was almost an exact replica of Ross's first attempt to meet a reluctant, suspicious Chavez, but Cesar, too, had learned the art of persistence. "[Cesar] came to the house in Corcoran, and he was talking to my wife and other people who had come to the house that night. I was out playing pool," recalls Hernandez, then a cotton-field supervisor and a CSO volunteer.

"I was a pretty good pool player, almost a professional, and I had played to support myself in Mexicali," Hernandez, now in his seventies, recalls. "I was told that

Chavez was at my house waiting for me to talk about a union, but I was not interested because we [unionists] had been blackballed earlier. We had worked with an organizer in 1951, and he just up and left us, stranding us in the fields as we were getting ready for a strike." Soon employers would no longer hire Hernandez, and his family had to move to Oregon in search of jobs. When he returned to California six years later, "I swore I wouldn't get involved again with a union."

So the day Cesar came to visit, Hernandez says, "I stayed out until two A.M. playing pool. I thought he'd be gone by then. . . . When I came home he was still there." Cesar kept Hernandez up until five in the morning, chipping away patiently at his stony resistance. Finally, Hernandez relented after his wife called him a fool. "He seems like good people," she said of Cesar, much as Helen Chavez had characterized Ross. "We found out that the harder a guy is to convince, the better leader or member he becomes," Cesar said, looking back on that summer.

"When we began, we were nothing," muses Hernandez, as if he was still surprised, even in memory. "But we got started by putting out leaflets, all day long. We'd begin before the roosters got up, around three in the morning. Then we'd go to work, and then we'd get back to passing out information, until well after dark."

Angie Hernandez Herrera, Julio Hernandez's daughter, was another early recruit. At sixteen, she was already bitter about farmwork, which she'd been doing since she was old enough to pick up a basket. "Always hated it. Always got blisters on my hands from the hoe," she remembers. "We'd hoe cotton, and clean it too. Pick grapes, melons. All for fifty cents an hour. I remember when my dad made it up to ninety cents, we were all happy because my brothers were working, too, and that was a big raise." Like the rest of her family, Angie was excited when her father later became one of the new farmworkers association's first vice presidents. Hernandez became the first "full-time convert," Cesar later said proudly. He pulled in more than three hundred members, more than any other recruiter in the valley.

The California Migrant Ministry, an interdenominational church group that had a history of charity work among laborers, was one of the association's early supporters; the ministry loaned Cesar a badly needed mimeograph machine, which was set up in Richard's garage. Chris Hartmire, the ministry's Los Angeles–based director, had met Cesar back in the CSO days, and he knew what Fred Ross's star organizer was up to. Most of the Migrant Ministry volunteers were convinced union organizing was the only way to really support farmworkers; Hartmire himself was shepherding as many as fifty college students who had volunteered to work with farmworkers in California. One of the ministers, Jim Drake, was eager to help, but had his doubts. How could Cesar support his family and build an independent union from virtually nothing, Drake wondered. One look at all those kids and the family's battered 1953 Mercury station wagon that devoured oil and burned gas like a tank, and Drake had to think the man was crazy.

But the volunteers were naturally attracted to Cesar, whose approach and conviction appealed to the ministry activists. Chavez accompanied the pastors and their aides to weekend retreats, where they prayed and talked about their work. "He was organizing us," Hartmire says. "He didn't come to our retreats just because he loved us so much. He came really because he knew that the day was going to come when he would need outside support. So we were the closest group other than his own family."

■ *Cesar Chavez and Gilbert Padilla (to his right) at a house meeting with farmworkers in Fresno.*

Still working for the CSO in Stockton, Dolores Huerta carved out time on weekends and evenings to pack her children into her own battered car and drive from camp to camp, distributing the farmworker survey cards in the northern valley counties: Stanislaus, Merced, and San Joaquin. Gilbert Padilla, who was called *Flaco,* "skinny," by his friends, was also a CSO staffer in Stockton. He had grown up working California's fields, and like Huerta, he jumped on board immediately. He took jobs picking cherries and peeling peaches so he could pass out cards on the sly and quietly talk up the idea of an association. The three of them carried each other through the early days of the union.

From his base in Delano, Cesar logged hundreds of miles, traveling the length and breadth of the valley, leaving stacks of questionnaires in grocery stores, visiting homes in every farmworker community he could. His obsession led to grueling sixteen-hour workdays. "I left Stockton at 2:30 A.M. and arrived in Delano at 6:30 A.M.," he wrote on May 22. "Saw a small crew on the road in Tipton and stopped off long enough to register them—and drop off more cards." Sometimes he'd squeeze his eight children into the station wagon and take them on his long rides. His daughter Linda remembers that Cesar made such days festive, often treating them to ice cream after the long hot rides.

"We are now really in the swing of things," Chavez wrote enthusiastically to Ross on May 16. "I have fifteen young bucks on the drive, including two Negroes and one Filipino. Wish I could tell how many have been registered but don't know myself. . . . The leaflet committee, all my kids plus my nephews and nieces, covered part of Delano last night and immediately we got some reaction from the workers."

In spite of his exhausting schedule, Cesar was beginning to enjoy himself. "The local office-supply store is running in circles trying to keep me supplied with white cards," he wrote to Ross. "I know the guy [in the store] wants very badly to find out what in the hell I'm doing but I'm just letting him suffer day in and day out. He won't ask me directly and I can't take any hints, either." He was also outgrowing the security of a regular salary. "When I missed the fourth paycheck and things were still going, the moon was still there and the sky and everything . . . I began to laugh, you know. And I began to feel free."

Despite his enthusiasm, Cesar confided to Ross that the campaign was more difficult than he had ever imagined. Farmworkers and other contacts tended to drop out of the cause as quickly as they had agreed to help out. Cesar would sometimes drive for hours to get to meetings he'd arranged, only to find an empty house. Initial optimism often gave way to skepticism and was blunted by the consuming task of making ends meet. "They are working every day and trying to pay off the winter debts, so that they can go right back and pile up more debts come this winter," Cesar wrote Ross, during a moment of acute discouragement.

On his long drives, Chavez had time to think about how the threads of his network were starting to come together—yet how easily it could all fall apart. "Anyone who comes in with the idea that farmworkers are free of sin and that the growers are all bastards," Chavez said of that time, "either has never dealt with the situation or is an idealist of the first order. Things don't work that way." Early in his campaign that summer, Cesar discovered he would face more obstacles than apathy and the growers' might.

CHAVEZ WANTED to build his union from the fields themselves, rather than listen to what some remote labor chief thought was best. He was liberal in his criticism, for example, of the Agricultural Workers Organizing Committee's methods, so much so that he had earlier turned down a job the group had offered him. Cesar's attitude ruffled some of AWOC's leaders, who were beginning to see him as a competitor. Ironically, AWOC had been reluctantly formed as a response to pressure from Dolores Huerta and a Stockton priest, Thomas McCullough, who had complained to the AFL-CIO that the American labor movement had abandoned farmworkers. McCullough had even traveled east and tried to shame AFL-CIO directors into doing something about the terrible wages and conditions in California's fields. When his appeal failed, the priest started his own organization, sponsored by the Catholic Church. Dolores Huerta had immediately immersed herself in the group, the Agricultural Workers Association, even though McCullough told her disapprovingly that rough-and-tumble "farm-labor organizing was no place for a woman."

After only a year, an embarrassed AFL-CIO finally told McCullough it would invest seed money and commit a handful of professional organizers to farm labor. And in 1959, AWOC was born. Huerta joined up and was eventually elected secretary-treasurer. She recruited Larry Itliong, a Filipino who'd immigrated to California as a boy, to work as a paid organizer. Itliong, a tough veteran of the fields who had lost fingers as a railroad worker, proved an admirable advocate for the many Filipino crews in the state.

While AWOC won pay raises for some farmworkers, Huerta recalls it had particular trouble organizing Mexican and Chicano workers, who formed the majority of the state's workforce. AWOC's top organizers, who were mostly white and had never done farmwork, were arrogant and used strategies that didn't work, Huerta says. One of AWOC's biggest mistakes was to fraternize with labor contractors. Instead of working directly with the Mexican workers, AWOC organizers used Spanish-speaking contractors as middlemen, asking them to collect union dues from workers. The workers saw it as merely another form of extortion: They received little in return, but the payoff for contractors was that AWOC wouldn't interfere with their thriving enterprises.

Huerta got angry when she went into AWOC's office and found union organizers chatting amiably with contractors. In 1959, she tried vainly to persuade AWOC to invite Cesar up from Oxnard to help train the organizers because he had been so successful in Ventura County. Huerta now believes AWOC leaders were "very threatened and jealous." Within a year, she quit the organization, although she remained on good terms with Itliong after she teamed up with Chavez.

By late August, Chavez had set up an office in his brother Richard's garage, where it was so hot the ink ran and the mimeograph machine often overheated as he and Helen churned out association flyers. Farmworkers from all over the valley began to hear Chavez was a man who knew how to get things done: They'd show up at the garage, asking his help when they were cheated out of wages or had an accident on the job; when an elderly parent died and there was no money for a burial; or even when their car broke down or they needed someone to translate at a child's school.

AWOC's suspicions of Cesar still ran strong, however. As Chavez's association gained strength, AWOC complained it was undermining its efforts and demanded a meeting with Cesar. In case the showdown got out of hand, Cesar recalled, "I went down and got my own boys and told them what to do in case [AWOC] tried to act smart. We had them outnumbered about five to one," Chavez wrote to Ross. "I told the people there that the drive was to get information directly from the workers and find out how they feel on some of those questions that outsiders have been deciding for the workers for too damn long now." The meeting ended in a draw, and the relationship between Chavez and AWOC remained frosty for three more years. AWOC director Al Green added to the chill by regularly referring to Chavez as "that Mexican."

Looming behind the Chavez-AWOC conflict was another potential competitor: Jimmy Hoffa's Teamsters union. When San Francisco Teamster official John Goldberger demanded a meeting, Cesar was sure that Goldberger was "fishing around" to figure out if Chavez really thought farmworkers could be organized. "I told this guy Goldberger that we didn't have money," Cesar wrote to Ross, describing the meeting that included Huerta, Padilla, and another volunteer organizer, Antonio Orendain. "But we have the other ingredient, which is more important than money, and that is time."

S IX MONTHS after he arrived in Delano, Chavez believed that it was time to cement his gains with a founding convention, a meeting that would bring together as many association members as possible from around the valley. He wanted to show the workers this organization was real, and that they, Spanish-speaking farmworkers, would have a voice in deciding its goals. He went in search of a hall in Fresno to hold the convention at the end of September. He immediately encountered resistance. Priests at St. Alphonsus Church were reluctant to help because they'd already been rebuked by the mostly Italian-American parishioners for allowing AWOC to hold a meeting in the church. Cesar turned to his reluctant cousin, Manuel, to secure a place for the big meeting. Manuel, who had left farmwork to sell cars in San Diego, had come to Delano grudgingly, promising only to stay a couple of months. Cesar sent him to Fresno with no money, but the fast-talking Manuel convinced the owner of an old theater that the union president would pay up after he flew into town for the gathering. When the NFWA members finally arrived, Manuel quickly passed the hat and collected enough money to make good on his promise.

Cesar, meanwhile, broke out the official stationery—paid for by his relatives—and a convention with about two hundred farmworkers and their families was set for Sunday, September 30, in Fresno. There was also a grand surprise, one that Manuel and Richard Chavez, Cesar's brother, had been diligently working on for some time: a union emblem.

Like Manuel, Richard was a reluctant volunteer to the farmworker group. He had become a journeyman carpenter and had little time for social work. Yet there he was, talked into helping design the symbol for his brother's group.

Cesar said "maybe an eagle of some type, that shows strength, that shows power," Richard remembers. He drew a version of the eagle that appears on the Mexican flag, substituting a cluster of grapes for the snake and cactus. Cesar liked it, but cautioned that people wouldn't be able to reproduce it. Richard and Andy Zermeño, a Los Angeles graphic artist and CSO friend who had worked on Gallo Winery advertisements, then came up with another design—one with straight lines and corners.

"So I come up with the idea of the thunderbird," says Richard. "The idea of straight lines and straight corners and all that was fine, but what [Cesar] didn't like was that I had two little feet on the thunderbird—they looked like chicken feet. It just didn't look good." The inelegant bird feet were replaced by a straight bar across the bottom. Cesar came up with the colors: white for hope, black for the plight of the workers, and red for the sacrifice that would be required of them. (Zermeño points out that the eagle, turned upside down, forms an Aztec pyramid.)

The farmworkers gathered at the theater, where Chavez and Huerta took to the stage. Chavez read a proposed plan of action: The new National Farm Workers Association would lobby the governor's

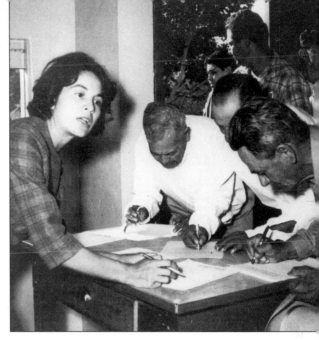

■ *Dolores Huerta signing up National Farm Workers Association members during the founding convention in Fresno in 1962.*

office in Sacramento for a $1.50 an hour minimum wage for farmworkers and the right to unemployment insurance. The association would still avoid the term "union," but it would advocate for the right to collective bargaining, something bound to raise the ire of growers. Equally ambitious was Chavez's plan to establish a union-run credit union for members, and a hiring hall.

The members embraced the ideas with an affirmative vote, but then the convention's real debate began. It was time to unveil the association's banner: a square-edged black eagle in a white circle against a red background. Manuel pulled a cord, tearing aside a sheet of paper that had been covering the flag—and gasps rippled through the crowd. Some people immediately left in protest because they thought the flag looked "Communist." Some thought the eagle design was inspired, half in jest, by the label on Gallo's cheapest wine, Thunderbird. To Richard's astonishment, still others thought the colors echoed that of the Nazi flag of the Third Reich.

But Manuel improvised a vision that calmed the gathering: "When that damn eagle flies, we'll have a union." The enthusiastic delegates voted to adopt the flag and *"Viva La Causa!"* was coined as its official motto. They elected Chavez president, and Dolores Huerta, Gil Padilla, and Julio Hernandez as vice presidents. And it was also decided to collect $3.50 a month in dues, money that would help finance a life insurance plan. In honor of Chavez and his mission, a loyal recruit from Madera named

Rosa Gloria had written a *corrido,* or Mexican "ballad," and delegates cheered as a recording of it rang out through the hall.

Despite the success of the convention, the nascent farmworker group had next to nothing in its treasury. Shortly afterward, in the fall of 1962, a new leaflet began to circulate through the valley, lightheartedly asking for donations. "Who finances the organization of the NFWA? Nobody! Nobody that is, except two or three of Cesar's buddies who send him a buck occasionally, and Cesar and his wife, Helena, who are now working in the fields from 6:00 A.M. to 2:00 P.M. picking cotton to support themselves and their eight kids. The rest of the day and half the night Cesar devotes to organizing."

Gil and Dolores, meanwhile, found a lawyer in Stockton who agreed to write articles of incorporation and hold off on collecting his fee. "I told him this might be many, many months," Dolores wrote to Cesar just days after the Fresno convention, adding with excitement: "Give me the word on the next line of tactic. The troops are restless."

Cesar cooped himself up in his garage in October to draft the NFWA's first constitution, which he'd promised would be put up for a vote at the association's next meeting. The document's most important passages called for the formation of "ranch committees" at each company, where farmworkers would represent the union. These committees would report to the association's directors, who would train workers to negotiate with managers and, eventually, manage contracts. Chavez told Ross that for his unconventional constitution, he was considering a passage from Pope Leo XIII: "Rich men and masters should remember this . . . To defraud anyone of wages that are his due is a crime which cries to the avenging anger of heaven."

In that same note to Ross, he added proudly that he had "put the bite" on a Catholic bishop's committee for a one hundred–dollar donation. By mid-December, Cesar, Dolores, and Manuel were in Sacramento lobbying for the association program, talking to legislators Huerta had gotten to know through CSO, and arguing bitterly with the director of the Department of Employment over the ongoing bracero program.

In the meantime, setting up the crucial life insurance plan for farmworkers was like "chasing the rat," according to Cesar, who was juggling skeptical insurance agents and unaffordable premiums. Getting the credit union started proved simpler, thanks to Richard Chavez. Richard had built himself a small one-bedroom home in Delano, and Cesar discovered that the house could be used to finance the credit union. "One day he came up to me and says, 'Richard, how much do you think your house is worth?' I says, 'I don't know, Cesar, probably twenty-five hundred dollars, three thousand dollars at the most.' Cesar said, 'Why don't we go to the bank and find out?'"

The Bank of America determined the house was worth $3,700, and Cesar persuaded Richard to let him borrow the money, using his house as collateral. "Before you knew it, my house was in hock," Richard recalls. "We somehow publicized that we had money to lend. The next day there were about fifty farmworkers outside the door—before we opened the door—who wanted money."

"I didn't even know what the word 'collateral' meant," Richard admits. "We just looked at them, and Cesar would look and say, 'Yeah, he's OK.' And so we'd lend him the money. Later on those loans became known as 'face loans,' and I would make fun of it." Many of the first loans were hard to collect, Richard says today, laughing at the memory. The union lost money on half of their first loans, but it provided a bedrock service to farmworkers who had been spurned by banks, and it helped establish the union's reputation and win more converts. Helen Chavez took control of the credit union, and it flourished, loaning out more than $5.5 million to farmworker families by the mid-1980s.

F OR THE NEXT two years, members dropped in and out of the new association, but it struggled along. Delano's clannish grower community was starting to realize that something was afoot. They weren't the only ones. Around this time, one of Cesar's daughters rushed into the house, crying, after she'd gotten into a fight with a boy up the street. "Your mother and your father are running an illegal house," the little boy had told Cesar's daughter. "She didn't know quite what that was," Cesar said. "She was very small—but she knew it was bad."

Not surprisingly, the Chavez family was barely scraping by financially. Although Cesar was able to get some unemployment insurance money owed him from his CSO work, that didn't last long. That first summer, Helen picked grapes and cotton, and the Chavez's eldest son, Fernando, whose family nickname is Polly, earned a dollar an hour doing his first field work. Chavez recalled Fernando's pride in a letter to Ross in July of 1962: "He came home and gave Mama all of the money and told her how it should be spent. He then waited up for me until I came in and wanted to know how old I was when I earned my first $36 in one week."

Cesar learned to accept food from laborers only after relentless pressure from his cousin Manuel. What he found, Chavez said, was that "people who give you their food give you their hearts." But he often turned down money, albeit reluctantly. "Met this fairly young Negro late last night," he told Ross in a letter, describing a labor camp he visited when he first arrived in the valley. "He was in bed, but we got him up. After some discussion he reached for his wallet and offered to help. He pulled out a five skin. Wouldn't take it, but it represented almost 2,000 [questionnaire] cards. . . . I think I impressed him more by not taking the money."

Chavez had fewer qualms about asking his former CSO colleagues for support. "Both Helen and I can't quite accept the idea of begging, but it looks as if we will have to," Cesar told Ross, wondering if he would put in a word for him in various quarters to seek donations. Checks for ten and twenty dollars began to trickle in from CSO acquaintances and Migrant Ministry supporters, but the money was quickly eaten up in payments for gas and other supplies the association needed. After the September convention, when dues started arriving, Cesar hoped to go on a $75-a-week salary. But by the summer of 1963, he was once again heading to the fields to make ends meet: chopping cotton, digging ditches, and picking peas. "Found out that I can still pull the vines as before," he remarked ruefully.

During one particularly lean week, Helen was overjoyed to discover that she had won $100 in a Safeway grocery store contest. "Oh, boy, what a lot of food for the kids!" she thought triumphantly. Cesar reminded her that it was urgent they pay the $180 they owed on a credit card bill for gas he had used to scour the valley for members. "I don't know if she cried," Cesar said of Helen, "but I think she did."

ALL THE VOLUNTEERS who started the union in 1962 made tremendous financial sacrifices, perhaps none more so than Dolores Huerta. The mother of seven was going through a painful separation from her second husband, Ventura Huerta, an ex-farmworker who shared her passions but disagreed with her on child rearing and the role of a wife. Huerta had patched together child care with a network of babysitters and relatives so she could work as a teacher and an organizer for CSO. As a part-time CSO lobbyist in Sacramento from 1960 to 1962, she excelled at her job, speaking out boldly before state legislative committees debating labor laws. Her political skills helped push through changes that, for the first time, extended retirement benefits to legal immigrant workers who'd lived in the United States most of their lives. Huerta was also the only woman to appear before a 1960 California senate committee hearing on labor and welfare, where she decried the unfair treatment of farmworkers.

Even while working as a CSO lobbyist, Huerta was already hard at work forging political contacts in Sacramento and Washington, D.C., for the Farm Workers Association. "Being a now (ahem) experienced lobbyist, I am able to speak on a man-to-man basis with other lobbyists," she joked to Cesar in one of her letters. "We have a hell of a task in front of us, but I do not think the task is impossible. . . . But if you think for one minute that the FWA will be kept a dark secret any longer you can forget about that. You know once the word gets around up here where all the representatives of organizations are, there will be not a ghost of a chance of keeping our infant under its protective garments."

Huerta was ready to take on male politicians, but some male farmworkers were not ready for her. There were times, she recalls, when she would show up to conduct association meetings, and stunned men wouldn't talk to her. Even some friends thought she was foolish when she began to prepare for her move to Delano. "I had been work-ing as a teacher, and people thought I was a little loony because I was also going through a divorce and I had seven children and was going to quit my job teaching to come organize the union."

Still in Stockton, she scraped by on some unemployment insurance and child-support checks from her former husbands. She was willing to sacrifice but was also vocal about her concerns: "I am not the quiet long-suffering type," she told Cesar in another letter. She was frank about her financial woes, reminding Chavez that she needed gas money—and might need to borrow his car—if she was going to continue making her rounds.

"If I can make it through August, and I know the Good Lord will not let us starve, then in September I can apply for my substitute teacher credential," she wrote in one letter, making sure, as always, to ask about the welfare of Helen and Cesar's eight children. "By the way, you guys should have another baby. They are so sweet and make you feel like you are close to the angels. Until next time, *Viva la Causa . . .*"

In November 1963, she wrote Cesar that California congressman Phil Burton, whom Huerta had befriended, was so impressed with her that he tried to get her a job working for the state. She wasn't interested, however. In 1964, Huerta made the move to Delano, where she and her kids would subsist on about thirty dollars a week, child-support checks, and the government surplus food that union volunteers would come across.

"When I made the decision I was going to do it, somebody left a big box of gro-ceries on my front porch in Stockton, and I thought that was just like a sign to me," Huerta remembers. "I knew what it was like to send my kids to school with shoes that had holes. So it was a very, very hard time for us. But again, this is what farmworker families go through every day of their lives."

SEVERAL YEARS EARLIER, Chavez had met Gilbert Padilla, by now one of his closest partners, when Cesar showed up for a house meeting in Hanford, the valley town where Gil was living. At the meeting, Cesar and the residents discussed how CSO might help the thousands of workers who were threatened with displace-ment by mechanical cotton harvesters. Padilla wasn't interested in charity or tempo-rary fixes, so when Cesar talked about getting Chicanos elected to office and their right to the minimum wage and unemployment insurance, he was excited. He joined CSO, and by 1961, was the group's third paid staffer. He also began to roam the val-ley as an organizer for the NFWA in conjunction with the Migrant Ministry. "In the eyes of the public it was slow," Padilla says of the birth of the union, "but in our eyes it was a very fast, involved thing."

Gilbert Padilla, who was born and raised in California migrant camps, also joined Dolores Huerta and Cesar Chavez to organize farmworkers.

Like Cesar, Gilbert Padilla was a World War II veteran who was fed up with discrimination and the powerlessness of farmworkers. He was born in a labor camp and raised out in the fields, and his mother, a supporter of the Mexican Revolution, taught him it was not right to stand by and do nothing about injustice. Still bristling with indignation fifty years later, Padilla remembers when he and his five brothers were discharged from the army and returned to a farm to get their jobs back. The foreman, he says, "wanted to pay us less than he was paying the braceros, who we trained to work in the fields, doing the tractor work and all that. He wanted to pay us less!"

Furious, the brothers refused, and went off to look for other work. A friend offered to get them a job thinning cotton plants for eighty-five cents an hour, ten cents more than the other workers were getting. It was a particularly blistering day, Padilla recalls, and he and his brothers dared to stop for a cigarette break. The foreman, incensed when he saw them, started shouting, so the men threw down their hoes and walked five miles back to town. "That was my first strike," Padilla says, smiling. "I was about nineteen or twenty years old."

Disgusted with the abuse of both braceros and Chicanos in the fields, Padilla found a job in a dry-cleaning business for a few years. His first venture into community organizing was in the town of Los Banos, on the parched western edge of the San Joaquin Valley, where he started a civil rights group called Club Mexico. In 1955, he returned to the fields, where he found more of the same: oppressive work for bosses he considered less than honest. One foreman who was pocketing Social Security deductions from the checks of laborers tried to offer Padilla a twenty-dollar bribe to keep him quiet.

In early 1965, Padilla was at the center of a pivotal event that made a name for the National Farm Workers Association and helped set the stage for the historic grape strike that started later that year. He and Jim Drake of the Migrant Ministry challenged the state of California for profiting from substandard farm labor camps in Tulare County, where workers were jammed into shacks slapped together during the depression. The state was planning to raise the rent on the shacks, flimsy structures that were never meant to be permanent and had never been connected to running water. Padilla and Drake organized protest marches and a summer-long rent strike that stunned officials and forced renovations of the camps. The black eagle banner flew for the first time that summer, and the action's success won new hearts and minds, including that of an accidental convert, a young Anglo student from the University of California at Berkeley.

Doug Adair, who happened to be in the valley that same summer, had been a Young Republican for the Free Speech movement at Berkeley in 1964. Except for his membership in the Republican Party, he was typical of the hundreds of students who would soon flock from urban colleges to the fight looming in the fields. A scholarly youth with a dry wit, Adair had grown up in the Southern California college town of Claremont, and was enrolled in a graduate program in history at UC Berkeley. One day, during spring semester, he was out on Sproul Plaza, the heart of the school's Free Speech movement, passing out literature on behalf of a Republican candidate for office. (The movement fought for the right to free expression for all students, regardless of political affiliation.)

At a literature table nearby, a thin young student named Marion Moses grew indignant. Didn't Adair know that his candidate supported terrible things, like bringing in Mexican braceros to be exploited out in the fields? Moses, who was involved with a university farmworker support group, told Adair he should go see for himself how bad conditions were in the valley. The more Adair thought about the idea, the less far-fetched it seemed. He loved the outdoors, and he was a talented gardener. Besides, other students had gone to pick fruit during the summer. So when school let out, he joined a group that traveled to the San Joaquin Valley, to see what farm life was really like.

For the young Republican, it was an awakening. Adair found a room at a dilapidated labor camp, where he bunked with Okie and Mexican workers and picked peaches and plums. He enjoyed their company immensely and respected their hard work; and he was angry when he discovered that contractors were deducting workers' money for Social Security and pocketing it. When he tried to stop the fraud, he found out that state officials were not legally required to investigate. It was up to the workers to confront the thieves, and most were too afraid to do that.

Adair's coworkers complained bitterly that there were no bathrooms in the fields; that a cup of ice water cost a nickel, a soda, a quarter or even fifty cents—far higher than the usual store price. Some families with five or six children were crammed into tiny huts with tin roofs. "No glass windows, and no screens," he says. "And oh, they were like ovens in the summer."

In July, some of Adair's friends from the camp reported to their seasonal jobs at a fruit dehydrating company, where they learned they would be paid five cents less an hour than the previous year. If they didn't like it, they could leave. Furious, the workers struck. Adair was asked to talk with the police to find out where the strikers could picket. Afterward, in the company of some strikers and David Havens of the Migrant Ministry, Adair followed a strikebreaker home to try to persuade him to change sides. "His dad came to the door, opened the door and took the glasses off my nose and—*pow!*—and just flattened me right out," Adair recalls.

"I'm sorry you hit me. But gosh, we're not here to threaten you or anything," Adair remembers saying, getting on his feet. They started talking, and eventually the father invited Adair into the house for coffee. He told Adair he'd been a union man before, but he'd been laid off, and his son needed money for school clothes. "It wasn't this vicious scab trying to steal someone else's wages. In every strike you have very human stories and very poor people that need extra work."

The strike fizzled anyway, as most did, and the workers dispersed bitterly, but Adair knew things were heating up. During the rent strike, he met Gil Padilla, who, in turn, took him over to Delano to meet Huerta and Chavez. Noticing his truck, the pair asked him if they could use it to deliver their new newspaper, *El Malcriado, The Voice of the Farm Worker.* Adair agreed, and neither he nor his truck ever went back to Berkeley. He accepted an invitation from Bill Esher, *El Malcriado*'s editor, to stay in Delano and work on the newspaper's English-language edition. Instead of studying history, Adair was living it.

THE UNDERGROUND farmworker newspaper made its debut in December of 1964. Originally published only in Spanish, *El Malcriado* was Cesar's labor of love. Its witty name means "ill-bred," or "children who speak back to their parents," as Chavez told Fred Ross in a letter. Its editorials called for living wages, lambasted indignities on the job, and took on growers for complaining about losing the Mexican braceros. (The program was about to be terminated by the federal government after years of intense pressure by churches, unions, and farm labor activist Ernesto Galarza, as well as bad publicity following several fatal labor-bus accidents involving braceros.) Cesar explained to Ross that the paper's name also had other resonance for Chicanos: "During the [Mexican] *Revolución* one of the people's papers was called *El Malcriado.* . . . The name is really the best we could find for the paper. It means many more things for the people."

Bill Esher, who belonged to the Catholic Workers Organization, came down from the San Francisco Bay Area to Delano to help Chavez paste up and print the newspaper. What made *El Malcriado* so appealing to farmworkers was its humor and irreverence. One of its first covers sported cartoons of a hapless farmworker, Don Sotaco, who delighted workers because they could see so much of themselves in him. "The cartoons were very important because a lot of the farmworkers couldn't read," recalls Andy Zermeño, the artist who drew Don Sotaco. "Cesar and I talked and decided we had to establish a way to use graphics to communicate with them about their rights. So he came up with Don Sotaco, a farmworker who didn't know anything. We wanted [farmworkers] to identify with this character and show that if you didn't know your rights, you would get into a lot of trouble." In one edition, Don Sotaco and his wife, Remedios,

visit the University of California at Davis, the state's top farm-research institute, where researchers were studying the productivity rates of different ethnic groups in the fields and devising genetically altered plants that could be picked by machines. Sotaco and Remedios are shown farmworkers of the future: one with very short legs to pick tomatoes, one with very long legs to harvest dates from palm trees.

Farmworkers scooped up *El Malcriado* as fast as Chavez and Esher could put it out and drop it off every two weeks at barrio grocery stores. Chavez and other association leaders also sold copies of the ten-cent newspaper when they threw giant barbecues that doubled as association fundraisers. A list of paying subscribers grew longer with each month, and the paper became the talk—in Spanish—of hamlets all over the valley.

Early one morning in Delano, Eliseo Medina plucked a copy of *El Malcriado* out of a stack at People's Market. The eighteen-year-old Chicano couldn't believe what he was reading: a story about a labor contractor, a white guy no less, who kept getting hauled before the labor commissioner by the Farm Workers Association for failure to pay wages. What

■ *Statues of the Virgin keeping watch over stacks of* El Malcriado *newspapers at the offices of the National Farm Workers Association in 1966.*

El Malcriado

15¢

"Don Sotaco"

■ *Farmworkers who read* El Malcriado *saw themselves in Don Sotaco, the hapless but amusing cartoon character popularized in the newspaper. Chavez started* El Malcriado, *which means "the ill-bred one."*

impressed Medina was how resoundingly familiar the story was: He had been working for a similarly unscrupulous Mexican contractor who habitually cheated workers in order to cover bad bets. "Every Saturday we'd have to get up at six in the morning and the whole crew would go and stake out his house and not leave until he paid us," Medina recalls. "Here, this group of people had taken this guy and made him pay back wages. This guy was the biggest labor contractor around, and he got dinged! So I really read that thing from cover to cover."

It wasn't long before the FBI would be scrutinizing *El Malcriado* as well, after it was brought to agents' attention by irate Delano townsfolk who demanded an investigation of the farmworkers group in 1965. Unknown to Chavez, at first, the agency began looking into his background. The name of the FBI file was "COMINFIL: Communist Infiltration of the National Farm Workers Association."

I N M A R C H 1965, up Highway 99 from Delano, more trouble was brewing in the valley, this time among the large and profitable tracts of young rosebushes growing near the town of McFarland. A thick-chested Mexican immigrant named Epifanio Camacho threw down his work tools and decided he was finally going to do something about the broken promises of better pay. A friend told Camacho, "Go see Cesar Chavez, who lives over in Delano; he wants to organize farmworkers," Camacho remembers. The blustery laborer was disappointed when Cesar told him he wanted to continue organizing workers in McFarland quietly, making sure they were signed up as members of the association before doing anything rash. "What's that going to resolve now?" Camacho demanded to know.

He returned to Chavez, telling him that some of the workers had resolved to strike with or without help from the association. They labored in one of the biggest flower industries in the country, crawling for hours, as they grafted rosebuds onto mature bushes. It was precision work that had to be performed at top speed. They had been promised $9.00 per thousand plants, but were only getting paid between $6.50 and $7.00.

Cesar told Camacho to get the workers together for a discreet house meeting; Camacho already had a reputation as a troublemaker, and any rose worker who teamed up with him had been threatened with firing. Only four workers came the first time, but thirty showed up for a second meeting. Some were seasonal workers who came from Mexico every year, and they were terrified. "We can't get mixed up in politics," they pleaded. An elderly woman who had come to the meeting scoffed at the men, boasting that she wasn't afraid and would be "worth ten men" in a walkout.

Emboldened by the woman's bravado, the men chose to strike the biggest company, Mount Arbor, which had about eighty-five employees in the fields. There wouldn't be any picket line—that would be too much, Cesar thought—but everyone was expected to honor the pledge to strike. Huerta held out a crucifix at the final meeting, and the men placed their hands on it and swore to stick together. "Things are getting exciting," Cesar wrote to Ross. "The most exciting thing is our drive to get a contract this summer. . . . If we are successful we will have something to crow about."

The first day of the would-be strike, Huerta and others began to make predawn rounds to check on the men. To her dismay, she found some were already up and dressed. It looked to her as if they were planning to report for work. In a flash, she moved her truck into the driveway to block their car and hid her key. Later, when she was sent to the company office as a union representative, she was called a Communist and promptly ejected.

A group of men, all from one small town in Mexico, were brought in to break the strike, angering Chavez so much that he dispatched a stern letter to the mayor of the workers' hometown. The note was posted on a town bulletin board, and word began to spread that the men had betrayed their countrymen by breaking a strike. After three days, desperate to win back its experienced workers, the company caved in and granted the workers a pay raise.

The workers never did get a contract with the company, however, and Camacho was blacklisted so he would never again get a job in the rose industry. But word of the strike spread quickly through the barrios of the valley's little towns, as did a *corrido* about the workers' bravery. Even though this "War of the Roses" was trumpeted in *El Malcriado* and gave the association its first strike, Chavez was convinced the NFWA would never win a real victory without a larger membership. That was at least a good three years off, he thought.

But events in California that summer began to outpace the union's cautious plan. Strike fever was sweeping the state, faster than Chavez and others imagined possible. Following the death of the bracero program, another key labor dispute had erupted

My Crime Is Being Filipino in America

by Carlos Bulosan

During the prewar years, Filipino men who worked the fields of California were subject to continual harassment and discrimination. Among them was Carlos Bulosan, who arrived in the United States from the Philippines when he was seventeen years old and worked as a dishwasher, fruit packer, and field worker on the West Coast. He helped form a new international cannery workers union at a time when an AFL-CIO drive excluded Filipinos from long-established unions. Later he signed on with Filipino Workers Association, an independent union, and risked his life organizing farm workers in California. He went on to write America Is in the Heart, *the classic account of the Filipino immigrant experience. In this excerpt, Bulosan describes the dangers he and his close friend José faced as farmworkers and organizers in the early 1940s.*

I know deep down in my heart that I am an exile in America. . . . I feel like a criminal running away from a crime I did not commit. And this crime is that I am a Filipino in America.

It was now the year of the great hatred [against Filipinos]. We walked to Holtville, where we found a Japanese farmer who hired us to pick winter peas.

It was a safe place and it was far from the surveillance of the vigilantes. Then from nearby El Centro, the center of the Filipino population in the Imperial Valley, news came that a Filipino labor organizer had been found dead in a ditch. . . .

Upon my return to the Santa Maria Valley, I found that the Filipino Workers Association, an independent union, was disintegrating. I rushed to join José in Lompoc, where he had gone with Gazaman to see if there was a possibility of establishing a workers' newspaper. The three of us decided to form a branch there and to make it the center of Filipino union activities in Central California. . . .

Persecutions against [union members] were sporadic at first, then concerted. In Salinas, for instance, the general headquarters were burned after a successful strike of the lettuce workers, and the president of the association thrown in jail.

It was during the membership campaign that I came into contact with fascism in California. The sugar beet season was in full swing in Oxnard, but the Mexican and Filipino workers were split. . . . They had not recognized an important part: that the sugar beet companies conspired against their unity.

I contacted a Filipino labor contractor and a prominent Mexican and José, who planned a meeting in the town park. I felt a little elated; harmony was in the offing. But in the evening, when we were starting the program, deputy sheriffs came to the park and told us that our right to hold a meeting had been revoked. I did not know what to do. I was still a novice. An elderly Mexican man told us that we could hold a meeting outside the city limits.

There was a large empty barn somewhere in the south end of Oxnard. A truck came and carried some of the men, but most of them walked with us on the highway. They were very serious. I glanced at José, who was talking to three Filipinos ahead of me, and felt something powerful growing inside me. . . . I walked silently with the men, listening to their angry voices and the sound of their marching feet.

I was frightened. But I felt brave, too. The Mexicans wanted a more inclusive union, but that would take time. We were debating the issue when I heard several cars drive into the yard. I signaled to the men to put out the lights and to take cover. They fanned out and broke through the four walls, escaping into the wide beet fields.

I rushed upon the improvised stage and grabbed José, whose wooden leg had become entangled in some rope and wires.

"This is it!"

"Yeah!"

"Follow me!"

"Right!"

I jumped off the stage, José following me. Then there was the sudden patter of many feet outside, and shooting. I found a pile of dry horse manure in a corner of the barn. I told José to lie down; then I covered him with it, exposing only his nose. I lay down beside him and covered myself with it, too. When I tried to talk, the manure went inside my mouth and choked me. I lay still, waiting for the noises outside to subside.

A man with a flashlight came inside and stabbed the darkness with the steely light, cutting swiftly from corner to corner. He came near the pile of ma-

■ *California growers imported Filipino farmworkers like these men starting in the 1920s. Only a limited number of Filipina women were allowed to immigrate.*

nure, spat on it, and searched the ceiling. A piece of manure tickled my throat, and I held my breath, bringing tears to my eyes. The man went outside, joined his companions, and drove off to town.

"Did you see his face?"

"No!"

"I saw it. He is a white man, all right."

"Let's run. There is still time."

I crept to the wall and crouched in the darkness. I wanted to be sure that every man had gone. The way was clear. José followed me outside. Then we were running across a beet field, our feet slapping against the broad leaves that got in our way. The moon came up and shone brightly in the night. As I ran, I looked up to see it sailing across the sky.

Then my fear was gone. I stopped running and sat down among the tall beets. José sat beside me. There were no words to describe the feelings in our minds and hearts. There was only our closeness and the dark years ahead.

farther to the south, in the Coachella Valley. The bracero program had been officially killed in December 1964, and California growers—who had imported sixty-five thousand braceros the year before—were convinced there would be labor shortages. They demanded that Governor Pat Brown pressure President Lyndon B. Johnson and his labor secretary to revive the program. The administration agreed, and a limited number of braceros were to be let in during the summer of 1965, as long as they were paid at least $1.40 an hour.

That decision backfired on the growers, who decided to pay their domestic workers in the Coachella Valley at a lower rate. When California's first grape harvest started that spring, the area's resident Filipino workers discovered they would be making $1.25 an hour, fifteen cents less than the braceros. AWOC agreed to lead the Filipinos out on strike. After ten days, their pay was increased. As the harvest moved north, other growers repeated the mistake of not raising the domestic workers' pay, inviting the strikes that rolled up the San Joaquin Valley. *El Malcriado* faithfully chronicled the developments, and that summer, an editorial probably written by Cesar acknowledged the momentum: "We who are picking the grapes and the peaches and tomatoes, which are the lifeblood of California, are soon going to share in the richness we have made. The little fights against the little grower and the contracts that you read about today are only the beginning."

ON SEPTEMBER 8, NFWA members Manuel Uranday and his wife, Esther, rushed into the association's office with news that Filipinos from nine labor camps—nine vineyards—had gone out on strike in Delano under the AWOC banner. Cesar was momentarily caught off guard; he had intended to slowly build the Mexican farm-workers movement, hoping to launch a strike, if necessary, only when he knew workers had the resolve to force growers to sign contracts. But with the Filipinos on strike, he and the other NFWA leaders would have to make some quick decisions. Chavez's union was rich in spirit, but unlike AWOC, cash poor: it had seventy dollars in its treasury, a volunteer lawyer, and no backing from organized labor.

The striking Filipinos around Delano, many of whom lived in grower-owned labor camps near town, had made an agonizing choice. When the harvest started that fall, they discovered that the same growers who raised wages to $1.40 an hour in Coachella were only willing to pay them $1.00. The workers, who needed enough money to last through the winter when jobs were scarce, appealed to AWOC's Larry Itliong. He sent two sets of registered letters to nine growers. None responded.

Itliong had suffered his share of hardship in the fields since he had arrived from the Philippines at age fifteen in 1929. He warned the young Filipino workers, who were most eager to walk off the job, that a strike in Delano could get brutal. For one thing, there were plenty of Mexican and Chicano workers around who could step into the vineyards to break the strike. "We told them, you're going to suffer a lot of hardship, maybe you're going to get hungry, maybe you're going to lose your car, maybe you're going to lose your house," Itliong recalled. "They said, 'We don't care.' They feel that they're not being treated fairly by their employers so they took a strike vote."

There was a catch, though. "They said, 'We don't want to picket our boss,'" Itliong said. " 'We've been working for him for ten, fifteen, twenty years. We don't want him to get mad at us.' So I said, 'Well, forget about the strike.' Couple of days later, they come back again and said, 'We will picket.'"

As a compromise, Itliong agreed to let the Filipino workers picket each other's bosses instead of their own. The growers thought the strikers would cave in after a

■ *Filipino grape workers started the 1965 Delano grape strike and were joined by Mexicans and Chicanos under the leadership of Cesar Chavez.*

few days, when they became hungry. But the workers hadn't budged when, after five days, the growers started shutting off the electricity and gas in their bunkhouses and barricading some of the strikers inside. Strikebreakers—called scabs—were trucked in, and fistfights broke out. "We knew that we had to support it," Huerta says.

Chavez was still worried, however, and for good reason. The association had about twelve hundred members at that time, but only two hundred were paying dues. While the young leaders huddled to discuss what to do, Cesar asked Helen if she was concerned about the kids. His wife made herself perfectly clear, Huerta recalls. "Well, what are we? Aren't we a union?" she asked Cesar. "That's what we're a union for, right?"

Itliong, for his part, knew the strike could fail if he didn't get the Chicanos and Mexicans to back him up. He talked to Padilla and met with Cesar and the rest of the Chicano leadership. Chavez and the union board offered unconditional support to AWOC. To join the strike, however, they would have to call a general meeting and convince their members. Equally urgent, they told Itliong, was the need to publicize the strike outside of the grower-dominated valley. If news of the strike stayed in Delano, the police and judges would surely crush it. "Cesar's genius and Dolores's genius and Helen's genius," Doug Adair says, "was to see that at this moment in American history there was a chance to finesse this local power structure and appeal to the wider community."

Furiously preparing for the mass vote, *El Malcriado* added a special leaflet in the latest edition urging workers to come to Delano for a general meeting in four days— September 16, Mexican Independence Day. "Now is when every worker, without regard to race, color, and nationality, should support the strike and under no circumstances work in those ranches that have been struck," the newspaper exhorted.

When the Filipinos walked out on strike, Eliseo Medina was sitting at home in Delano nursing a broken leg and watching *I Love Lucy.* His mother and sister, also field workers, came in shouting, *"Estámos en huelga! Estámos en huelga!* We're on strike!" Everyone knew it was coming. "We had heard about the strike in Coachella, and everybody thought it was so great," Medina recalls. "You know how it is when there are certain times in history when you can feel something in the air?"

On September 14, among the news accounts of car accidents and the deepening war in Vietnam, the *Bakersfield Californian* reported on the Filipinos' strike: "A grower spokesman predicted a back-to-work movement tomorrow while other growers expressed fears the strike may spread. . . . Growers said they are standing pat against the wage boost."

On September 16, Medina hobbled over on crutches to Our Lady of Guadalupe Church, where the size of the crowd stunned him: The parish hall was absolutely jammed, its balconies filled, and some of the fifteen hundred people were spilling out of the doors. The black eagle flag had been hoisted, along with posters of Emiliano Zapata and copies of Jack London's biting pro-union tract, "Definition of a Strikebreaker." The priest was visibly nervous: after all, the church's stained-glass windows had been paid for by some of Delano's most prominent growers—the Pavich, Zaninovich, and Pandol families, whose immigrant forefathers had come to the valley in the 1920s and built small fortunes with hard work and the help of cheap labor and federal water subsidies. (In 1951, valley growers received a boost from the federal government when they were charged only $123 for every acre-foot of irrigation water it cost the government $700 to deliver.)

At the church, Medina remembers a tall, distinguished-looking man with a commanding presence, who rose and spoke forcefully. "He must be Cesar," Medina thought. He was surprised when Gilbert Padilla introduced a much smaller man as the union president. Chavez's words were as strong, however, as his manner was calm. "One hundred and fifty-five years ago in the state of Guanajuato in Mexico, a padre proclaimed the struggle for liberty. He was killed, but ten years later Mexico won its independence," Cesar told the crowd. "We are engaged in another struggle for the freedom and dignity which poverty denies us. But it must not be a violent struggle, even if violence is used against us. Violence can only hurt us and our cause." That same night, knowing he had to convey the justice of the strike to other farmworkers and to people in cities, Chavez called on the workers to pledge to remain peaceful.

The church's priest was comforted by Cesar's words. But the audience, including Doug Adair, who was struggling to understand the Spanish ricocheting around him, was also motivated by the emotional testimony of the workers. Cesar had forewarned the militant Epifanio Camacho he would be sought out at the right time to rise and speak. Camacho complied, delivering a fiery exhortation laced with slogans from the Mexican Revolution. "It's better to die on our feet than live on our knees," Adair remembers Camacho shouting. "My Spanish wasn't very good, but, gosh, it swept me off my feet." Cesar also asked farmworkers from the area's various ranches and from different states in Mexico to rise and speak, and Mexicans from their part of the homeland cheered them on with cries of *"Viva Chihuahua! Viva Nuevo León!"*

An older farmworker, Felipe Navarro, answered the call. "I saw two strikers murdered before my eyes by a rancher," he said, recalling a strike in the thirties. Tears came to his eyes. "There was nothing to eat in those days, there was nothing. And we're still in the same place today, still submerged, still drowned. . . ." After solemn applause, a man from Tamaulipas stood and said he wanted to show that Mexican immigrants were willing to support farmworkers from other backgrounds. "It is an honor to come here and we must not abuse it. We are all humans, and we have to aid our brothers the Filipinos in this just cause. Let's go out on strike!" he concluded, as the crowd cheered wildly.

As cries of "Strike!" rocked the hall, union leaders called for a show of hands. The crowd voted to demand the same terms as the Filipinos, whose strike by then had spread to twenty labor camps. After the meeting, the union counted some twenty-seven hundred signed cards authorizing NFWA representation. With the Mexicans joining the Filipinos, forty-eight ranches would be struck.

Eliseo Medina was beside himself: This was the most exciting thing he'd seen since coming to Delano from a little village in Mexico with his family a decade before. That night he shook out all the money he had in a piggy bank and went down to the union office and paid three months' worth of dues—$10.50—all at once.

■ FOLLOWING PAGE: *Cesar Chavez debating with a police officer in Delano before leading a three-hundred-mile march to Sacramento called the* peregrinación, *the "pilgrimage."*

The experience of working in the civil rights movement gave us what we called "Mississippi Eyes," because you'd go back to where you came from and suddenly it would look different. . . . It hadn't changed, but you'd changed. . . . And when I went back to Bakersfield, it looked different. It looked a lot more like Mississippi.

—MARSHALL GANZ

CHAPTER FOUR

HUELGA!

NO ONE KNEW what lay ahead the morning that a caravan of cars arrived at a Delano vineyard under cover of darkness. Several weeks after the town's farmworkers leaped into history with their vote to walk out of the vineyards, Cesar Chavez and eighty others had risen before dawn to picket their most formidable target yet: a 4,500-acre vineyard—the region's second largest—owned by Schenley Industries, a national corporation known for distributing such brands of liquor as Cutty Sark whiskey and Roma wines. The strikers, so far, had been remarkably peaceful. But Chavez had seen ranchers throw punches, and he wanted to avoid an escalation of violence. He invited prominent clergymen to Delano to bear witness, hoping their presence would keep both sides calm. The day strikers arrived to picket the Schenley vineyard for the first time, they were accompanied by Bishop Sumner Walters of the Episcopal Church's San Joaquin Diocese.

As the sun rose over the Sierra Nevada to the east, strikers took up their positions at each of the ranch's twenty entrances. Cars of strikebreakers reporting for work streamed toward the sprawling vineyard, which lay still and ghostly in the early morning. As the workers approached, pickets began to shout through bullhorns and rolled-up copies of *El Malcriado:* "Don't work here!" *"No pasen!"* "Strike!" The action worked: Some of the strikebreakers drove off or stopped and got out of their cars to join the picket line. Jubilant union members applauded the new recruits, promising them a hot meal at AWOC's meeting place, Filipino Hall, and bags of donated food. But others sped past the picket lines, disappearing into the canopy of leaves. Chants followed them: "Don't betray your brothers! Think things through!"

Suddenly, a blast of engines from an empty field behind the pickets silenced the crowd. Two men poised atop tractors jerked their machines into gear and began lumbering toward the pickets, dragging sharp-edged plow disks behind them. Bearing down as if they were driving tanks into battle, they drove within a few feet of the strikers, churning up choking clouds of powdery dust. The pickets and Bishop Sumner were soon covered in grit. But the Schenley foremen didn't stop. They seemed to relish this mission, driving back and forth continuously, showering the crowd with fine Delano dirt. Finally, after watching the spectacle from a distance, the grape pickers who'd crossed the picket line were so disgusted they, too, decided to leave the field.

"After almost an hour, when we are unrecognizable with dust, and our throats are hoarse with yelling and choking, we see the thrilling sight of a worker's car heading toward us through the vines," recalls union volunteer Eugene Nelson, who wrote a memoir about the grape strike. "Their admiring looks seem to say: After seeing what you went through, your persistence in the face of all that, it was impossible not to join you."

Against all odds, the Schenley harvest had stopped, at least for the day. The bishop smiled, and wiped dust from his brow. "The growers' spite has backfired on them," a satisfied Nelson wrote.

E VEN THOUGH he and his union had joined the AWOC strike, a still-cautious Chavez hoped to avert a protracted walkout that could prove disastrous to workers and growers alike. Immediately after the vote in Guadalupe church, he rushed off registered letters to the growers whose workers planned to strike, calling on them to negotiate. Chavez also alerted state labor officials, hoping they would step in and mediate a deal. But the growers refused to talk. California farmers had a long history of ignoring letters from labor leaders, and Chavez's letters were returned unopened. The growers, it seemed, preferred to throw down the gauntlet. They were confident—especially the most powerful and corporate among them—that they could outlast the Filipino and Mexican unionists.

But by September 25, five days after the Mexicans joined the Filipino workers, labor reporter Harry Bernstein of the *Los Angeles Times* was in Delano reporting on what was shaping up as the largest farmworker strike in recent memory. Despite the growers' contention that only a fraction of their five thousand workers were involved, the California State Church Council estimated that at least two thousand employees had walked out. The growers' recalcitrance, the council told Bernstein, was fostering a "dangerous potential for violence."

The Great Delano Grape Strike, as it later came to be known, cemented the National Farm Workers Association as a group and vaulted the determined Chavez into a public role. Chavez and Itliong emerged as important leaders, as did Dolores Huerta and Gilbert Padilla. Every morning, they rose before dawn and told strikers where to go to set up the picket lines that covered an estimated 450 square miles of vineyards in Kern and Tulare Counties. Chavez, who had been observing the success of Martin Luther King Jr.'s nonviolent civil rights movement in the South, took the lead in repeatedly telling pickets not to fight with strikebreakers or ranchers; any show of violence by strikers, he said, would give the police and sheriff's deputies an easy excuse to descend on them with clubs and guns. The strike would be lost, and people might die.

Not content to seek only the support of California's labor unions, Chavez began to travel to university campuses and churches, speaking to students and worshipers who were curious about the strike.

■ *Young actor Luis Valdez returned to his birthplace, Delano, to found the Teatro Campesino—the Farmworkers Theater—and cheer on strikers in 1965.*

He asked them for desperately needed donations of food, money, and clothing—and he asked for more. He asked them to come to Delano to see for themselves what life was like, and to join the farmworkers in *la causa*—the cause.

"He wasn't a good speaker at all, but there was something about him. I always say that word 'charisma,' which is often used, doesn't begin to describe it," says Marion Moses, who first heard Chavez at the Unitarian Church in Berkeley in October 1965, just a month after the strike began. The speech provoked Moses, then a nurse, into driving to Delano to volunteer. She later became a UFW doctor.

When the first groups of students and clergymen began to show up on picket lines in Delano, townsfolk started to realize this was no ordinary strike. The grapes were ripening, and some of the ranchers, especially local residents who viewed the smooth functioning of their vineyards as a birthright, grew increasingly antagonistic. A few sought immediate court injunctions to limit roadside picketing outside their ranches. Smarting with injured pride, they hired armed security guards, equipped foremen with shotguns, and began to patrol the edges of their property menacingly. The ranchers wanted to get their grapes off the vines, and they sought out strikebreakers

Zeferino's Fathers

by Alfredo Véa, Jr.

A *criminal defense attorney and novelist living in San Francisco, Alfredo Véa, Jr. was born in the desert near Phoenix. After leaving Arizona, he entered the migratory farmworker circuit as a child in California, where Filipinos in Stockton gave him a set of encyclopedias and taught him to read and write; he also met a young man named Cesar Chavez who was stirring some farmworkers with his demands for change. Véa went to Vietnam when he was eighteen, and after his discharge, put himself through law school, later working as a volunteer with the United Farm Workers union. In his recent autobiographical novel,* The Silver Cloud Café, *he describes an incident in which a group of older Hindu, Filipino, and Mexican farmworkers saved him from humiliation.*

At the edge of the hot, shadeless fields, Zeferino the boy silently worked the furrow closest to the highway. Squash were as hard to pick as cotton, he thought. His hands were getting raw again, almost as raw as his back, which was covered now with a layer of white salt. The petroleum jelly that one of his *tíos* had applied there to prevent sweat rash had been completely washed away by midmorning.

Zefe paused to look to his right. What he saw there soothed and comforted him. Señor Chavez was out there as usual, passing out handbills that protested the miserable working conditions and the slave wages. He was young and unknown and picketing all by himself, stooping now and again to carefully smell the leaves for the telltale scent of insecticides.

One day a young welfare worker came out to the fields where Zeferino was working. She looked totally out of place in her suit and heels. Someone had told her that there was a boy laboring out there who should've been in school. Zeferino remembered that she had asked where his mother was and that he had told her that he didn't know, that she never spent much time in the camps. When she asked him about his father he told her that he had no idea who he was and that he had never seen him.

"But I have lots of uncles, señora. In fact, I have over thirty of them," the boy had explained. The welfare worker didn't seem very impressed by what he had said, so she dragged the boy—kicking and screaming—out of the fields. His *tíos* had protested, but they all knew that she was right. They knew that a childhood should not be spent laboring in the furrows. Though the boy hated her at the time, he understood the woman's heart was in the right place.

"I remember that she bought me some brand-new school clothes and made me take off all of my muddy gear," Zeferino said excitedly to an unconscious King and to Anatoly, who was parked in front of Radio City Music Hall. "Then she placed me into a classroom in some town that I can no longer recall. I can still see the faces of those children staring at my battered work boots. I do remember clearly that the teacher in that classroom ended the first day of class by reminding the children that in three days the school would be having their yearly Father's Day celebration. All of the children were expected to bring their fathers to school on that day."

"*Pobrecito,*" said Raphael from above.

"I remember how miserable I felt as I walked back to the camp. I imagined myself at Father's Day, the only kid there without a father." It bothered the boy so much that he couldn't eat and even the sweet music from the Mexican Quonset couldn't cheer him up. He told one of his *tíos* that he wanted to run away, that he just couldn't go back to school on Father's Day.

"Well, I must have told my *tío* the wrong day, because, on the day before Father's Day, thirty farmworkers dressed in their work clothes walked slowly, single file into the classroom. Each one had carefully combed his hair and put on his cleanest dirty workshirt. The entire classroom smelled of pomade and brilliantine and sweat.

"How could I have ever forgotten that morning?" Zeferino said, exhaling his emotions onto the rusted hood of the cab and onto the windshield. Anatoly suddenly ceased his incessant chatter as Zeferino's voice dropped to a whisper.

"There they were, forming a circle around the class, all of them shifting nervously and smiling shyly at the teacher and so proudly at me. They had their sharpened paring knives and long machetes hanging from their belts. They had taken the time to brush the mud from their knee pads and each man had carefully folded his bandanna into his shirt pocket. Each of my uncles held his straw hat respectfully in his hands.

"They were Mexicans, Pinoys, and Hindus, all missing a day's pay . . . for me." Zeferino paused while the power of the memory washed over him. Anatoly said nothing and King moaned inconsolably in the backseat. Above them all, a silent Raphael watched and listened. Zeferino's eyes glistened with emotion.

"These were strangers in America, men who had been told again and again that their own lives were a poor imitation of the lives around them. But that

■ *Here, in the Delano grape strike, children used as strikebreakers pack grapes while a sheriff's deputy looks on.*

morning, they all braved the other world for my sake. I remember that, finally, one of them stepped forward and with a heavy accent spoke a single sentence to the class: 'We are Zeferino's father.'"

from in town, from other parts of California, from Mexico, and even from Los Angeles ghettoes, where winos were recruited to cut grapes.

Some of the growers tried to appeal to pickets' sense of loyalty. "They were like members of our family. We gave them a place to live all year long," one grower said of his Filipino workers. His kinship apparently had its limits. When pickets suggested to him that he could have averted the strike with a twenty-cent raise, the man grew angry. "We'll let the grapes rot first," he snapped, speeding off in his truck.

As days turned to weeks, and picket lines swelled and became more animated, some of the security guards and ranchers antagonized strikers, sometimes driving their trucks toward them at high speeds. Police officers and sheriff's deputies failed to discourage the intimidation, and strikers wondered darkly: *Who will be the first to die?*

One of the first displays of true menace erupted at the C. J. Lyons Ranch, where two Chicano pickets held up *Huelga* signs and exhorted workers to join the strike. Although they were standing on the shoulder of a public road, a ranch foreman stormed toward them with a shotgun, shouting, "You can't picket here! Get the hell out!" In one seamless movement, he seized a picket sign, threw it down, and blew it apart with a round of buckshot. "Now get the hell out of here!" screamed the foreman, who was also the owner's son-in-law. Punctuating his order, he fired another deafening round into the sky.

Grape growers Bruno and Charles Dispoto also became notorious. The Dispoto brothers sprayed pickets with pesticides, threatened them with dogs, and showered them with obscenities. They reserved their most venomous barbs for clergymen, whom they denounced as "creeps," "Communists," and meddling "fairies." Bruno backed his pickup truck into Eugene Nelson, knocking him violently to the ground. As Chavez predicted, law enforcement was not eager to pursue charges against members of the area's farming establishment: A Kern County prosecutor promised to look into the assault, but nothing came of his pledge.

After Charles Dispoto beat up an AWOC volunteer—drawing blood with his punches—it took a week and the intervention of Governor Pat Brown's office to get Dispoto arrested for battery. The victim, Hector Abeytia, had been a member of the governor's committee on farm labor.

Other ranchers sometimes tested strikers by walking up to picket lines and stomping on toes, punching men in the ribs, and issuing racist taunts: "Oh, pardon me, Mex! Why don't you watch where you're going, stupid!" After Chavez witnessed such an incident, he walked up to Al Espinosa, a Delano police captain—the rare Chicano on the force—and demanded he do something. But nothing happened until Espinosa saw grower Milan Caratan knock a Mexican striker to the ground with a punch to the stomach. Caratan was taken into "protective custody."

"Those early days were really mean," recalled Cesar. "We had cops stationed at our office and at our homes around the clock. Every time I got in the car in Delano, they'd follow me around, and I wouldn't shake them until I was well past several towns." Right from the start, some of the strikers grumbled about Chavez's insistence that they shouldn't retaliate against the growers or the police; others argued that damage to property shouldn't be considered a violent act. Chavez responded that the growers' petty behavior would backfire on them in the end. As more supporters arrived and empathetic news accounts of the strike began to appear, the strikers started seeing the effectiveness of Chavez's nonviolent philosophy. "We took every case of violence and publicized what they were doing to us," Chavez said later. "By some strange chemistry, every time the opposition commits an unjust act against our hopes and aspirations, we get tenfold paid back in benefits."

But with his own ribs bruised by punches on the picket lines, Chavez knew the discipline it took to keep hostility in check: "Love is the most important ingredient in nonviolent work—love the opponent—but we haven't really learned yet how to love the growers. . . . Maybe love comes in stages. If we're full of hatred, we can't really do our work. Hatred saps all that strength and energy we need to plan."

K N O W N   F O R its placidity, and more than content with the status quo, Delano wasn't accustomed to reporters from Los Angeles and San Francisco coming into town to interview Mexican and Filipino workers who claimed they were being mistreated. All this attention quickly enraged local agribusiness, which grew to hate Chavez, the short Chicano with the long hair who had brought in what they called "outside agitators," students from Berkeley, peaceniks, and ministers who'd never touched soil in their lives. The growers retaliated by forming ad hoc groups in defense of their town—Mothers Against Chavez and Citizens for Facts—and counterpicketed with signs demanding that outsiders go home. "I don't like the people that are here in our town, they don't belong here, and I wish they'd all go back to where they came from. . . . We have no labor troubles," a woman told a television newsman during a protest. The stridently anti-Communist John Birch Society took an early interest in the Delano strike, dispatching investigators to the town to search for "subversives."

Some of the locals had dubbed Delano, a town of fourteen thousand, "the Little United Nations," because they believed the races lived in harmony, sharing in the prosperity of the agribusiness industry. If the dark-skinned field workers weren't as successful as the white-skinned growers, the prevailing attitude seemed to be, it was probably because they didn't work as hard. But basically, they argued, everyone got along. The town's image of itself as a peaceful melting pot may have stemmed, in part, from the fact that its mix of highly successful Yugoslavian, Italian, and Armenian resident growers employed laborers from every race. Its public schools were integrated. The local Cinco de Mayo celebration was covered by the *Delano Record*; at least one Chicano labor contractor was on a school board; and a black man who was a local pastor had been admitted to the Kiwanis Club. Delano even had a Mexican-American police captain, Al Espinosa, who was familiar to many farmworkers—not for his law enforcement work, but because he moonlighted as a labor contractor.

The strike, however, uncovered festering resentments among minority families all over the San Joaquin Valley. Their children, they reported, were simply not treated as well as white children. A Stanford student, observing valley schools over a forty-day period for a doctoral thesis completed in 1966, recorded numerous examples of discrimination. For instance, when a local teacher was asked why she appointed an Anglo boy to form a line and lead five Mexican students out of the classroom, she innocently replied, "His father owns one of the big farms in the area and one day he will have to know how to handle Mexicans."

Growers resented the depiction of their town as mean-spirited and were even more angered by stories about the miserable living conditions and low pay of the workers. "These men are extremely happy. If they weren't, they would not be coming here from other states, coming here from Puerto Rico and all over the state of California," insisted grower Jack Pandol, talking to a reporter about the strikebreakers. "These men have chosen this type of work. . . . We think that we have tried to better their lot by some of the actions that we've done in this area, and by higher wages, which are the highest in the United States, by the way." Grape pickers in Delano earned about $2,400 a year—more than many farmworkers, but well below the nation's poverty line in 1965 of $3,223 per family of four.

In editorials and letters protesting the strike, growers painted a curiously slanted picture of themselves: They were the weak underdogs oppressed by the all-powerful farmworkers. "We just cannot in our own mind see this complete power of destruction being granted to the type of union organizers we have observed in action. These men are cold, hard, and brutal," wrote Jack Pickett, editor of *California Farmer* magazine in January 1966. "They have preached hate against the farmer. Even under con-

■ *Grape workers, whose earnings depended on speed, rushing to pick and deliver wine grapes.*

ditions of greatest restraint the strike would be too lethal a weapon to use against the man with one payday a year."

Reverend Chris Hartmire, director of the Migrant Ministry and a native of Pennsylvania, couldn't understand why corporate growers wouldn't join the modern age and accept unionization of the industry. He decided, finally, that it was because the growers had worked so hard to insulate themselves from their field workers. "They had developed this mythology to justify the misery of the farmworkers. And the mythology was that this was all they deserved, and this is what they liked," Hartmire recalls, recounting how, in his years of work among migrant farmworkers, he had tried to persuade growers to increase wages or make repairs at the ramshackle labor camps. Instead, he heard such responses as, "If I gave them more money, what would they spend it on? They'd just spend it on booze and waste it on gambling. . . . If I fix the windows, the kids will break them."

George Zaninovich's family owned a four-hundred-acre grape ranch in Delano which was struck and picketed during the grape strike, when the young Zaninovich was at Stanford studying political science. "I could not talk to my family about this back then," he says. "They were so convinced that it was a Communist conspiracy that was behind the labor organizing. They couldn't see that I was more of a Marxist than Chavez ever thought of being. All they could see were those signs saying *Huelga!* and they were afraid."

George Zaninovich was part of a Delano clan that included distant cousins whose ancestors had left Croatia's Adriatic coast at the turn of the century. His father had been lured to the San Joaquin Valley in 1908 with promises of cheap land; after years of sharecropping and buying property bit by bit, he developed a prosperous farm. Zaninovich explains that as immigrant farmers, his hardworking parents were afraid of losing what they had. "They had peasant mentalities that affected the way they did business. They figured they worked hard for what they got and hated like hell to let it go."

But some of the other Zaninoviches in Delano were far wealthier and more powerful, and, like the corporate growers with boards of directors in big cities, they were the ones who would decide whether the industry would fight the strike or capitulate. And in 1965, there was still enough bigotry among the growers to dismiss Chavez and his followers, according to George Zaninovich. "A lot of these growers, especially those who built their farms from nothing, took the view that the Mexicans did not deserve any more money because they just didn't work as hard for it."

Eugene Nelson, a union volunteer who, ironically, was the son of a Modesto grape grower, was still more critical. The growers, he said, would not face up to their own hypocrisy. They were "only too willing to accept socialist Big Government subsidies such as braceros and free irrigation water, but unwilling to accept the older and less socialistic institution of the union." Their reaction to the strike, he said, was "an anachronism: the last-ditch stand for a feudalistic way of life which in other industries disappeared long ago."

O VER AT FILIPINO HALL in Delano, not far from the Elk's Club where Delano's farmers sometimes gathered, tacos and tamales were starting to show up next to platters of the traditional lumpia and rich adobo dishes favored by the Filipinos. The Filipino-dominated AWOC had union money to spare for now. Union officials had invited the Mexicans and other pickets to dine for free at their mess hall, and vats of steaming rice and hot coffee were always at the ready to fortify hungry pickets.

The National Farm Workers Association kept its own closet-sized office and rented an adjoining "*huelga* house" to contend with all the activity. But the Filipinos and Mexicans mingled and got to know each other at Filipino Hall. "It became like a brotherhood, I think, for the first time," Gilbert Padilla remembers. In the past, Filipinos and Mexicans had often been segregated into different picking crews; this separation was often exploited by ranchers to pit one group against another in a labor dispute. The current mood of exhilaration and unity changed that dramatically. Pete Velasco, a Filipino striker, experienced the newfound sense of brotherhood as he walked into the mess hall to get some dinner. One evening, as he passed by a cluster of workers, Velasco heard a mellow voice saying, " 'Hello, brother.' I stopped, looked back, and then walked back to shake hands with this Mexican who was talking to a Filipino. He said, 'Cesar Chavez.' " Velasco did a double take when he realized who was talking to him, but introduced himself gamely: "Pete Velasco, glad to meet you." The two would call each other brother for decades to come.

Many such spontaneous friendships were formed in the strike's early days, and overnight, farmworkers were transformed into committed activists. One young field worker eager to get involved but unsure how to go about it was Eliseo Medina. He had heard rumors since the night of the strike vote that the NFWA would pay $1.40 an hour to picket—the same base wage the unions were demanding the growers pay. Although the idealistic teenager didn't know exactly what the term "picketing" meant, he was more than willing to find out. On his first assignment he and a friend accompanied a fifty-year-old Mexican to confront strikebreakers out in a field not far from Delano. As they drove down a highway that bordered an expanse of tangled vines, their enthusiasm quickly turned to fear when they realized three patrol cars were following them. The older man slammed on the brakes of his car and jumped out, shouting epithets in Spanish at the *esquiroles*—the strikebreaking "scabs"—working the vines. Eliseo and his friend looked on aghast as he insulted the deputies as well. "I was convinced they wanted to shoot us," he says.

■ *Tulare County law enforcement guarding the fields in 1965. They were accused of favoring the growers during the Delano grape strike.*

But the deputies shrugged and left, and the strikebreakers slunk farther into the field, visibly shamed. Eliseo was impressed. All of a sudden he didn't care so much about the money, which he later found out didn't exist anyway. He was overcome by the romance of the cause. "We were standing tall," he says. "That was my baptism into the union."

Many of the strikers with families to support sent their men outside Delano to find other work to help feed their children. That put women, who were on the picket lines from the beginning, into an even more prominent role. Like the men, they first tried to reason with workers who crossed the line, calling out *"Huelga!"* in a tone more inviting than angry. If this didn't work, the women would turn to shouting and trading insults with strikebreakers, trying to shame them into joining the strike. They'd brandish Mexican flags alongside Filipino men who chanted in Tagalog, *"Mag labas kayo, kabayan!"* "Come out of there, countrymen!"

One particularly effective Mexican striker would bark at strikebreakers hiding in the vines, "You—you with the stringy hair—come over here and hear what I've got to say! *Sí, usted!* You want to be a slave all your life?" she shouted. "We just walked off a vineyard down the road—come out here and listen to us! What has the grower ever done for you—has he invited you to his house?"

Jessie De La Cruz remembered how the double standard for men and women came under fire during those tumultuous times. "At first it was hard to organize those that believed that women should stay home and do the washing and the cooking," she says. "But they never stopped to think about them working out in the fields. So why shouldn't they attend meetings and be involved with the union? It got to the point where most of us out on the picket lines were women." *El Malcriado* went so far as to publish a photo of Cesar washing dishes, then an unconventional pose for a Mexican man.

Chavez also changed the culture of the farm-labor community by encouraging so many students active in the Free Speech movement or the civil rights movement to come to Delano. "At the beginning, I was warned not to take the volunteers, but I was never afraid of the students," Chavez said. "If it were nothing but farmworkers in the union now, just Mexican farmworkers, we'd only have about 30 percent of all the ideas that we have. There would be no cross-fertilization, no growing. It's beautiful to work with other groups, other ideas, and other customs. It's like the wood is laminated."

Student volunteers from the black-led Congress of Racial Equality and the Student Nonviolent Coordinating Committee, fresh from Freedom Summer in segregated Mississippi, were quick to respond to Chavez's call. In Freedom Summer of 1964, whites and blacks in SNCC worked side by side in dangerous territory to help register black voters; others risked lives to integrate lunchrooms and bus lines in other southern states. Three young civil rights volunteers—two whites, one black—were murdered within days of their arrival in Mississippi. For many of the white college students, the South was the first place they saw police as brutal aggressors rather than trusted protectors.

The first SNCC organizers to come to Delano arrived about a month into the strike, offering short-wave radios and help with nonviolent organizing. One of the volunteers was Marshall Ganz, a student who'd grown up in Bakersfield—not far from Delano. Ganz's father was a rabbi, and his mother a teacher. While attending Harvard University in the early 1960s, he got involved in the civil rights movement and lived in Mississippi during Freedom Summer. While there, Ganz read an article in

---

■ *Dolores Huerta, three of her children, Julio Hernandez, and Rev. Jim Drake singing* "De Colores" *at a 1966 meeting of the National Farm Workers Association.*

an SNCC newspaper about a farmworkers' march in the San Joaquin Valley. It was a revelation: Instead of returning to Harvard in the fall, he headed back to California to see his family, and then up to Delano to see what was going on.

"I hadn't really made the connection between what was happening in Mississippi and where I'd grown up," Ganz remembers. "But some of us talked about how the experience of working in the civil rights movement gave us what we call 'Mississippi Eyes,' because you'd go back to where you came from and suddenly it would look different. It had been changed. I mean, it hadn't changed, but you'd changed. You'd been changed by that experience where in the South every conflict was so clearly drawn and the power relations were so clear. . . . And when I went back to Bakersfield it looked far different. It looked a lot more like Mississippi."

Before jumping into the thick of strike organizing, Ganz first approached Chavez, to make sure he was wanted. "He seemed very interested in people who had experience in the civil rights movement," Marshall said of Cesar. "That seemed very important to him. It also seemed very important that he was constructing a movement that, [while] rooted in a very strong sense of cultural identity and pride, was not to be exclusive.

"People of diverse backgrounds and experiences enrich the movement a lot, but [diversity] also had its problematic aspects," he adds. "Of course, again, it never stopped being something of a tension, often a very constructive tension."

Chavez conceded that some problems were inevitable. He told biographer Jacques Levy that two young free-spirited volunteers had scandalized the farmworkers by strolling through a nearby cornfield in the nude. The mix of morals and ideals generated sparks within the movement. "Some of the volunteers were for ending the Vietnam War above all else, and that shocked the workers because they thought that was unpatriotic," Cesar recalled. "Once, when there was a group more interested in ending the war, I let them have a session with the farmworkers. After a real battle, the volunteers came to me astounded. 'But they support the war!' they said. 'How come?' I told them farmworkers are ordinary people, not saints."

Whatever their political views, all were welcome in the first heady days of the strike. Union volunteer Doug Adair remembers strikers were happy with anyone who showed up on the picket lines during the first couple of months. Indeed, when the strike began, members of the Communist Party, which only had a couple thousand members in California at the time, were among the first to donate food. "There came a point when there were people who were very much opposed to Communists in the union," says Adair. (*El Malcriado* took a tongue-in-cheek approach to red-baiting, running a short blurb offering a "cash reward" for the "arrest and conviction" of anyone referring to the union as Communist-inspired. "Such references are false and illegal, and we intend to punish anyone saying these things to the full extent of the law," advised the notice.)

The NFWA didn't discriminate at the onset. Catholics, Communists, trade unionists, and antiwar students were all welcome as long as they were nonviolent, promoted the strike, and appeared to have no hidden agendas.

"I—a Republican—could be picketing next to a Communist, and we would be brothers and sisters," says Adair. "There is a job to get done and there is a big tent, and boy, everyone that can help, get your ass down here and help."

Most Chicanos and Filipinos involved in the new farmworkers movement had tasted bigotry all their lives, and they were quick to compare the Delano uprising to the South. The modern civil rights movement, they wrote in *El Malcriado,* "began in the hot summer of Alabama ten years ago when a Negro woman refused to be pushed to the back of the bus. The Negro is willing to fight for what is his, an equal place in the sun. Sometime in the future they will say that in the hot summer of California in 1965, the movement of the farmworkers began."

THE SIGHT OF farmworkers in league with students wearing SNCC buttons was too much for many to bear in Delano. To this day, it's not known whether it was an influential grower, a resident, or a local congressman, but somebody placed a call to the FBI in early October and demanded a meeting with agents to discuss the "subversive connections" of Cesar Chavez, Dolores Huerta, and David Havens of the Migrant Ministry, as well as Larry Itliong and three other Filipinos.

The caller was indignant that the federal Poverty Program's Office of Economic Opportunity was giving the National Farm Workers Association a grant of $267,887 to teach workers about U.S. citizenship and money management. The NFWA had applied for the money in Washington, D.C., in March, long before the grape strike. In early October, the program's director, Sargent Shriver, announced that the association had been awarded the grant. Chavez, overwhelmed, asked that the award be postponed until the strike was over. Shriver agreed, but Delano was already up in arms.

The town council condemned the grant, and locals derided Chavez as a suspected Communist. "The general public in Delano," an FBI agent wrote, "believes that this money will be used, not for its intended purpose, but to further the political beliefs of [identity blacked out] and other 'left wing' type groups, such as the Student Nonviolent Coordinating Committee and Congress of Race Equality (CORE) [sic], which may attempt to affiliate with NFWA if this grant is actually received." Another source, reported the FBI, said that Chavez "is not qualified to manage this large a sum. He has heard that Chavez has only a grammar school education."

So began the FBI's probe of alleged Communist infiltration of the farmworkers union. The surveillance lasted more than a decade and generated nearly two thousand pages of documents filed by agents and informants around the nation. The COMIN-

FIL, or "Communist infiltration," documents, which were released in 1992 under the Freedom of Information Act, provide a remarkable portrait of the union's meteoric rise from obscurity to the national political scene. With names of informants blacked out, the memos dutifully report Chavez's whereabouts, suspected plots against his life, boycott activities across the country, break-ins and bombings of union offices, and every possible link, however remote, to groups on the FBI's very wide Cold War radar screen.

Within weeks of the investigation's inception, agents assigned to watch the union seemed satisfied that Chavez had a "clean background." An early memo noted that "We have conducted limited inquiries concerning NFWA for the purpose of determining whether it has been infiltrated or controlled by the Communists. Our information indicates that it is not controlled or guided by the Communists but that several individuals that are allegedly members of this organization have extremely liberal views." Nonetheless, marching orders from J. Edgar Hoover's headquarters in Washington, D.C., were to continue surveillance.

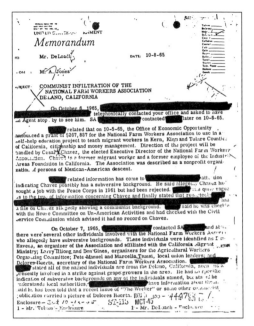

■ *Beginning in 1965, the Federal Bureau of Investigation kept extensive files on Cesar Chavez and his farmworker organizing.*

In the view of the FBI, Dolores Huerta was prominent among the individuals with liberal views. As one FBI agent put it, Huerta was "the driving force on the picket lines of Delano and Tulare County and daily inspires the pickets and their cause." Huerta also piqued the FBI's interest when her picture appeared in *People's World,* the Communist Party newspaper, in October. Sam Kushner, a reporter for the *World,* was in Delano to cover the strike before most major newspapers decided it was big news. His presence, and the fact that Chavez gave him interviews, seemed evidence enough to the FBI that the union might be a subversive entity.

Even the word *huelga*—"strike"—seemed suspicious to the FBI. A source told the agency that *huelga* means "revolt" to some Mexicans. The agency had to resort to expert help to determine if this was true, and was informed that it was not: "Bureau translator [identity blacked out] advises the word *huelga* means 'strike' or to 'leave the place vacant.'"

While the FBI was hard at work, the Delano police department and the Kern County sheriff's department conducted their own surveillance of strikers. They snapped photos of everyone on the picket lines and recorded automobile license plates around strike headquarters, so they could be checked against records at the state Department of Motor Vehicles. The Kern County sheriff's department compiled some five thousand dossiers on pickets with this information. Police officers stopped pickets and asked them for their names and backgrounds, filling out cards that would be turned over to the sheriff.

Harassment did not stop with the police; it extended to the fire department and other agencies. "More than once, the authorities threatened to close our office down. Some of the people that were sent to do a job on us obviously didn't like it at all," Chavez said. "I think it was the fire marshal who hated the politics of his assignment and told us so on more than one occasion. But they kept up their investigations under the health and building codes and the fire code."

On October 16, Kern County sheriff Roy Galyen issued a special directive to his deputies that Chavez knew couldn't go unchallenged: Strikers would henceforth be forbidden to "disturb the peace" by rallying and shouting on roadsides. The abridgment of free speech threatened to cripple the strike. As growers avoided pickets by moving workers deep inside vineyards, away from public roads, or by stringing up thick sheets of plastic to shield them from the pickets, shouting had become the strikers' key means of communication.

The day after the sheriff's order, pickets drove over to a ranch just outside Delano, to put the injunction to test. Lining up to confront a group of thirty Anglo and black migrants who'd been brought in from outside the area, striker George Gonzalez held up a copy of *El Malcriado* and began to read a reprint of Jack London's "Definition of a Strikebreaker."

Sergeant Gerald Dodd of the sheriff's department had heard "this piece of trash" before, and arbitrarily decided it was not to be read to the strikebreakers. He ordered Gonzalez to stop and vowed to arrest the next man who dared read the essay out loud. A few minutes later, Reverend David Havens, dressed in a crisp black jacket and tie, climbed up onto the back of a flatbed truck parked next to the field and calmly began to violate Dodd's order: "After God had finished the rattlesnake, the toad, and the vampire, he had some awful substance left with which he made a strikebreaker. . . ." Reading in a loud, steady voice, Havens paused occasionally to glance meaningfully toward the field. "A strikebreaker is a two-legged animal with a corkscrew soul, a waterlogged brain, and a combination backbone made of jelly and glue. Where others have hearts, he carries a tumor of rotten principles. . . ."

The workers were compelled to stop and stare back. Dodd motioned for a squad car to move in. He let Havens finish the essay, then charged him with disturbing the peace and whisked him off to Kern County jail. Dodd was disappointed to learn that the author of "Definition of a Strikebreaker" was the same man whose books he had read and admired as a boy.

Two days later, the sheriff upped the ante once again, forbidding the use of the word *"huelga."* Chavez decided to challenge this edict under the glare of cameras. "More than a victory issue was the morale issue," Chris Hartmire recalls. "You have people standing all day long and not even speaking, not even trying to get people to come out. So that's when Cesar cooked up this idea." On October 19, Cesar's wife, Helen; thirteen farmworkers; and several clergymen from the Bay Area and Los Angeles were dispatched to shout the forbidden word at the W. B. Camp Ranch outside Delano.

Helen Chavez and the other pickets took up their positions across from a harvest crew, just far enough away to justify shouting. With reporters from several television and newspaper outlets watching, the pickets suddenly burst into repeated cries: *"Huelga! Huelga!"* Sergeant Dodd immediately declared the gathering an unlawful assembly, ordering deputies to round up the accused and place them under arrest. Reverend Hartmire, whose mission was not to chant but merely to talk to reporters, was grabbed from behind and shoved hard into the back of a squad car—cutting short his interview with Harry Bernstein of the *Los Angeles Times*.

Forty-four pickets were transported to Bakersfield in paddy wagons and police cars, and booked into the county jail. Bail was set at $276 each, and Helen Chavez and twelve other women spent three days in jail. Like her husband, Helen tended toward stoicism: "Being in jail didn't scare me because I know that what I'm doing is right."

At the political level, the arrests were a coup: As planned, Cesar was at the University of California at Berkeley that same day, where he was speaking to hundreds of students who had just been through the fight for free speech on their campus. When Chavez announced that his people had been arrested for shouting *"huelga,"* the students broke into choruses of the word and took up a collection for the cause; they also deluged Delano with telegrams protesting the arrests. After stops at Stanford University, Mills College in Oakland, and San Francisco State University, Chavez returned to Delano with $6,700 in cash donations. Soon after his return, 350 people gathered to picket and sing protest songs outside the Kern County courthouse, and Helen and the others were released. The sheriff's order was later declared unconstitutional.

The union didn't know it, but the incident inspired more than student sympathy and good press. FBI agents in Los Angeles sent an "urgent" teletype to Washington on October 22 to inform the mother office about the arrests. "These individuals and others expected to arrive in Delano from San Francisco, Berkeley, and Los Angeles, California, tonight and tomorrow, will make themselves subject to additional arrest this weekend. New arrivals expected to be college professors, students, and ministers. . . . Officials in area fully cognizant of situation, and state it is not possible to predict potential of situation at present."

O NE OF THE youths inspired to head toward Delano in October was a young college graduate, Luis Valdez. Chavez and Dolores Huerta, who was also becoming a popular speaker on college campuses, were developing a following among urban Chicanos eager to help a new Mexican American civil rights movement just a few hours away from them. Many urban Chicanos had relatives who were farmworkers, or they had picked crops themselves and knew the sweat and toil it involved.

■ *Using masks, picket signs, and ingenuity, the bilingual Teatro Campesino regaled strikers with clever skits and used comedy to teach farmworkers their rights.*

Valdez had been born, coincidentally, to a Delano farmworker family that moved to San Jose when he was fourteen. It was there that he discovered he had a talent for comedy and theatrics. He studied drama at San Jose State University, and by the time the strike hit, had moved to San Francisco, where he was working with the irreverent San Francisco Mime Troupe. Already a passionate critic of the draft and the war in Vietnam, the twenty-five-year-old Valdez was eager to celebrate his Mexican-American heritage and thrilled that someone was finally sticking up for the Mexican farmworker.

In June 1964, the radical Valdez had journeyed to Cuba, a trip that earned him the attention of the FBI and, later, proved fodder for scandalized growers. Returning from Havana that August, on an airplane flight from Chicago to San Francisco, the innocent Valdez regaled the person sitting next to him with lively tales of his visit to Cuba, never suspecting that the passenger would soon report their conversation to the FBI. The FBI informant claimed that Valdez, fresh out of college, was reading a book by Lenin. "Valdez exhibited a photograph of Valdez and Fidel Castro taken at a baseball game," the FBI report added.

In October of that year, Valdez penned a strident letter to the "señores" at the Selective Service in San Jose, bitterly denouncing the draft and American support for dictatorships in Nicaragua, Spain, and the Dominican Republic. On fire with youthful

■ *Cesar Chavez and Reverend Jim Drake, Chavez's assistant, appreciating the Teatro Campesino's brand of humor at a performance in 1966.*

indignation, he asked how they expected him to respond to their call for "assassins" when "a stupid, racist U.S. senator such as George Murphy can praise the use of bracero labor because Mexicans are 'built close to the ground' "; and when "reactionary, fascist gringo farmers through the California Growers Association refuse to pay Mexican Americans slaving in the fields the minimum wage necessary to the survival of their families." This letter to the Selective Service was added to his growing FBI file and, after he arrived in Delano, to the expanding collection of memos on the farmworkers union.

Valdez would go on to become a well-known and stubbornly independent director, producing such evocative Chicano films as *Zoot Suit* and *La Bamba*. At the time, however, he wanted nothing more than to follow *la causa* back to his roots in Delano. "Cesar came to San Francisco—and Dolores, too—in the first weeks of the strike, and that's when I approached him." Valdez's previous attempt to talk with the distracted Chavez in Delano had been unsuccessful. "So I took this opportunity," he says, "and literally chased him all the way from the Mission District to Oakland over the Bay Bridge."

Chavez and El Teatro Campesino

by Max Benavidez

One of the most exuberant and unexpected out-growths of the farmworkers movement was Teatro Campesino, whose phantasmagoric dramas have captivated audiences not only in the United States but in Mexico, Central America, and Europe. The Teatro "is somewhere between Brecht and Cantinflas," a bilingual theater that borrows heavily from Mexican folk humor and is "salted with a wariness for human caprice," according to founder Luis Valdez. In this essay, poet Max Benavidez examines Teatro's genesis in Cesar Chavez's movement and farmworker culture.

El Teatro Campesino, The Farmworkers Theater, recently celebrated its thirtieth anniversary. In 1967 the troupe left its spiritual home in Delano to take up residence in the small California town of San Juan Bautista, where it opened its "Packing Shed Playhouse." As Teatro director and impresario Luis Valdez put it, the actors were still migrant workers in search of a permanent home. Evolving into the most famous Chicano theater group in the world, it toured thousands of miles across the United States with such plays as *Corridos, Soldier Boys,* and *Zoot Suit.* The troupe has also traveled to Europe, where it participated in the World Theatre Festival in Nancy, France. Yet, for all its fame and success, Teatro never forgot its true roots: the strike-agitated fields of Delano and their mentor, the legendary union organizer, Cesar Chavez.

"Without Cesar," said Valdez, "there would have been no Teatro. When I asked him if I could put together a theater company, Cesar told me: 'There is no money. Nothing. Just workers on strike.' But he

■ *Luis Valdez and Agustín Lira during a Teatro performance in 1974.*

also told me that if I could put something together, it was fine with him. And that was all we needed—a chance. We jumped on top of a truck and started performing. Then something great happened. Our work raised the spirits of everybody on the picket line and Cesar saw that.

"Cesar was supportive of our work," Valdez said, "until the day he died. He understood what we were all about. 1967 was a turning point. El Teatro went its own way. We moved from Delano to Del Rey and there we established an art center. That year, Antonio Bernal painted the first outdoor Chicano mural. Teatro also made the first Chicano film, an adaptation of Corky Gonzales's poem, 'I am Joaquin.' We shot the film in a kitchen in Fresno. That's how it was."

UFW organizer Doug Adair recalls that the "hilarious" Teatro skits took strikers' minds off their financial problems, especially when the "Children's Teatro" performed: "People would laugh so hard they cried." Richard Montoya, actor and founding member of Culture Clash, remembers that, as a child, he was intoxicated by the sight and sound of Teatro actors performing on an outdoor stage lighted by car headlights in the fields of the San Joaquin Valley. "It was all so vital and alive," he recalls, "but there was also a sense of danger that was exciting, because you never knew when some Teamster goons might show up and start busting heads."

According to Valdez, the late sixties marked a turning point for El Teatro and other Chicano artists. Their scope expanded beyond the farmworkers to more global issues such as Chicano identity, racism in education, the Vietnam war, and police brutality.

"But always," Valdez pointed out, "the cultural root is the campesino, the farmworker. I don't care how sophisticated we get in the city, we share the communal remembrance of the earth. This goes for Chicanos as well as anyone else."

He concluded: "Like many Chicano artists, Cesar was self-taught. What amazed me was that he could completely absorb everything around him. He was brilliant, a genius. He didn't just read about Gandhi, he became a living late-twentieth-century version of him transposed to the American Southwest. He didn't just read about labor movements, he started one. He didn't just read about the arts, he became them."

Once he caught up with Chavez on the other side of the bay, the young actor asked Cesar if he would object to him coming to Delano to start a street theater group, in the satirical agitprop tradition. Together he and the farmworkers would come up with ideas and stage the skits, boosting morale in the process and maybe persuading a scab or two that the strike was so much fun they'd have to join. "He said, 'Sure, come along,'" Valdez remembers. "The only thing is that you got to know, there's no money, there's no actors, there's no time for you to rehearse even. Do you still want to do it? And I said, 'Absolutely! What an opportunity. Let me join you, Cesar.'"

Chomping a trademark cigar with great aplomb and sporting a handlebar mustache, Valdez arrived back in Delano in October, to found the Teatro Campesino, or Farmworkers Theater, in collaboration with striker Felipe Cantu and singer and farmworker Agustín Lira, who'd already been entertaining picket lines. One of the more imaginative and enduring symbols of the UFW, the Teatro has become a California institution, playing in opera houses and barrios alike. The *huelga* was the spark igniting an explosion of Mexican-American mural and visual artistry, song, and theater. "The urge to create, to express had been there for some time," Valdez explains. "It was just waiting for the *huelga* to come along and set it off."

When he got to union headquarters, however, Valdez found theater was the last thing on people's minds. Strikers were scrambling to cover the vast area they had to picket, struggling to feed themselves and pay bills and car payments and rent. Only after nine o'clock at night could Valdez finally gather a group of farmworkers to get the *teatro* started. He began by working on acting and improvisation techniques, but a woman raised her hand and asked him when the show would begin.

"So I said, 'Well, you're the show. You're the actors,'" he explained. A dozen or so faces frowned and heads shook in front of him, and he realized he hadn't done all his homework. A couple of days later he collected some more picket signs and gathered the workers in the kitchen of the pink *huelga* house. "I started to improvise with the workers. I asked them to do what they did on the picket line. At first nobody wanted to play a scab, but we got a couple of people who were inclined to be clowns to get out and do the scabs, shouting epithets and stuff. The strikers began to get into it. It was an amazing evening because it was like an explosion."

Working with cartoonists at *El Malcriado,* the *teatro* developed characters drenched in irony, like the grower, the *Patroncito,* a well-fed boss in sunglasses struggling to learn Spanish so he might enjoy the good life of the Mexican field worker. There were characters who explained abstract ideas, like a person dressed as a union contract, brandishing a huge pencil in defense of the workers. Another was the grower who told a worker he was going to take away his seniority. "*Mi señora?* No, you're not going to take away *mi señora,*" the wide-eyed worker responds. During "intermissions," musicians like Agustín

Lira would lead crowds in *"De Colores,"* a religious hymn in Spanish that became the union's anthem.

Looking back, Luis says, "I'll never forget a moment on the picket line early on when someone was on the bullhorn talking about *dignidad* ('dignity'), and a campesino came up to me and said, 'What's *dignidad?*' He didn't understand the word dignity. And it was a good question because it stopped me for a moment. How do you define dignity?" Teatro turned the question into a skit. Most of the plays were performed on the back of flatbed trucks, entertaining strikers and convincing many reluctant workers to join the cause. "The farmworker who has never said anything is now speaking," Valdez said after the theater group started attracting press coverage.

Explaining why he considered the strike long overdue, Valdez brings up an issue that, perhaps even more than poor wages, symbolized how farmworkers were deprived of simple human dignity. "There were no bathrooms out there, not even for the women," he recalls. "You consider you're out there, miles out in the middle of the field. . . . Where do women go to the bathroom? They would be forced to go out behind the parked cars. And walk a mile or two to get there and a mile or two to get back because they were out in the middle of the fields—if they were allowed to get permission to leave their hoes. . . . And in most cases, they were not even allowed to do that."

Chavez "knew the only way you could resolve those problems would be by having some kind of mediated exchange between employers and workers," says Valdez. "That's the union contract. Those are the negotiations. He wasn't talking about anything so far-fetched."

Cesar and Valdez soon developed a strong bond, one that would endure throughout Chavez's life, even though the two men were very different. At first Luis was worried Cesar wouldn't like his brand of political satire. But Cesar, although shy in public, had a healthy and very dry sense of humor. (He once pointed at field workers with hoes and told biographer Jacques Levy, "We call that Mexican golf." Another time, at the suggestion he run for office, Chavez held aloft a can of Diet-Rite, declaring imperiously "Down with the Grape Society!") He also liked the young Valdez and stuck by him, even though accusations that Luis was a Communist grew more shrill every day. The source for many of these charges was a widely circulated 1966 John Birch Society pamphlet called "The Grapes: Communist Wrath in Delano." It referred to Valdez as "Cuban-trained" and to almost everyone else prominent in the strike as a Communist or revolutionary. "From Selma, to Watts, to Berkeley, to Delano may look like a circuitous route on your road map, but it is a straight line on the road to revolution," the pamphlet intoned ominously.

Also featured in a book spun off from the pamphlet was a mug shot of Marshall Ganz, courtesy of the Mississippi police. Underneath Ganz's photo, the Birchers wrote: "Southern law enforcement authorities, who found it necessary to jail him on several occasions for his revolutionary activities there, are doubtless glad that he has returned to California."

Amid the Cold War paranoia that pervaded Delano, Valdez found that many moderate Americans outside of the valley could empathize with the strikers. "Up until the 1920s and 1930s, people lived on farms in great numbers, so it was a recent memory we were evoking," he says. "In 1965 there was still an opportunity to relate to the dust bowlers, there was still an opportunity to relate to the people who had grown up picking cotton. . . . It was possible to tap into that sense of labor justice in America."

For Chicanos, *la causa* was to become an almost religious mission, and Cesar its prophet. "Here was Cesar," Valdez would write, "burning with a patient fire, poor like us, dark like us, talking quietly, moving people to talk about their problems, attacking the little problems first and suggesting, always suggesting—never more than that— solutions which seemed attainable. We didn't know it until we met him, but he was the leader we had been waiting for."

WITH AUTUMN'S passing, the temperature fell in Delano along with the leaves from the vines. It had become increasingly difficult to get picket crews out of their homes, let alone bundle them into jackets and gloves for their daily assignments. There wasn't much of a harvest left—the work had turned to stripping vines and clearing fields, and many of the strikebreakers had gone back to Mexico, where they had been recruited originally by labor contractors, or to the California or Arizona desert to pick vegetables.

But there was still plenty of work ahead for the strikers. In winter's first chill, Cesar directed volunteers to follow grapes from cold-storage to warehouse areas and loading docks in Los Angeles and San Francisco. Longshoremen in San Francisco refused to cross a farmworker picket line, and a behind-the-scenes fight broke out when the companies called to complain that the unions were violating their own contract. Al Green, the AWOC chief, attacked Chavez fiercely, threatening to break ranks if Cesar didn't call off the picketing and stop embarrassing other unions. Chavez lashed back, pointing out that volunteers had stopped grapes from being loaded—and that's what counted. It was an important showdown: "Don't you ever threaten me," Cesar said, "You're not big enough to threaten me, and you're not even big enough to begin to carry out your threats." He realized he'd have to fight Green to be treated as an equal.

In early December, the union took another step in trying to enlist public support for the farmworkers strike: it launched a boycott of Schenley Industries, whose various liquor brands and other products drew at least $250 million in annual sales. SNCC supporters around the country offered to help, and enterprising volunteers from California hitchhiked their way to East Coast cities to try to get the campaign going. The DiGiorgio Fruit Corporation, which operated more than thirteen thousand acres of vineyards and fruit orchards in California, would also become a primary target. Both companies already had contracts with other labor unions covering truckers and packers, giving them little excuse for ignoring field workers, and both had recognizable products that could be boycotted. In more than a hundred cities, students and other volunteers picketed and gave away cards that supporters could send to the company telling them they wouldn't buy their brands without the union imprimatur.

Meanwhile, in San Francisco, the AFL-CIO was about to convene its annual meeting. Paul Schrade, western states director for the United Auto Workers, asked powerful UAW president Walter Reuther to pay the Delano strikers a visit to boost morale. On December 16, fresh from an AFL-CIO vote endorsing the strike, Reuther joined Chavez and Itliong at the head of a march through the streets of Delano. It was the fledgling movement's first public recognition from the giant union, but Reuther seemed right at home. He carried a new plywood National Farm Workers Association picket sign, and told hundreds of cheering farmworkers: "There is no power in the world like the power of free men working together in a just cause. If General Motors had to change its mind because of the autoworkers, then the growers have to change their minds; and the sooner they do, the better for them, the better for you, and the better for the community."

Police had warned that any protesters who marched without a permit would be arrested. But Reuther's presence, and the dozens of journalists it attracted, dissolved the threat. Reuther met with the mayor and the city manager and even with a contingent of growers. His advice to them was simple: "Sooner or later these guys are going to win."

Afterward, the UAW president spoke at Filipino Hall, taking the stage with Itliong and Chavez. "This is not your strike, this is *our* strike!" he declared to wild cheers. Then Reuther made a stunning announcement: the UAW was pledging to give $5,000 a month each to both unions for the duration of the strike. It was an incredible victory for the independent National Farm Workers Association, although not as easy to achieve as it may have appeared. Cesar had worked behind the scenes to secure the money.

Encouraged by this infusion of support, the union leaders returned to their picket lines with renewed vigor. Singing and chanting, Teatro Campesino performed next to vineyards where only a handful of strikebreakers were now pruning naked stems in preparation for the next season. The Teatro became a regular presence in the valley's dense midwinter fog, lightening the days' load with wit and humor. Strikers were further cheered when a couple of trailer trucks loaded with toys and turkeys arrived to rescue Christmas for the families of the *huelgistas.*

Delano was also becoming a familiar place to journalists and policymakers. As winter warmed into spring, politicians of every stripe descended on the small town. The California State Senate Factfinding Subcommittee on Un-American Activities arrived in town to conduct a "preliminary investigation" of the strike, according to a March 9, 1966, FBI memo. But on March 14, strikers won the upper hand again in the media battle when reporters converged on Delano for a series of hearings by the United States Senate Subcommit-

■ *Larry Itliong and Cesar Chavez of the Agricultural Workers Organizing Committee flanking Walter Reuther, president of the United Auto Workers, during a visit to Delano in December 1965. The UAW donated $5,000 a month to the strikers.*

tee on Migratory Labor. Senator Robert Kennedy was on the panel, as was Senator George Murphy of California, a conservative friend of agribusiness. (A Murphy aide had called the FBI on March 7, hoping the agency would help him develop advance information that might be useful in his interrogation of the committee's witnesses. The agency demurred, and suggested Murphy contact an "appropriate liaison at the Seat of Government.")

As the hearing got underway, Murphy squirmed noticeably as Chavez offered convincing examples of police bias. The farmworker leader had also brought sworn affidavits that accused growers of illegally importing strikebreakers from Mexico and other areas without informing them that they would be walking into a strike. Chavez reminded the senators that Congress had considered bills that would have guaranteed American farmworkers the minimum wage and collective bargaining rights and abolished child labor— rights taken for granted in other industries. There the bills, introduced by one of the panel's own members, Senator Harrison Williams, had failed, he added.

"All that these bills do is say that people who work on farms should have the same human rights as people who work in construction crews, or in factories, or in offices," declared an impassioned Chavez. "All these bills do is to overcome the farm lobby that Franklin Delano Roosevelt's administration was subjected to in the thirties which forced them to decide that farmwork and farmworkers were somehow different from everyone else." Winding down his testimony, Chavez ended on a satirical note: "Ranchers in Delano say that the farmworkers are happy living the way they are—just like the southern plantation owner used to say about the Negroes."

The growers lost another round when Bishop Hugh Donohoe of Stockton testified, representing the Catholic Bishops of California, a group that until then had remained officially neutral. Breaking the Catholic hierarchy's silence, Donohoe avoided endorsing the grape strike, but told the senators that the bishops had concluded there was "no compelling reason" for excluding farmworkers from the National Labor Relations Act, or forbidding them from resorting to a strike when all other attempts at compromise failed.

But a lively exchange that took place two days later, and stung the valley's establishment to the quick, would have the greatest impact on the hearings. The historic give-and-take took place when Kern County sheriff Roy Galyen agreed to discuss the police harassment that Chavez alleged during his testimony. The senators had been disturbed when Chavez told them that the sheriff had ordered deputies to photograph and interrogate all pickets for the department's dossiers. He went on to describe the *huelga* arrests in October, along with a mass arrest of pickets that occurred later—allegedly to prevent angry vineyard employees from assaulting the strikers.

Kennedy seized the opportunity: "Do you take pictures of everyone in the city?"

GALYEN: "Well, if he is on strike, or something like that."

Kennedy then politely inquired why the pickets were arrested as a preventive measure.

GALYEN: "Well, if I have reason to believe that there's going to be a riot started, and somebody tells me that there's going to be trouble if you don't stop them, then it's my duty to stop them."

KENNEDY: "You go out there and arrest them?"

GALYEN: "Absolutely."

KENNEDY: "Who told you that they were going to riot?"

GALYEN: "The men right out there in the field that they were talking to said if you don't get them out of here we're going to cut their hearts out."

KENNEDY: "This is a most interesting concept, I think, that you suddenly hear talk . . . about somebody's going to get out of order, perhaps violate the law, and you go in and arrest them, and they haven't done anything wrong. How do you go arrest somebody if they haven't violated the law?"

GALYEN: "They are ready to violate the law, in other words—"

KENNEDY: "Could I suggest that in the interim period of time, in the luncheon period of time, that the sheriff and the district attorney read the Constitution of the United States?"

The room rocked with laughter at Kennedy's retort, with the exception of the sullen growers. It was a quip they would not soon forgive. Nor would they forget Kennedy's subsequent, unexpected jump into the farmworkers' camp: After the hearings Kennedy headed for Filipino Hall, where incredulous farmworkers, who revered his late brother, heard him declare support for the grape strike. He even went so far as to march on a picket line at the DiGiorgio Fruit Corporation's 4,400-acre ranch, where Helen Chavez had worked. Paul Schrade of the UAW, who later became a key Kennedy supporter, says of the historic show of support that day: "I'd never seen any Democrat, much less a Republican, politician do that before."

IT WAS DURING a union brainstorming retreat in the winter of 1966 that the notion of a massive farmworkers march was first raised. The union needed something dramatic to lift spirits, and to continue to push the strike beyond the borders of Delano. How about a farmworkers cross-country bus trip to New York, where strikers could camp out in front of the headquarters of Schenley Industries? What about a march to San Francisco, ending with a rally in front of the company office there? Union strategists settled on the idea of a march to Sacramento, with the goal of pressuring Governor Pat Brown to lean on Schenley and DiGiorgio to negotiate.

It was decided: on March 17, a group of farmworkers would begin to walk three hundred miles from Delano to Sacramento, dramatizing the six-month-old grape strike with the longest protest march ever in the United States. Cesar suggested the march be called a *perigrinación*, or "pilgrimage," which would light up farmworker towns across central California as marchers stopped to pray and rejuvenate themselves in the small campesino barrios that stretched like a chain up the valley. The theme of the march was to be *Perigrinación, Penitencia, and Revolución,* "Pilgrimage, Penitence, and Revolution"—appropriate messages during the holy season of Lent. Arriving in the state capital on April 10, Easter Sunday, the marchers would present their grievances to Governor Brown. Cesar also proclaimed that the march would help prepare farmworkers for "the long, long struggle" ahead. "We wanted to be fit not only physically but also spiritually, and we wanted to stress nonviolence even more, build confidence, and have more visible nonviolent tactics."

Angie Hernandez Herrera, Julio's daughter, was twenty at the time of the march. She remembers the excitement that pervaded the Mexican barrio in Delano the morning of March 17, when about one hundred disorganized union supporters gathered to commence their walk. "We were right there on Albany Street, with the organizers trying to get us in line," she said of the march's start. "But as soon as they did, others would fall out. It was like a big party."

Diary of a Strikebreaker—1974

by Demetrio Diaz

What about the faces on the other side of the picket lines—the strikebreakers, scabs, esquiroles? Seventeen-year-old Demetrio Diaz was among them. An "illegal" worker in the United States picking lemons at Arrowhead Ranch in Phoenix, Arizona, which the UFW struck in 1974, he was one of an estimated 100,000 Mexican workers being smuggled into the country to break UFW strikes in Arizona, according to the union. In this account, published in El Malcriado, he explains why he joined the farmworkers union.

Five months ago a man came to our village of Cuamil, Michoacán, and contacted us. He said, "Don't you want to go North? Up there you will earn lots of money, hundreds of dollars. All you have to do is pay me and walk for two days." We believed him and I went back home and said, "Mama, there is a man who will take me to the North. Lend me money." So she borrowed 1,500 pesos from my grandfather and I went with him.

We crossed the border frontier and began to walk for four days in the desert. There were twenty of us on this crossing with a "coyote" who called himself Alberto. He never got tired of running, pushing us along. He wouldn't let us sleep. There were spines and tough thorns that pierced my skin, and we got our water from the nopal cactus. As we walked along we saw cadavers, human bones. [We] continued the 150 mile march to Casa Grandes, where we were met by the foreman of Arrowhead Ranch. He brought us to some fields near the ranch and there we slept.

The following day he took us to work in the lemons. We tried to tell [him] that we couldn't work, that we were too tired and our feet were swollen. "Whatever you say," he said, "but *la migra* [INS] will be here for you right away," and since we didn't know any better, we had to go.

They paid us thirty cents a sack of lemons, and if you were really fast, you could make twenty boxes a day—six dollars. Many times when the [Chicano] foreman paid us our check it was zero, because he would charge us Social Security and the ride to the orchard—two dollars per person. And if he took us to the store, he would charge us, too.

Every day we asked him to buy food for us. Then he would deduct as much as $15 per week for a little sack of flour, two dozen eggs, and some lard.

There were many days when it rained and we wouldn't be able to work, so the foreman didn't bring us any food. We had to make it through somehow so we ate oranges. We would go inside our little box houses, but the rain came in and we were wet all the time.

We went two days without drinking water, without eating, and with such hunger. . . . But we would not leave the camp because the foreman told us the *migra* was outside. He had us scared.

I was not used to sleeping under trees and many of us got sick. The other day a boy broke his foot running behind the tractor. He slipped and his foot went under the tractor and was crushed. He fainted and we took him from the orchard to the foreman, but he just said, "You boys should be more careful." The boy was older, almost twenty-five, but he was crying and afterwards, he couldn't walk right.

The boss does all the work with illegals. He never hires any others because we can't do anything to him. He kept us against our will. There's not one of us who doesn't have debts to the contractors. . . . They treated us like their slaves and I am no one's slave.

One day I finally told him I was going to leave. I was covered with filth—there was no place to bathe, or wash clothes, or buy new ones. He said he was going to call immigration but I thought, if they come, it doesn't matter. Then I saw the strikers. My boss had told us, "Don't believe them, they're crazy." Pay no attention to them: you're here to work." But the next day I was going to work, a girl from the *huelga* came close to me and said, "Come," and I went with her.

Cesar led the parade, which, before heading out to Highway 99, cut through downtown Delano—without a parade permit. The police took the bait, and after a few minutes of public negotiations, the chief of police, who by then had become more savvy about the press, relented. The exuberant marchers proceeded. Invisible to the marchers, but shadowing the protest like a disembodied spirit, was a clandestine network of FBI informants, who already had obtained a detailed itinerary of the pilgrimage and had shared it with U.S. Army commanders in Pasadena. As soon as the march left Delano, an FBI agent promptly filed an urgent memo to Washington: "March group began with about one hundred persons, about 75 percent Mexican Americans or Filipinos, and remainder Anglo Americans except for two or three Negroes. . . . Affair was given publicity by all news media and will probably receive nationwide attention."

Filing past the vineyards they had struck six months earlier, the farmworkers trudged north as the hot sun poured down upon them. They held aloft portraits of the Virgin of Guadalupe, the matron saint of Mexico and the movement's adopted symbol of hope. Others shouldered large crosses to represent the final journey of Christ, and carried a Star of David, Mexican and American flags, and cloth banners representing AWOC or emblazoned with the NFWA's black eagle, a particularly suspicious symbol to townsfolk in Delano. Some of the Chicano war veterans wore VFW caps, while others sported broad-brimmed straw hats. Luis Valdez carried the farmworker manifesto he'd been asked to write and read in each of the pilgrimage's stops. His "Plan de Delano" was inspired by Mexican revolutionary hero Emiliano Zapata's "Plan de Ayala," a declaration for land, liberty, and bread.

Not everyone was comfortable with the emphasis on Catholic imagery. Epifanio Camacho, the caustic hero of the first rose strike, refused to carry any images of virgins, crosses, or saints. (He'd already scandalized the more religious among the strikers with a bumper sticker on his truck that read "I too was a virgin once.") By the time the group reached Sacramento, however, Camacho would be there as well, wearing an expansive Mexican sombrero and waving a union flag.

Newsreels captured Chavez leading the march, limping on blistered feet, sometimes with Itliong at his side. Cesar invited Angie Hernandez Herrera, who used to baby-sit his children, to walk near him throughout most of the parade. Most of the pilgrims dropped in and out along the march, but Hernandez Herrera was among the handful of *originales*—those who walked every step of the way. "Some people had bloody feet," she recalls with a touch of awe. "Some would keep on walking and you'd see blood coming out of their shoes."

With the pain came joy, she remembers. The march was a revolution on the move, and for many, it was one of the most exciting events of their lives. As the walkers approached each new town, their ranks would swell, sometimes stretching for two miles. In the small towns and cities, the march became a feast, with excited farmworkers and community supporters grilling hundreds of tortillas and cooking up barbecued beef and traditional bean and rice dishes to feed the marchers. In each town, the tired pilgrims were assigned to different local families, where they would have a chance to shower and get a good night's sleep. Often, Hernandez recalls, walkers would be so overwhelmed with the hospitality that despite being worn down by the daily hikes of up to fifteen miles, they'd stay up late into the night, talking about the march and their dreams with host families.

The union used the subscription list for *El Malcriado* to scout out places where people could sleep at night in each town along the way. "We walked into Parlier, that's the night I remember," says Doug Adair, "walking through this little town in Fresno County, and every citizen of Parlier was welcoming us. In those days, the barrio part of Parlier, the streets weren't even paved. The women had prepared a big feast, and then in the community hall or the Catholic church we had a big dinner. Then we had a candlelight parade all through the barrio, back and forth, back and forth, and then back in this hall, and Luis reading the 'Plan of Delano,' and the Teatro, and the singing. It was just overwhelming, it was such a beautiful experience. That whole town was solid."

Each night, Valdez unfurled "El Plan de Delano" to thrilled crowds who would gather for the festive rallies. The vintage sixties manuscript, laced with quotes from Pope Leo XIII and tribute to the Virgin, so disturbed the establishment that it was reprinted a year later in a report by an alarmed California Senate Factfinding Subcommittee on Un-American Activities. "We are sons of the Mexican Revolution," the essay declared. "Along this same road, in this very same valley, the Mexican race has sacrificed itself for the last hundred years. Our sweat and our blood have fallen on this land to make other men rich. . . . We seek our basic, God-given rights as human beings. . . . We do not want the paternalism of the ranchers; we do not want the contractor; we do not want charity at the price of our dignity. We want to be equal with all the workingmen in the nation. . . . Wherever there are Mexican people, wherever there are farmworkers, our movement is spreading like flames across a dry plain. . . . *Viva la Causa!*" (Luis Valdez's writing turned out to be prophetic. On April 4, the FBI's Chicago office reported that more than 250 Chicagoans had marched through a Mexican American and Puerto Rican barrio in support of the California migrant workers, bearing placards reading *Huelga!* And a year later, farmworkers in Texas joined in a massive four-hundred-mile march from the Mexican border to the state capital.)

At each stop, *la causa* attracted new followers. The *perigrinación* so moved Hijinio Rangel that he quit a coveted job as a tractor operator at a fruit ranch to sign on as a union volunteer. When he first heard the march was headed toward his home in Dinuba, Rangel had no thoughts of changing his life. He enjoyed his work and had no complaints about it, but he had seen how badly the field workers at his own company were treated. When the parade finally arrived, Rangel remembers working his way to the front of the procession so he could ask Chavez a few questions. After several minutes of listening, he found himself converted. Soon the tractor operator was out rustling up food for the marchers, and despite the sacrifice his large family had to make, he volunteered in Delano.

As the procession attracted new followers, the FBI, with its informants and agents, recorded the twists and turns of the march in journalistic detail. One informant noted almost admiringly, "The marchers have a field kitchen for emergencies, as well as chemical toilets, and they are in possession of a car equipped with a two-way car radio, bearing the call number KFZ-478."

As the pilgrimage snaked northward, opponents of the strike protested by setting out card tables with bottles of Schenley liquors and signs declaring their support for the company. The tactics failed to provoke the marchers. The farmworkers, some of whom had never expressed their feelings so boldly, were surprised, then delighted, when strangers would honk horns in support. "White people would honk, too," Hernandez Herrera remembers with amazement.

Midway through the march, on March 25, Cesar received word that Governor Brown was going to Palm Springs to vacation at Frank Sinatra's house over Easter weekend. A special contingent of strikers formed to head south to the desert, and with the help of a hastily formed Riverside County committee that supported the grape strikers, sent a fruitless telegram to Brown asking to meet him in Palm Springs. Since former president Eisenhower was also in Palm Springs golfing, the FBI put out an alert to army personnel and the Secret Service to watch for disturbances.

On the final leg of the pilgrims' journey, a few days before Easter, Gil Padilla rushed up to Cesar before a rally in Stockton and told him a guy who said he was from Schenley Industries was on the phone—and that he wanted to sign a contract. "Oh, the hell with him," a disbelieving Cesar said. After more lobbying from Padilla, Chavez finally agreed to talk to the caller. It was Sidney Korshak, a lawyer who handled the corporation's affairs with other labor unions. Cesar was incredulous. The boycott was working, or maybe Kennedy had exerted a little pressure, no one knew for sure. But Schenley, whose products a Teamsters local in San Francisco had refused to load, was starting to find the strike and the boycott annoying.

It turned out a false rumor was the last straw for Schenley. The company heard that bartenders affiliated with a California local of the Bartenders International Union were considering promoting a boycott of the company's liquors. It wasn't

■ *Cesar Chavez limping on blistered feet during the farmworker pilgrimage to Sacramento that he led in March and April of 1966. The march made national headlines, attracted thousands more supporters, and resulted in a contract with a major grape grower.*

true, but a sympathetic union officer and a union secretary, who had earlier met Chavez and found him "a nice little guy," circulated a phony memo hinting at such a boycott. The company flipped. The chairman of the board of Schenley was so angered by the imagined boycott threat that he almost sold the company's five thousand acres of grapes. Korshak, however, advised him to settle instead and be done with it.

After the phone conversation, Chris Hartmire drove Cesar and a group of workers all night to Korshak's mansion in Beverly Hills. Bill Kircher, the AFL-CIO's national organizing director, was there, along with a Teamsters representative. When Kircher and the Teamsters rep started arguing, Cesar went off in disgust to play billiards on the lawyer's pool table. A Los Angeles AFL-CIO representative tried to talk Cesar into turning the Farm Workers Association over to AWOC, and Cesar refused: "When we fought for it, bled for it, and sweat for it? You must be out of your mind!" Kircher took Chavez's side, sensing a historic opportunity for the labor movement.

Schenley agreed to recognize the National Farm Workers Association, and a preliminary agreement was signed that night. Huerta was put in charge of drawing up a full contract that the two sides would negotiate within ninety days. It was a miracle: the workers would get a raise of thirty-five cents an hour and a hiring hall immediately.

Chavez rushed back to the pilgrimage, and on Easter Sunday an exultant throng entered Sacramento, headed by several Chicano flag bearers on horses, wearing black sombreros and dressed in mariachi garb. The crowd swarmed the capitol steps, where Dolores, wearing a rakish cowboy hat, her face radiant with joy, took the microphone: "On behalf of all the farmworkers of this state, we unconditionally demand that the governor of this state, Edmund Brown, call a special session of the legislature to enact a collective bargaining law for the state of California. We will be satisfied with nothing less. . . . You cannot close your eyes and your ears to us any longer. You cannot pretend that we do not exist. You cannot plead ignorance to our problems because we are here and we embody our needs for you. And we are not alone."

Cesar, fresh from a triumph he had been dreaming of most of his life, chose to be conciliatory when he addressed the crowd. He thanked the Teamsters, the Longshoremen, and the AFL-CIO, the church, and all the students and civil rights workers who had helped them win this one victory. In Spanish he told the farmworkers: "It is well to remember there must be courage, but also, that in victory there must be humility."

This first victorious moment for the new farmworkers union would be sweet, but short. Schenley was just one of dozens of grape growers who had been holding out against the union. The real work lay ahead.

■ FOLLOWING PAGE: *Union men joined students and others in publicizing a boycott of all California table grapes that began in 1968. It was the most ambitious consumer boycott in American history.*

I saw these two young white men walking toward me. . . . I made myself strong and went up to them and said, "Excuse me, could I ask you to help farmworkers by not buying grapes?" As if on cue, they both turned around and showed me their jackets, which had giant United Auto Workers emblems on them. And they said, "We're all for you."

—JESSICA GOVEA

CHAPTER FIVE

BOYCOTT

IN JULY 1966, three months after Cesar Chavez and the farmworkers marched into Sacramento, Adelina Gurola was at her job with
the DiGiorgio Corporation, a farming empire whose produce and
canned vegetables and juices lined shelves in grocery stores across the
country. The fifty-year-old farmworker was sorting plums as fast as her
hands could fly when the supervisor called a lunch break. Relieved,
Gurola headed for the railroad tracks with her friend Josefa when she
saw a young man solemnly passing out leaflets. Josefa snapped at the
boy, asking him why he even bothered with them when "burros" like
her couldn't read. But Adelina scolded her for being rude. "He's probably just one of those Jehovah's Witnesses," she whispered. She was
Catholic, but wanted to be polite. Smiling at the boy, she took a leaflet
and slipped it inside her purse without so much as a glance.

After the break, Gurola reported back to work at DiGiorgio. She
was just starting in on the plums again when her supervisor, Mariana,
approached. "Adelina, I don't got no place for you anymore," she told
her coldly.

Gurola was almost speechless. "But I haven't done anything wrong!"
she cried out. She'd been in the San Joaquin Valley town of Arvin since
1948 and had worked for DiGiorgio for years. But Mariana was adamant:
She was fired. With no one to appeal to, Adelina left, bewildered and furious. Later that day, Josefa came by her home with some urgent news.

"Do you know why Mariana fired you, Adelina?" she said.

"Why?" Adelina answered angrily. "Because she's a *cabrona*?"

"No! She fired you because you took a paper from one of those
strikers."

"What! What strikers?"

The leaflet was buried, still unread, deep in Gurola's purse. She pulled it out, noticing it had a return address in Bakersfield and asked interested farmworkers to come join the union. Armed with this information, Gurola persuaded her reluctant boyfriend, Teofilo Garcia, to drive her to Bakersfield, where both were taken aback by the sight of many other DiGiorgio workers milling around outside the union office. At first Gurola was afraid to go inside, but she realized she had nothing to lose. And that, she remembers, was the day the DiGiorgio Corporation, thanks to its own vindictiveness, unwittingly converted Adelina Gurola into another union recruit as *la causa* rolled down the valley.

AFTER SCHENLEY VINEYARDS signed with the union in April 1966, shock waves reverberated through the San Joaquin's grower grapevine. The next to break ranks was the DiGiorgio Corporation, which floated a hint over the radio just as Chavez was marching into Sacramento that it might be willing to hold a union election for its workers—even though the law didn't require it to do so. Chavez jumped at the opportunity to talk.

As he trudged into Sacramento on the final day of the pilgrimage, Chavez offered conciliatory words in the soft Chicano accent that was becoming ever more familiar to Californians. "We go back to continue the strike that's lasted now almost seven months," he said, "with great hopes and expectations that it will not last much longer, and that we, as civilized men, can sit down together with employers."

A round of talks began in Fresno with DiGiorgio executives. Chavez broke them off abruptly, however, after union members who were trying to talk to workers in the company's Delano vineyard were met with violence. DiGiorgio guards threatened a woman organizer with a gun and struck a male unionist so hard he needed thirteen stitches. Chavez retaliated with a weapon that was proving highly effective: He declared a boycott against the company, including its ubiquitous TreeSweet brand juices and S&W canned foods. The NFWA turned to its student supporters and old-time trade unionists to start picketing stores and warehouses as well as the company's headquarters on San Francisco's Market Street.

Going into the boycott, the union was bolstered by the Schenley victory and the high-profile exposure that the pilgrimage had bestowed upon it. But Chavez knew that taking on DiGiorgio was like wrestling with a titan. One of California's farming giants, DiGiorgio had annual sales of more than $230 million from its vast fruit orchards and processing plants. Robert DiGiorgio, the son of the company's founder, was on the board of the powerful Bank of America. Arvin, where the company farmed 9,000 acres of fruits and vegetables, was a company town. DiGiorgio also owned a 4,700-acre ranch in Delano—where Helen Chavez had worked—as well as thousands more acres of grapes, fruit trees, and Florida citrus.

More ominously, DiGiorgio also was notorious for refusing to allow organized labor among its field workers, so much so that John Steinbeck had modeled the violent Gregorio ranchers after the DiGiorgios in *The Grapes of Wrath*. In 1939, DiGiorgio broke a strike in Yuba City with help from a sheriff's posse that destroyed a strikers' soup kitchen, pummeled strikers, and drove its leaders out of town. In 1948, Adelina remembers, the company pitted Mexicans against striking Anglo and African American harvest crews by illegally bringing in braceros as strikebreakers. When the braceros stopped working in sympathy, DiGiorgio went so far as to contact Mexican officials, who forced the men to go back to work.

■ *Chavez asked his brother Richard to build an altar on the back of his station wagon so strikers could pray outside the DiGiorgio Corporation for elections and contracts.*

Chavez had already come up against DiGiorgio once before. In 1964, a year before the grape strike, DiGiorgio tried to replace women workers from its Sierra Vista ranch in Delano with braceros. Unknown to the company, Chavez convinced state labor officials to investigate. To the anger of DiGiorgio officials, the state officials ordered the company to lay off the braceros and rehire the women.

"We were facing a giant whose policy was to break legitimate unions," Chavez recalled. "They had done it before, and they were very comfortable at it. But they met with a very different brand of unionism when they met us." His plan was to slowly loosen DiGiorgio's grip on local judges and communities by pressuring the company with methods it had never even dreamed of. He would hit DiGiorgio from all sides—in the cities, where the boycott promoted by students and housewives and unionists would scare the company back into negotiating, and out in the fields, where the spirit of the strikers had begun to flag.

After seven months, many union members had found themselves broke and desperate for money. Some strikers, including those who had walked out of a DiGiorgio ranch in Delano, drifted back across the picket lines; they couldn't realistically get by on what little food and supplies were available at Filipino Hall. Chavez tried another tactic: in secret meetings, he asked workers to strike from within the ranch by persuading others to back the union and engaging in slowdowns, what the laborers called, in Spanish, *planes de tortuga,* or "turtle plans." Chavez's imaginative strategies included "anything that was legal and moral, but that would cost the grower more money." The union also developed an inside network of informants: foremen who acted as *submarinos,* or "submarines," who would relay company plans to the union.

DiGiorgio had its own battle plan: The company was importing a steady stream of fresh migrants from Texas and Mexico, who would travel thousands of miles in buses and arrive, unsuspecting, in the middle of Delano's tumultuous strike. Dolores Huerta was dispatched to Texas to try to stem the flow by picketing contractors who were rounding up the migrants. Talking with migrants as they arrived in Delano, the union uncovered the most alarming development yet: DiGiorgio was actually requiring strikebreakers to sign cards that authorized the Teamsters union as their representatives. Then, in late May of 1967, DiGiorgio obtained an injunction limiting the number of pickets who could circle its Sierra Vista ranch in Delano, another blow that sapped strikers' energy and hope.

Not long after the injunction, three women approached Cesar privately and offered an idea that was intriguing in its simplicity. Instead of picketing, why not pray across the road from the ranch entrance? "My mind just flashed to all the possibilities," Chavez recalled. That same night, he asked his brother Richard to fashion a wooden altar on the back of his station wagon. By dawn it was ready. The union rushed out a stack of leaflets announcing not a picket line but a prayer meeting at the DiGiorgio ranch in Delano. Local Spanish-language radio stations also announced the meeting.

The idea of praying for DiGiorgio's workers was a smashing success as well as an ingenious organizing technique. For months, Mexican farmworkers flooded into Delano by the hundreds to pray at the shrine in the station wagon. The handcrafted mobile altar, adorned nightly with flickering candles, flowers, and images of the Virgin of Guadalupe, was such an attraction that some migrant strikebreakers braved supervisors and came out from the DiGiorgio camps to look at it. More than a few knelt and prayed, and they were embraced by the strikers, whether they returned to work or not.

WHILE STRIKERS PRAYED, the DiGiorgio boycott spread farther across America, with strongholds of support in San Francisco, New York, and Chicago. The union also unveiled another tactic: blocking distribution of grapes at the source. That May, as the FBI looked on, a peaceful but lively march of about sixty trade unionists and students in Chicago blocked an S&W Foods distribution center, which served the entire Midwest. Many members of the Teamsters union refused to cross the picket line, the FBI noted, and the Catholic bishop of Chicago allowed union organizers to use his office.

These unexpected tactics wore down DiGiorgio's resistance. In June, Chavez and Bill Kircher, the AFL-CIO's chief of organizing, sat down for a new round of negotiations with the company at its headquarters in San Francisco. After several days of intense talks, both sides agreed to take a short break. Chavez and Kircher weren't gone but a day when they heard stinging news: DiGiorgio, without consulting either of them, had decided to hold elections in just three days at the Sierra Vista ranch in Delano and at a Borrego Springs ranch. The ballots had already been printed. There would be four choices: the Teamsters union, the NFWA, AWOC, and "no union." DiGiorgio planned to bring in a public relations firm, not a government arbiter, to monitor the election.

Kircher stormed a subsequent DiGiorgio press conference in San Francisco, interrupting Robert DiGiorgio in midsentence. Kircher denounced the sudden election as a setup, rigged to deliver the DiGiorgio workers to the Teamsters union. Soon thereafter, Kircher obtained an injunction to have AWOC and the NFWA removed from the election ballots they had never approved.

It was also clear now that the Teamsters union, notorious for its corruption and its connections to organized crime, was declaring war on the new farmworkers union. "The decision was made to test Chavez's resolve and learn more about his true intentions on down the line," says Bill Grami, a Teamsters director of organizing who sparred with Chavez for many years. The Teamsters, who had been expelled from the AFL-CIO in 1957 for corruption, represented many truckers and cannery workers in the agricultural industry, but had largely ignored the Chicano and Mexican field hands at the bottom of the agribusiness hierarchy. They had only one contract covering field workers, a deal Chavez considered a sweetheart agreement with a grower. The contract had been signed in 1961 with the world's largest lettuce company, Bud Antle of Salinas, which had faced a serious organizing threat from AWOC lettuce strikes in the Imperial Valley. AWOC had previously picketed Antle for using braceros instead of unemployed local workers, but the Teamsters had no objections to braceros—in fact, its contract didn't cover them at all. Instead, it explicitly promised to support Antle's need for "supplemental foreign workers." As a bonus, the Teamsters agreed to loan Antle $1 million from its pension fund.

Even though he felt that the Teamsters could be beaten, Chavez knew they were dangerous opponents. Moving quickly, he urged workers to boycott DiGiorgio's June 24 elections at Sierra Vista and Borrego Springs, and called on his closest friends for help. Fred Ross, Chavez's mentor, put aside the book he was writing and raced down to the valley to help organize protests. Soon afterward, his son abandoned Syracuse University to pitch in.

With only hours until the election, Cesar traveled to the scorching desert community of Borrego Springs near San Diego to urge workers not to vote. Many workers refused to go to the polling stations, but the Teamsters union, which had campaigned by doling out beer and branding Chavez a Communist, was declared the winner of elections at both ranches.

Chavez didn't give up. A few days after the Teamsters' tainted victory, Dolores Huerta won the support of the Mexican-American Political Association, a largely white-collar group, at its convention in Fresno, giving the union an enormous political advantage. When Governor Pat Brown showed up soon afterward for a scheduled meeting to seek MAPA's endorsement for his reelection, Chavez was there to greet him. A surprised and embarrassed Brown shook Cesar's hand, paused to gather his wits, and exclaimed loudly, "This meeting is two months overdue."

Soon afterward, Brown agreed to ask DiGiorgio for a new election. He also appointed a team of veteran mediators from the American Arbitration Association and Wayne State University in Detroit to investigate the DiGiorgio vote. U.S. senators Robert Kennedy and Harrison Williams of the Senate Subcommittee on Migrant Labor also appealed to the company to postpone negotiations with the Teamsters until the mediators could finish their probe.

Chavez, leery of most politicians, kept pressure on DiGiorgio. He went to Borrego Springs to organize a strike, and summoned Luis Valdez and the Teatro Campesino, who were on the road, touring farm towns and college campuses. When the Teatro finally rolled into Borrego Springs late one night, Valdez says, "We couldn't find the *huelga* headquarters because there *was* no *huelga* headquarters. They were staying out at a state park, where I could see these bundles—*bultos*—these mounds in the desert. There were men sleeping there. So we got our sleeping bags and proceeded to go out there, and I bumped into somebody. And Cesar rolled over and said, 'OK, you can just lie down here.' He was one of these bunches of people sleeping out in the desert sand!"

Valdez and company were up before dawn, only to find Chavez already out on the picket line, Valdez remembers, coaxing workers to come out of the vineyards. Ten strikebreakers were convinced quickly and walked out of the field. The union ran into trouble later that day, however, when some of the converts asked Chavez if he would help them retrieve a few possessions from the company labor camp. Chavez accompanied them onto the ranch in his station wagon, with Chris Hartmire behind the wheel and Victor Salandini, a Catholic priest, riding along for moral support.

Half a dozen nervous security guards, trembling and barking orders, trained their guns on Chavez's old wagon as it rolled through the entrance. Chavez asked if the men could get their belongings and leave; instead, he and the others were forced at gunpoint to climb inside a transport truck with small slits for windows. The prisoners were held there for hours under the desert sun, and workers, beginning to fear they might suffocate, started chanting for their release. At about ten that night, sheriff's deputies arrived, shackled the men together by their ankles and wrists with heavy chains, and put them in a patrol wagon. "In the backseat I fell asleep right away," Cesar recalled. "Although I was happy because of the confrontation, I was disgusted with the so-called justice we were getting. When I woke up, we were in San Diego. There were all kinds of bright lights in my face and a TV camera poking through the window taking my picture." In the city jail, the guards stripped each of the men naked and searched them.

After a night in jail, Chavez and the others were released, to be convicted later of criminal trespassing and placed on three years' probation. But the damage to DiGiorgio was done: When word spread that Chavez had been shackled and strip-searched, more outraged farmworkers flocked to the cause.

Two weeks after the arrests, Brown's investigators recommended that DiGiorgio hold another election to resolve the labor dispute under "fair and equitable" circumstances, with the American Arbitration Association supervising the vote. After being assured employees who'd walked out would be allowed to vote, Chavez agreed to suspend the strike. But DiGiorgio couldn't resist the last shot: It laid off two hundred employees just before the August 30 election.

This was the upstart union's first election, and Chavez knew two things for certain. If he didn't win, the NFWA might never recover. He also knew he needed more help to mount an effective election campaign. For months, Bill Kircher had been urging Chavez to merge with AWOC, creating one powerful organizing tool. In July, with the vote just weeks away, the membership voted to merge. Kircher offered a $10,000-a-month organizing budget for the new group, which was called the United Farm Workers Organizing Committee—a union that would later become the UFW. The executive board, composed of Chicano and Filipino leaders, selected Chavez to be director and Itliong assistant director.

TWENTY-YEAR-OLD Eliseo Medina was at the union's new hiring hall at the old Azteca Tortilleria, where farmworkers could sign up for a job picking Schenley's wine grapes. He had come too early, however. Dolores Huerta reminded Medina that the season hadn't started yet, and asked him if he'd be willing to help out a bit on the DiGiorgio election. He agreed, little guessing what was in store.

"Don't leave *anything* to chance," Fred Ross repeatedly warned the young DiGiorgio organizers, many of whom weren't old enough that summer to drink or vote for president. Medina had never witnessed such meticulous record-keeping: Ross had a card file on every voter, and for every visit with potential recruits, organizers would record the name, date, and jot down any questions the worker had about any imaginable concern. Ross told organizers to keep going back to those workers until all their questions were answered. In the end, he predicted, those cards would be added to the file of farmworkers who would vote for the union.

Gil Padilla, one of the movement's most persuasive organizers, was assigned a herculean task: He would go to El Paso, Texas, and Juárez, Mexico, to round up a group of eligible migrants who'd been laid off or had moved on since the strike. "I was looking for people that I had no addresses for," Padilla recalls. "I had to go into the *colonias* to find them, and a lot of them were not known by their names but by their nicknames." Nonetheless, after following word-of-mouth leads house by house through the barrios, Padilla and ten other organizers managed to fill a bus with DiGiorgio workers and drive them hundreds of miles back to California for the historic election. As election day drew closer, the Teamsters, fearing that they might lose, became increasingly worried—and increasingly violent.

"One day I was in my car. We had mounted the sound system on the top," Eliseo Medina recalls, "and another organizer and I were going down the DiGiorgio path announcing a rally. About thirty Teamsters came and blocked us. They piled out of their cars, and without saying anything, they reached in and started beating on us. My friend was trying to beat them back with a microphone, so I stepped on the gas and took off. I got about ten stitches in my mouth. It was very scary, but it was also very exciting."

On the day of the election, the farmworkers union made sure all their supporters had a ride to the polls. At 8:00 P.M. the ballots were put in a box, then in the trunk of a California Highway Patrol car to be rushed to San Francisco for the count. Representatives from all sides rode with the patrolman, and Huerta followed in a separate car.

Cesar remained behind in Delano, waiting for the news. Everybody in the union was gathered in Delano at Filipino Hall. "That was an emotional day, more than any other I've seen. People were crying . . . ," Cesar recalled. "I was . . . making plans on what to do in case we lost, how to deal with the strikers. Then Dolores called from San Francisco."

"We were all worried as hell," Medina remembers. After hanging up the phone, Cesar quieted the crowd by saying he had an announcement. Cutting through the silence, Chavez revealed that the Teamsters had won the packing sheds 97 to 45. Medina's heart sank, and it took a few seconds for him to realize that there was more: The NFWA had won among field workers, 530 to 331. "Everyone just exploded. People were jumping up and hugging. It was such a feeling of euphoria and happiness," Medina says, remembering that the victory party soon migrated to People's pool hall to continue the celebration. The next day, "as soon as they got the word in Delano, the merchants began to close down," Cesar recalled. "For them it was a day of mourning." News of the victory, however, prompted a September 22, 1966, telegram of congratulations from Martin Luther King, in one of his few formal contacts with Chavez:

"I extend the hand of fellowship and good will and wish continuing success to you and your members. The fight for equality must be fought on many fronts. . . ."

Now the union demanded an election at DiGiorgio's third property, the Arvin ranch. Adelina Gurola and Teofilo Garcia were sent to San Francisco with Marshall Ganz and a group of other farmworkers to confront DiGiorgio about why the company was dragging its feet on setting an election date. Workers had sent petitions to the company, appealed to the governor, and now they wanted an answer. When they refused to leave the DiGiorgio office, police descended on Gurola and the other workers and dragged them out of the building.

Faced with the threat of even more embarrassing protests, DiGiorgio agreed to an election in early November. The biggest challenge would be to unite the ranch's segregated crews—blacks, Filipinos, Latinos, and Anglos—and win a cross-section of votes. Many migrant workers had already left Arvin by then, but the union was once again victorious, winning 285 out of 377 votes.

Mack Lyons, an African American union leader in Arvin, realized the fight was just beginning after negotiations began when the company balked at basic requests: "They didn't see why they should be required to put ice in the water in the summertime, or to buy cups, so we would have individual drinking cups. Just some of the smallest things." Even after contracts were signed, supervisors made a point of resisting the new work rules. Teofilo Garcia remembers a foreman laughing scornfully when his crew tried to take the ten-minute break guaranteed by the contract. A Puerto Rican shop steward had to produce a copy of the rules, his hand trembling in fear. Garcia and the other workers were impressed by his courage. The resentful foreman finally acknowledged the contract, and the crew took what for some of them was the first workday break of their lives. "Before the union, we were *rancheros del rancho*—'hicks from the sticks,'" shrugs Garcia, who is now in his seventies. Gurola, at age eighty-one, says, "After that vote, everyone became brave, even me."

The DiGiorgio contract broke new ground by establishing the first employer-financed health and welfare fund for farmworkers, promotions and layoffs based on seniority, and vacation and holiday pay. (Within two years, however, this model contract would disappear. DiGiorgio sold off its properties because federal agriculture regulators declared it too large to qualify for generous water subsidies that most other valley farmers enjoyed, and the growers who bought the pieces of the DiGiorgio ranches refused to honor the union contract.)

At the height of the turmoil with DiGiorgio in 1966, the union and the Teamsters engaged in a briefer skirmish at another vineyard in Delano. In September 1966, workers at the Perelli-Minetti vineyard in Delano demanded their company negotiate with the NFWOC. But Perelli-Minetti, a winery, responded by replicating DiGiorgio's original

A Migrant Harvester's Letters Home

by Jane Kay and Bernabe Garay

In the early seventies, reporter Jane Kay of the Arizona Daily Star *became interested in letters that Mexican migrant workers regularly sent home to their faraway families. Of the following excerpts, written by migrant worker Bernabe Garay of Querétaro, Mexico, she wrote:*

No matter how far Bernabe Garay traveled from his home in Mexico to the fields and orchards of Michigan, Tennessee, Indiana, Texas, California, or Arizona, he never forgot to write. And his twenty years' worth of letters give an account of the life he lived while harvesting the nation's fruits and vegetables, always on the run. Now his two sons, Reginaldo and Francisco, are with him working in the orchards.

The following are only a few translations of excerpts from the dozens of letters he has sent over the years to his wife, Pilar, his mother, and his children. Of the letters, he says, smiling, "With us, there were no secrets."

La Mesa, Texas
October 21, 1963

Dear Little Regito:

I greet you and Celi and the other children. May you be happy and may you and Celi have a nice namesday. May God keep you well for a long time is what your dad wishes for you, and that you be very good and obedient with your mother and don't be bad so that the Most Holy Virgin will love you very much. And you take good care of the donkeys and the cows and let me know if we are going to have any small calves or piglets from the sow. I wonder if the horse has died; you don't tell me anything about him, if he's still alive. And be very obedient with your mother, because, according to Grandmother, you have been behaving badly with your mother. Don't be bad; be obedient, because I love you all very much, but I will not permit you to disobey your mother, because she and I are the same, so be very careful. That is all from your dad who blesses you and does not forget you, not even for a moment.

■ *Bernabe Garay, seen here taking a break with other farmworkers from his home state in Querétaro, Mexico, wrote letters regularly to keep in touch with his family during more than twenty years as a migrant worker in the United States.*

Paducah, Texas
(undated)

My beloved wife:

I write this letter to greet you, hoping that when you hold this letter in your hands you will be enjoying perfect health, as mine is, thanks be to God. About renewing the contract, I can tell you nothing right now, because the boss has not told us anything yet. But I believe that he will renew, and if not him, someone else. It would be better to have him renew, though, because he is very kind to us; there are only two of us and he gives us watermelons, small bags with peaches, a half-dozen eggs. The old man is very good, and above all, he has very good cotton. God willing we will stay here with him; I'll let you know in my next letter.

Vieja, I am afraid to send Mexican money through the mail, but am enclosing ten pesos in your letter and ten in mother's. I am going to save as much as I can here. All I have spent is on a pair of shoes for myself last Saturday, because I didn't have any to wear and I looked like a beggar without shoes. . . . And about what L. told you, please tell her I asked for her to "go visit her mother." It is true that I bought soft drinks for her and my *compadre* Claudio's girls, but that does not mean anything; I did not hide from anyone to do it; so be careful of gossip. And please pray to God to help me out, for you know that what I earn is for you and my little cubs. You are right in telling me these things, but don't worry. That is all for the moment.

Your husband—

Marion, Indiana
October 23, 1967

Dear honey,

I greet you lovingly, as I do my whole family. . . . Tell my mother that I am well, and please excuse me for not writing to her. It's because I have no stamps; I just have one. . . . I am enclosing my Social Security card so that I will not carry it with me and also that of my *compadre.* Please pray much to God and the Little Saint to take care of me so that we won't be caught. Your old man who loves you and wishes you the best.

Phoenix, Arizona
November 30, 1972

My dear wife,

. . . We arrived after eight days and then, on Thursday, we began to work; we earn low wages, because there are many people, but something is better than nothing. We sleep right on the field. We already bought a blanket, because it is very cold and only that way can we stand it. Immigration cars almost daily are after us, but here where we are staying nothing happens to us; only, pray to God and my Holy Mother to take care of us. I send you $60, $50 for you all to eat, and the remaining ten you give to the Holy Virgin for her birthday gift in the name of myself and my son, and you all pray to her for us all you can. When you answer this letter, in case they have not caught us yet, then I will send you something to pay some of all we owe; and you be sure to tell me the needs you have to see what we can do over here. Regards to all my children and my little mother-in-law, and tell me if Lollo has not left to come over here. . . . Don't fail to answer right away, please. . . . I love you and I don't ever forget you.

J. Bernabe Garay

strategy, ushering the Teamsters in for a sweetheart deal it hoped workers would accept. The farmworkers union declared a boycott against Perelli-Minetti's labels, then took an even bolder step: asking consumers not to shop at stores that sold Perelli-Minetti wines. Fred Ross sent busloads of farmworkers and volunteers into the Bay Area, Los Angeles, and New York to picket stores and leaflet, laying the foundation for more ambitious boycotts to come.

In July 1967, the winery relented and allowed the workers an election. When the NFWOC won, the Teamsters union relinquished its contract, and under pressure from a committee of clergymen, signed a peace pact with Chavez. Gallo, the world's biggest vintner, along with Christian Brothers, Almaden, and Paul Masson, also agreed to negotiate with the farmworkers. Dolores Huerta, who had trained herself in the fundamentals of contracts, was in charge of this blitz of negotiations; she worked so hard that she had to be admitted to a hospital for a few days after she fainted.

TEXAS WAS BECOMING another test case for the union. Farmworkers getting paid as little as forty cents an hour went on strike in the Rio Grande Valley, and union organizer Eugene Nelson shook up the state's politicians by leading field workers in a four-hundred-mile *perigrinación* through Texas in the summer of 1966, again with the FBI trailing the marchers every step of the way. (Many Chicanos in Texas said nothing had ever inspired them to break their silence as much as *la marcha*.) In October 1966, the FBI also scrutinized Nelson and others as they blocked a bridge on the border to prevent Mexicans recruited as strikebreakers from crossing into Texas.

In January 1967, Cesar sent Gilbert Padilla back to Texas because the Texas Rangers were intimidating and beating up strikers. Almost as soon as Padilla, with Reverend Jim Drake, arrived in Rio Grande City, they were ordered off the steps of the courthouse, where they had come to inquire about strikers who were arrested for using "abusive language." The pair ignored sheriff's deputies, and instead dropped to their knees in prayer. "I was in jail the second day I was in Texas for praying. Nonviolence wasn't taught there as much as we did in Delano," recalls Padilla, who didn't return to California for two years.

Despite Padilla's hard work, the union drive was nearly crippled by Texas's tough right-to-work laws and the close proximity of Mexican workers, whom the Texas Rangers would often escort across the border to break strikes. The real breakthrough in Texas would come several years later, thanks in large part to the efforts of a young Berkeley-educated lawyer who would become one of agribusiness's most formidable opponents.

The lawyer was Jerry Cohen, who in late 1966 was living in McFarland, working as a legal aid attorney and hanging around People's bar in Delano, shooting pool with union organizers. He had already met Gilbert Padilla, who, over bottles of Mountain Red, piqued Cohen's interest with stories about the farmworkers movement. The son of a navy doctor, Cohen was fresh on the job with California Rural Legal Assistance, a government-funded agency that helped the poor. Within months, the young attorney attracted Chavez's interest because Cohen loved a good fight. In early 1967, Chavez lured him away from CRLA with the promise of plenty of good fights to come.

Although Cohen knew next to nothing about labor law when he joined the union as a staff attorney, he was a quick study, and he loved constitutional law and working with Chavez. "Cesar had as much raw brain power as anyone I ever knew," Cohen recalls. "He knew the lay of the land that we were fighting was not land that we could win on. . . . I could go into court in Kern County court and lose, and his reaction was always, 'That's great!' "

There was an extraordinary lack of labor laws protecting farmworkers. Nor was there a government agency with which to file claims of bad-faith bargaining. All farmworkers had in the way of rights was the First Amendment, and the freedom to boycott growers and strike and picket. In agricultural areas, where the economy was based on "grower power," the courts often sought to limit, or even deny, those basic rights. Judges, for example, would routinely violate the law by slapping limits on picketing without first allowing the union to argue against such injunctions in court. The union's strategy was to expose unconstitutional injunctions by violating them, thus elevating them to higher courts and into the public view.

Sometimes the union resorted to sheer improvisation. One of Cohen's first assignments after he joined UFWOC was to get rid of injunctions prohibiting the union from "secondary boycotts": boycotts aimed at grocery stores that sold products the union wanted consumers not to buy. To Cohen's amazement, the injunctions were based on the National Labor Relations Act, which explicitly excluded farmworkers.

The case in point turned on nine union members employed in a peanut warehouse at one of the DiGiorgio ranches. Because the peanut packers worked in a warehouse, growers had argued that those employees were industrial workers; therefore, the whole union was subject to National Labor Relations Act restrictions. But Cohen counterattacked. Under his direction, the union set up an organization called the United Peanut Shelling Workers of America, and transferred the nine warehouse employees out of UFWOC, freeing farmworkers to pursue secondary boycotts. "We sort of practiced legal karate," Cohen recalls. "We were trying for some set of rights in an area where there wasn't a set of rights. It was the law of the jungle."

In 1968, Cohen and the union sued the Texas Rangers for interfering with labor disputes. In the 1970s, when the case went to the U.S. Supreme Court, the union finally won.

IN A MERE two years, the union had made tremendous strides. But its contracts, mostly in a handful of wineries, covered only about 5,000 of the state's 250,000 farmworkers. The original strike in Delano was still going on, as the growers formed enough skeleton crews of strikebreakers—disillusioned and hungry locals or Mexican migrants—to keep the grape harvest coming. Some of the growers had raised wages in response to the strike, but they allowed children as young as six to work as strikebreakers in the vineyards—an ugly spectacle that generated startling photographs of children struggling to lift boxes and stack fruit while sheriff's deputies stood by nonchalantly.

Chavez saw table-grape growers as the most exploitative ranchers in the valley, and he was determined to keep up the pressure. During the summer of 1967, he and the other UFWOC leaders chose their next prime target: Giumarra Vineyards, a family-run business that was the biggest grape concern in the state. The Bakersfield-based company farmed more than eleven thousand acres, and its employees were ripe for a fight. They complained to the union that their wages were lower than those of other big companies because Giumarra required that they harvest only the best grapes, a rule that slowed workers and drove down their piece-rate earnings. Dolores Huerta saw Giumarra as the domino that could make the rest of the grape growers tumble: "If we can crack Giumarra, we can crack them all."

The union sent registered letters inviting the company to the bargaining table, but the invitation fell on deaf ears. Joseph Giumarra, the patriarch who founded the company, was an ardent foe of all unions. He also refused to listen to the pleas of a State Conciliation Service representative, who urged him to at least attend to what Chavez had to say.

At a huge rally on August 3, 1967, Giumarra workers voted to strike, and two-thirds of the company's five thousand workers walked off the job during the harvest. Giumarra immediately brought in strikebreakers and obtained injunctions prohibiting the use of bullhorns and limiting the union to only three pickets at each entrance. Cohen attacked in the courtroom, but Chavez knew better than to wait for legal machinery. He activated the union's most effective weapon yet—a nationwide boycott of Giumarra grapes.

This time, farmworkers were sent out to bolster the union boycott efforts that had been started by students and SNCC activists. Dolores Huerta took off for New York—with some of her kids in tow. Fred Ross organized a busload of fifty farmworkers to join her that winter. "The first day we went out on the picket line, one of the Filipino women fell down and hit her head on some ice and had amnesia for about an hour. Everybody was slipping on the ice and falling. But they had a heck of a lot of spirit," Huerta remembers. Eliseo Medina, who'd never been east of Texas, was given a ticket for the first airplane ride of his life. "Where's Chicago?" he remembers asking. By then young Medina was used to urgent directives. He was watching a movie in Delano with his girlfriend one night when somebody tapped him on his shoulder in the dark theater: "Cesar needs to see you." He was immediately sent to Napa County to help another organizer—no time for clothes or to say good-bye to the family.

Giumarra fought the boycott shrewdly, shipping grapes under the labels of other vineyards in California and Arizona. All in all, with the cooperation of other growers, it used one hundred different names in an attempt to fool shoppers and union members. The result was mass confusion at stores and on picket lines. Huerta and Ross urged Chavez to widen the boycott to all table-grape growers, an action Cesar was at first reluctant to take because he thought boycotts would only be effective if they were linked to specific strikes. By January 1968, however, with the growers continuing to falsify the origins of their grapes, Chavez accepted the wisdom of Huerta and Ross's advice. That month, the union began its legendary boycott against all California table grapes—a campaign that cut across all age, class, and regional differences, and became the most ambitious and successful boycott in American history. The campaign put a human face on American farm labor, and, for the first time, sent a generation of Latino field laborers into a world beyond California.

WHILE VOLUNTEERS packed their bags to head east, farmworkers on the picket lines were becoming more restless and the mood more volatile. Union members voted on all key decisions, but as the growers' wrath increased, maintaining the strike required great sacrifice. Families were suffering under the pressure, hoping that in the end the fight would be worth losing what few possessions they had.

The pressure on the Chavez family had also intensified because of Cesar's increasingly high profile. The oldest son, Fernando, was subjected to ridicule in school, and other boys would pick fights with him; eventually Chavez felt forced to send him to his grandparents in San Jose to finish high school. Rumors also spread among growers that Chavez had amassed a small fortune through the union. Linda, the Chavez's third child, remembers, "I had one of the grower's kids—in fact, his name was Zaninovich—in one of my classes. One of the kids said, 'That's Cesar Chavez's daughter,' and he made a remark like, 'Oh, *yeah,* Cesar Chavez's daughter. If you were Cesar Chavez's daughter you wouldn't be here! All his kids go to private schools in Switzerland.'"

Delano was deeply divided, and by 1968, some farmworkers found that their pledge to remain nonviolent was being sorely tested. In October 1966, Manuel Rivera, one of the first rose workers in McFarland to support *la causa,* was seriously injured when a sales representative for a local grower plowed through a picket line, ran over Rivera, and crushed one of his legs.

Pickets surrounded the truck, threatening to kill the driver, who called out to Chavez in a panic. Helen quickly called Cesar, who had just left the line moments before. He ran back, pushing through the mob, crawled under the vehicle, and appeared on the truck's running board. Although the strikers cursed his pleas for nonviolence, later some conceded that his intervention may have saved the man's life. Chavez escorted the salesman into the company office, where Cesar upbraided the grower, blaming the incident on his greed. *El Malcriado* published a photo of Rivera, his leg wrapped in a cast and bandages, after he was able to walk with crutches. He smiled bravely, but he was disabled for life.

The assault on Rivera, coupled with other indignities, was apparently more than some union members could bear. An elderly Filipino, Alfonso Pereira, told Gilbert Padilla that he was tired of suffering and ready to offer his life to the strike. Padilla sent him home to cool off, but Pereira got behind the wheel of his car and drove it straight into three growers, clipping one of them and breaking his hips. He was sentenced to a year in jail.

As the strike neared a boiling point in late 1967, Chavez became increasingly concerned about the union's ability to keep people under control. Police had arrested some strikers for carrying marbles and ball bearings, considered potential weapons for sabotage. Epifanio Camacho remembers discovering that cells of two or three men had secretly blown up water irrigation pumps, scattered nails across roads to flatten the tires of squad cars and farm trucks, and even roughed up some people they suspected of being grower spies. Several packing sheds stocked with grapes had also gone up in flames, and to this day, both sides of the dispute blame each other.

Everyone from farmworkers and growers to politicians expected Chavez to resolve the bitter dispute. Paul, the Chavez's sixth child, says his father privately questioned how long he could shoulder the burden of leadership and sudden fame, and turned to his wife for counsel. Paul describes finding a book of his father's thoughts, written in spring 1967. Thumbing through the book, he came across a simple entry: " 'It's very tough. I don't know if I can continue.' That was it. And then I turned the pages and I saw a couple days later another real simple entry. 'I spoke to Helen. I'm ready to go.' That's the strength my mother gave my father."

Chavez would become furious when he heard about acts of violence, according to Al Rojas, who, ten years after Cesar helped him settle his small dispute with the Oxnard broccoli farmer, had joined the union as an organizer. Chavez confiscated a few guns from strikers and expelled those found provoking physical confrontations on picket lines. Chavez's fear that violence would soon overtake the movement was compounded by what was happening outside of Delano; on a swing through the Southwest in early 1968, he met Chicano dissidents who had lost faith in nonviolence. The war in

Vietnam was claiming thousands of lives, including a disproportionate number of Chicanos and blacks. Antiwar demonstrations were met with tear gas and clubs, and many activists wanted to fight back. Taking its cue from the Black Panthers, a militant brown power movement that gave birth to the Brown Berets was rumbling through the Chicano barrios which Cesar toured on his drive to promote *la causa*.

Although some Chicano dissidents accused him of "cowardice" for refusing to attack or harm growers, Chavez would not budge, repeating that patience was a surer path to victory than violence. "I despise exploitation and I want change," he

■ *Inspired in part by the farmworkers' fight, Chicano students in Los Angeles staged walkouts to protest abusive treatment and poor education of Mexican-American children.*

said, "but I'm willing to pay the price in terms of time. There's a Mexican saying, *Hay más tiempo que vida*—'There's more time than life.' "

Still, Chavez was not sure he had enough time to prevent a wave of violence in Kern County. In February 1968, Chavez was called to a confidential meeting with union directors who had urgent news: The Kern County district attorney said he was considering filing charges—serious charges—against union members alleged to have destroyed valuable property. The district attorney indicated he would pursue long prison sentences. "Goddamn it!" Chavez shouted despondently at a closed-door meeting of union leaders. "We'll never be able to get anywhere if we start using tactics of violence. . . . You have to believe in that!"

The meeting adjourned, and Chavez and others retreated for some soul-searching. It was a terrible time for the union: After two and a half years, it had only nine winery contracts, morale was unraveling, and picket captains were failing to show up at the fields. The strikers' patience had worn as thin as their wallets. When people talked about how it was time to start resorting to violence, Chavez realized, they really meant it. He had to do something drastic to make the union recommit to nonviolence and humility. He decided to simply stop eating. Moreover, Chavez resolved not to break his fast until union members renewed their pledges of nonviolence.

The inspiration for Chavez's protest was Gandhi, the Indian independence hero who had fasted for weeks and whose writings Chavez had read over and over. Chavez's own pronounced Catholic faith and its tradition of sacrifice—*sacrificio*—made Gandhi's concept of overcoming enemies through "moral jujitsu" appealing. Chavez had practiced fasting before and found it extremely painful. But this time he would make individual *sacrificio* the centerpiece of a political statement, one aimed squarely at his own people.

Chavez BEGAN his fast, privately, on February 15. He confided in aide Leroy Chatfield, a former Christian brother who had joined the cause, that he had stopped eating solid food. It was only the third day, Cesar said, and already he was having terrible cravings. Nor was his choice of liquid one which Gandhi would have necessarily approved: Chavez was drinking only Diet-Rite cola, one of the few junk foods for which he had a fondness. On the fourth day, Chavez resolved to drink only water and to continue the fast. He called for a meeting at Filipino Hall to break the news to the union.

■ *Chavez, who worked almost around the clock, sometimes found the burden of leadership daunting.*

Although already weakening, Chavez had enough energy to speak angrily to those who gathered at the hall. Union organizers were unaware of what Chavez was going to tell them, remembers Al Rojas. Cesar seemed unusually emotional, criticizing organizers for not working hard enough, asking them how they dared to condemn the war in Vietnam while advocating violence in the fields. He was also particularly harsh in his attack on saboteurs, who he said were endangering the union's credibility by resorting to "macho" antics that would get the movement nowhere. It was exactly the kind of machismo his mother warned him against. He preferred to show his toughness with his fast.

Some broke down and cried, Rojas recalls, and looked at each other in disbelief when Chavez announced he would not eat another bite of food until everyone once again pledged themselves to nonviolence. Chavez said he was walking back to the Forty Acres—an old labor camp outside Delano that had been converted into the new union headquarters—and that nobody should try to stop him. Chavez left before anyone could respond. After the initial shock dissipated, Filipino Hall exploded. "We didn't know what he was talking about. 'You're not going to eat?'" Rojas says. "People thought, He's just going to kill himself, what is he going to accomplish?"

When Chavez reached the Forty Acres, a flat, dusty yard surrounded by vineyards and orchards, he disappeared into a little room where he slept on a cot, read, and prayed with visitors. Helen had walked part of the way with him that first night, arguing that he was being ridiculous. Knowing how stubborn Cesar was, she relented.

The fast would later be viewed as a defining moment for the union, one that renewed its sense of hope and unity and restored the power of nonviolence. At the time, however, some of the union's other leaders refused to talk to Cesar because they thought the fast was an absurd waste of time. Tony Orendain, who thought the fast was religious folly, sat with his back toward Chavez when they discussed union business. Other volunteers found the mystical and Catholic character of the fast so offensive that they also quit, complaining that Chavez was developing a messiah complex.

But within a few days, Rojas and other union members rallied to Chavez's side, fanning out across the valley to tell members about the purpose of the fast and to invite them to the Forty Acres to show their support. The response was overwhelming. Thousands of farmworkers streamed to the Forty Acres with crucifixes and other offerings, pledging support and imploring Chavez to stay alive and healthy; some built shrines to the Virgin of Guadalupe at the union headquarters. Priests wore vestments cut from union flags and offered mass with union wine. People slept in tents they pitched in the yard and at night had festive prayer rallies with singing and hot chocolate. "The irony of the fast was that it turned out to be the greatest organizing tool in the history of the labor movement—at least in this country," says Leroy Chatfield. "Workers came from every sector of California and Arizona to meet with Cesar, to talk to him about the problems of their areas. . . . Cesar had more organizing going on while he was immobilized at the Forty Acres fasting than had ever happened before in the union."

There were even some would-be heroics that provided comic relief. Late one night, long after midnight, a worker who'd been drinking heavily sneaked into Chavez's room. His ranch committee had assigned him the mission of making Cesar eat. "He had tacos, you know, with meat, and all kinds of tempting things," Cesar recalled. "I tried to explain to him, but he opens up this lunch pail and gets out a taco, still warm." The inebriated farmworker pinned the weakened Chavez to his bed and tried to press the rolled tortilla into his mouth. Cesar squirmed and fought "like a girl who doesn't want to get kissed," until his muffled cries finally prompted his brother Richard and others to rush in. They thought the man was trying to kill Cesar. After struggling to sit up, Chavez implored them not to hurt the worker. Soaked in the melancholy of alcohol, the weeping farmworker was allowed to stay with Cesar, who tried to explain the logic of the fast to him.

■ *During a fast against violence, Chavez was ordered to Kern County court to answer charges that he had violated an injunction limiting picketing. Farmworkers flocked to the courthouse and lined the halls inside, while others waited outside, on their knees, praying.*

Delano growers watched in frustration and disbelief as the national news media poured into the Forty Acres to record the event. Maybe the fast was a hoax, they suggested hopefully, or at the very least, a cynical stunt to attract sympathetic coverage. Jerry Cohen, who appreciated Chavez's media savvy, says Cesar's fast was no publicity stunt: Chavez was sincerely trying to pull the union through a crisis. "No union movement," Chavez told the workers, "is worth the death of one farmworker or his child or one grower and his child."

On day thirteen of the fast, Chavez was forced out of bed by a Kern County Superior Court order to answer contempt charges stemming from violations of injunctions that limited pickets to three at each entrance to Giumarra's ranch and barred strikers from standing within three hundred feet of each other. "Kern County courthouse to me was enemy territory before that day," Cohen says. "I got up early that morning because here's a guy in the thirteenth day of his fast, and they might throw him in jail, and I frankly was pretty scared. And I went out to the Forty Acres and I could see out of the fog, the lights of farmworker cars, just hundreds of cars."

The show of support left authorities befuddled: More than three thousand farmworkers packed the corridors of the courthouse and knelt in quiet prayer outside. Cesar, who for two weeks had consumed nothing more than water and a few ounces of bouillon and eucharistic wafers, leaned on Cohen's shoulder as he entered the building. When flashes from cameras exploded in their faces, Cohen asked, "Are you OK?" Cesar looked up and winked.

Giumarra's lawyer asked the judge to remove the farmworkers because they were a disruptive presence. "The presiding judge said, 'Well, if I kick these farmworkers out of the courthouse, it'll be another example of goddamned gringo justice,'" Cohen recalls. From that day on, Cohen says, "that courthouse became our turf." Giumarra dropped the contempt charges.

Another two weeks passed, and Cesar had lost more than thirty pounds. He had already fasted longer than Gandhi's hunger strike in 1924, but his message about nonviolence seemed to have gotten across to the union's rank and file. Chavez decided it was time to publicly break the fast.

On March 10, the night before Cesar was to eat for the first time in twenty-five days, Delano was drenched by a terrific downpour. Al Rojas was worried the rain wouldn't stop and would ruin the planned celebration. But the next day, the air was crisp and the sky a brilliant blue, and hundreds of cars bearing red flags began to appear on the horizon, heading into town. The parade of cars had been diverted away from the Forty Acres because organizers saw the crowd would be larger than expected. The rally, six thousand strong, reconvened at a park in Delano, where chants of "*Viva Kennedy!*" greeted Robert Kennedy, who had just flown in to celebrate the end of Chavez's fast.

Kennedy sat next to Chavez, who was wrapped in a blanket, and Helen, who was wearing a lace mantilla. The two men, both Catholics, shared a piece of bread blessed by a priest.

"Well, how goes the boycott, Cesar?" Kennedy asked.

"How goes running for president, Bob?" Chavez replied, and both men laughed.

A week later, Kennedy announced he was indeed running for the Democratic nomination for president. The union was stretched to the limit with the boycott and strike, but its leaders felt they could not miss an opportunity to campaign for the man who might be the next president of the United States—a man who had become their friend in the early days of the Delano strike. For two months volunteers saturated Chicano barrios throughout the valley and in the big cities, mustering a critical number of votes that lifted Kennedy to victory in California's June primary. Dolores Huerta and Chavez, both in high spirits, were at the Ambassador Hotel in Los Angeles for the

■ *Robert Kennedy flew to Delano to celebrate the end of Chavez's fast. Helen Chavez and Cesar's mother, Juana, were also at his side.*

Kennedy victory party on election night. As Kennedy walked up to the podium, Cesar went to look for Helen and prepared to welcome the senator into a room where a mariachi band would play in Kennedy's honor.

Dolores Huerta accompanied Kennedy to the stage in the Ambassador's ballroom, where he made sure to thank her and Cesar for their help, and uttered his last public words: "And now it's on to Chicago, and let's win there!" In the crush of people in the room, Huerta was separated from Kennedy, who'd been diverted to an exit through the ballroom kitchen. Moments later, Robert Kennedy was shot down by assassin Sirhan Sirhan. Kennedy died the next day.

The farmworkers union was devastated. In April, Martin Luther King Jr. had been assassinated. Now the union had lost another inspirational ally and, with him, the hope that the cause of the farmworker would soon be espoused from the bully pulpit of the presidency. Instead of Kennedy, Republican Party nominee Richard Nixon, who had publicly declared opposition to the grape boycott, was elected to the White House. In Sacramento, California, Governor Ronald Reagan remained contemptuous of the union, showily eating grapes in public and labeling the boycott "immoral."

While the growers had powerful allies like these, the union believed it could still reach the hearts of consumers. They would redouble their boycott campaign by sending out even more farmworkers whose personal stories and tireless lobbying helped them seize the higher moral ground.

To MAKE the ambitious boycott work, the union needed more volunteers, so staffing was cut back at some of the service centers that had sprung up around the state. Picket captains were pulled back from the strike lines and field organizers redirected to the urban campaign, for which the union had targeted forty strategic cities across the United States and Canada.

With his work on the Kennedy campaign halted by tragedy, Al Rojas bundled his wife, Elena, and their three children into a union car and left the Forty Acres for Pittsburgh, Pennsylvania, in July. Neither he nor Elena had ever been out of California. He was born in a labor camp—with a peach box as his cradle—and he had no idea what it would be like on the East Coast. They were stuffing their clothes into grocery bags when Chavez came over to say good-bye. "Take care, good-bye, make sure you have all your contacts," Rojas remembers Chavez saying. "I stepped back and I said, this is my chance to ask him the real big question . . . 'Cesar,' I said, 'How do you do a boycott?'"

Chavez was speechless at first. After an uncomfortable pause, he told Rojas the boycott would be on-the-job training. "To be truthful," Cesar said, "what you and your family are going to do has never been done before. I don't know how you do a boycott. You just go out there and tell those people to stop eating grapes. Get them to stop eating grapes."

On the way to Pittsburgh, Anglo sympathizers—contacted at random from *El Malcriado*'s subscription list—fed the family, offered them money for gas, and helped them on their way. The Rojases were taken aback, however, when they drove through a tunnel into downtown Pittsburgh and found a metropolis of high-rises and bridges; Cesar had told them Pittsburgh was a "small mining town." They were put at ease, though, after an Irish Catholic priest welcomed them enthusiastically, and the Steelworkers union put them up for a few nights in a luxury hotel.

■ *Al and Elena Rojas and their children, seen here singing "De Colores" with Cesar Chavez, volunteered to go to Pittsburgh, PA, to publicize the boycott.*

For more than two years, the Rojases worked with all sides of Pittsburgh, from unions and churches and synagogues to the black power and antiwar movements. From the beginning, there was much to learn: A friendly steelworker took Rojas aside and advised him to get rid of his Volkswagen plastered with boycott stickers. "He says, 'Rojas, goddamn it, in this town you don't drive around pushing the union scene in a non-union car.' . . . I corrected that right away and we got a Plymouth." Rojas found that Jewish groups were among some of the cause's strongest supporters in Pittsburgh. (He also learned, after a gentle admonishment, to stop referring to them as "Anglos." In 1969, Al and Elena would name their fourth child, a "boycott baby" born in Pittsburgh, Shalom, Hebrew for "peace.")

The black community also backed the boycott, Rojas recalls. But his first meeting with activists was a near disaster: Young men in dashikis, mourning Martin Luther King Jr.'s assassination, told Rojas they would be happy to get rid of the grapes "by any means necessary." With Al trailing behind them, they marched into a store and ordered the manager to remove all California grapes immediately. When he hesitated, the militants carried all the boxes of grapes the store had outside, poured gasoline on them, and set them on fire.

By November 1968, the FBI was keeping close watch over Rojas and the boycott, with spies reporting on events as innocuous as the screening of an Andy Warhol movie at Carnegie Mellon University, where a student urged the group to attend a march in support of the boycott. When the march took place, an FBI source was there as well.

As recorded in bureau files, "Rojas distributed to each participant one grape, a plain envelope, and the address of President-elect Richard M. Nixon. . . . Each person receiving the above was instructed to mash the grape, place it in an envelope, and mail it to Nixon at the New York address."

In the spring of 1969, the union declared a boycott on the West's largest grocery chain: Safeway, where Giumarra sold 20 percent of its grapes. Safeway's directors, some of whom were also involved in agribusiness, fought back. They refused to take grapes off their shelves, and "freedom-to-work" committees shadowed boycotters to argue against union contracts. Growers hired the Whitaker & Baxter public relations firm to launch a $2 million campaign to churn out op-ed pieces, ads, and bumper stickers urging the public to protect "consumer rights" and to "Eat California Grapes, the Forbidden Fruit."

Meanwhile, the Rojas family branched out to New York, West Virginia, and Ohio, organizing parades and marches against the Kroger chain, where the couple was once arrested for using a loudspeaker. After months of work, Rojas finally had his big victory: The Pittsburgh boycott yielded an agreement that removed all California grapes from all inner-city grocery stores and supermarkets.

The disappearance of grapes from the shelves was not all due to the boycott. In January 1969, the union had started hearing rumors that the Defense Department was gobbling up reduced-price California grapes and sending them to troops in Vietnam. Rojas's brother, who was a navy cook on a destroyer off Vietnam, wrote him that he and some crewmates threw dozens of boxes of grapes over the side of the ship. "There's no way we're going to eat them," Rojas's brother said. "I told the guys here those are being boycotted."

By the middle of 1969, the Defense Department was buying 11 million pounds of fresh grapes a year, a staggering increase from the 6.9 million pounds it had bought the year before. Shipments to troops in Vietnam skyrocketed from .5 million pounds to 2.5 million pounds, an increase that translated into eight pounds of grapes for each serviceman. The union reacted by picketing military installations from California to Washington, D.C.

But the farmworkers' campaign had touched a nerve in America and abroad. It was no longer just college students and trade unionists boycotting grapes. Consumers all over the country were avoiding the fruit—something that struck fear into the hearts of agribusiness—and people were thinking about farmworkers in a new way.

One of the most effective young boycotters sent out to appeal to consumers was Jessica Govea, a bright twenty-one-year-old from the racially polarized town of Bakersfield, who, starting at age four, had picked everything from grapes to cotton. Govea volunteered to go to Toronto on the boycott, not realizing it was three thousand miles away. In July 1968, she left with Marshall Ganz, her boyfriend at the time,

■ *Jessica Govea, who had been a child farmworker, volunteered to go to Toronto and Montreal, where she successfully persuaded supermarkets to stop buying California grapes.*

and a Catholic priest, Mark Day, to set up a boycott office on a very important front: Toronto was the third largest export market for California table and wine grapes. Canada as a whole purchased about 20 percent of the state's crop. "On the one hand, it was tough; on the other, it was an incredibly liberating and wonderful experience," Govea remembers.

"My strongest memory of going out on an information picket line in Toronto was when I saw these two young white men walking toward me who reminded me of young white men in Bakersfield," Govea recalls. "I made myself strong and went up to them and said, 'Excuse me, could I ask you to help farmworkers by not buying grapes?' As if on cue, they both turned around and showed me their jackets, which had giant United Auto Workers emblems on them. And they turned around and said, 'We're all for you. We're all for you.'

"It turned out the world wasn't like Bakersfield," concludes Govea, who now teaches organizing at Rutgers University in New Jersey. "It turned out the world was much, much bigger and much kinder than Bakersfield."

Jessica's sincerity and youth impressed newspaper and television reporters curious about the boycott. Favorable stories rich in detail about "the girl from the south" as the *Toronto Star* called her, helped boost the boycott's profile. "The Mexican kids are shunted into school courses that make them mechanics and tractor drivers and

An Open Letter to the Grape Industry

by Cesar Chavez

E. L. Barr, Jr., President
California Grape & Tree Fruit League

Dear Mr. Barr,

I am sad to hear about your accusations in the press that our union movement and table grape boycott has been successful because we have used violence and terror tactics. If what you say is true, I have been a failure and should withdraw from the struggle. But you are left with the awesome moral responsibility, before God and man, to come forward with whatever information you have so that corrective action can begin at once.

If for any reason you fail to come forth to substantiate your charges then you must be held responsible for committing violence against us, albeit violence of the tongue. I am convinced that you as a human being did not mean what you said but rather acted hastily under pressure from the public relations firm that has been hired to try to counteract the tremendous moral force of our movement. How many times we ourselves have felt the need to lash out in anger and bitterness.

Today on Good Friday, 1969, we remember the life and sacrifice of Martin Luther King, Jr., who gave himself totally to the nonviolent struggle for peace and justice. In his letter from Birmingham Jail, Dr. King describes better than I could our hopes for the strike and boycott: "Injustice must be exposed, with all the tension its exposure creates, to the light of human conscience and the air of national opinion before it can be cured." For our part, I admit that we have seized upon every tactic and strategy consistent with the morality of our cause to expose that injustice and thus to heighten the sensitivity of the American conscience so that farmworkers will have without bloodshed their own union and the dignity of bargaining with their agribusiness employers.

By lying about the nature of our movement, Mr. Barr, you are working against nonviolent social change. Unwittingly perhaps, you may unleash that other force that our union by discipline and deed, censure and education has sought to avoid, that panacean short cut: that senseless violence that honors no color, class, or neighborhood.

YOU MUST understand, I must make you understand, that our membership—and the hopes and aspirations of hundreds of thousands of the poor and dispossessed that have been raised on our account—are, above all, human beings, no better no worse than any other cross section of human society; we are not saints because we are poor but by the same measure neither are we immoral. We are men and women who have suffered and endured much and not only because of our abject poverty but because we have been kept poor. The color of our skins, the languages of our cultural and native origins, the lack of formal education, the exclusion from the democratic process, the numbers of our slain in recent wars—all these burdens generation after generation have sought to demoralize us, to break our human spirit. But God knows we are not beasts of burden, we are not agricultural implements or rented slaves, we are men. And mark this well, Mr. Barr, we are men locked in a death struggle against man's inhumanity to man in the industry that you represent. And this struggle itself gives meaning to our life and ennobles our dying.

As your industry has experienced, our strikers here in Delano and those who represent us throughout the world are well trained for this struggle. They have been under the gun, they have been kicked and beaten and herded by dogs, they have been cursed and ridiculed, they have been stripped and chained and jailed, they have been sprayed with the poisons used in the vineyards. They have been taught not to lie down and die or to flee in shame, but to resist with every ounce of human endurance and spirit. To resist not with retaliation in kind but to overcome with love and compassion, with ingenuity and creativity, with hard work and longer hours, with stamina and patient tenacity, with truth and public appeal, with friends and allies, with mobility and discipline, with politics and law, and with prayer and fasting. They were not trained in a month or even a year; after all, this new harvest season will mark our fourth full year of strike and even now we continue to plan and prepare for the years to come. Time accomplishes for the poor what money does for the rich.

This is not to pretend that we have everywhere been successful enough or that we have not made mistakes. And while we do not belittle or underestimate our adversaries, for they are the rich and powerful and possess the land, we are not afraid nor do we cringe from the confrontation. We welcome it! We have planned for it. We know that our cause is just, that history is a story of social revolution, and that the poor shall inherit the land.

Once again, I appeal to you as the representative

of your industry and as a man. I ask you to recognize and bargain with our union before the economic pressure of the boycott and strike take an irrevocable toll; but if not, I ask you to at least sit down with us to discuss the safeguards necessary to keep our historical struggle free of violence. I make this appeal because as one of the leaders of our nonviolent movement, I know and accept my responsibility for preventing, if possible, the destruction of human life and property.

For these reasons and knowing of Gandhi's admonition that fasting is the last resort in place of the sword, during a most critical time in our movement last February, 1968, I undertook a 25-day fast. I repeat to you the principle enunciated to the membership at the start of the fast: if to build our union required the deliberate taking of life, either the life of a grower or his child, or the life of a farmworker or his child, then I choose not to see the union built.

MR. BARR, let me be painfully honest with you. You must understand these things. We advocate militant nonviolence as our means for social revolution and to achieve justice for our people, but we are not blind or deaf to the desperate and moody winds of human frustration, impatience, and rage that blow among us. Gandhi himself admitted that if his only choices were cowardice or violence, he would choose violence. Men are not angels and the time and tides wait for no man. Precisely because of these powerful human emotions, we have tried to involve masses of people in their own struggle. Participation and self-determination remain the best experience of freedom; and free men instinctively prefer democratic change and even protect the rights guaranteed to seek it. Only the enslaved in despair have need of violent overthrow.

This letter does not express all that is in my heart, Mr. Barr. But if it says nothing else, it says that we do not hate you or rejoice to see your industry destroyed; we hate the agribusiness system that seeks to keep us enslaved and we shall overcome and change it not by retaliation or bloodshed but by a determined nonviolent struggle carried on by those masses of farmworkers who intend to be free and human.

Sincerely yours, CESAR E. CHAVEZ
1969

■ *Cesar Chavez, who had a passion for amateur photography, snaps a picture of young UFW organizers leaving California in 1969 to lead the grape boycott in far-flung cities.*

housemaids and other menials," Govea told the *Star.* "I was lucky. I spoke English better than most and both my parents spoke English so they went to the school and insisted I get the liberal arts courses."

Govea dazzled the Ontario Federation of Labor, which gave her a standing ovation when she spoke at its convention, and the United Churches of Canada responded to her entreaties by declaring support for the boycott. Eventually, three chain stores in Toronto removed grapes from their shelves. The work wasn't easy, though, and sometimes spirits sagged: Govea was sometimes mistaken for a Canadian Indian, and the racist remarks she overheard stung the way they did in her childhood. She missed her family so much that she cried herself to sleep at night after she transferred to French-speaking Montreal to direct the boycott there in January 1969. With the help of a bilingual staff, she organized creative picket lines around stores that featured skits and all-night vigils. In June 1970, she was victorious again when one of the largest chain stores in Montreal agreed to stop selling California grapes.

Labor union contributions poured in and helped Govea maintain the boycott office independently. As required by Delano, she always sent any surplus back to the home office to help support strikers. "We all worked for five dollars a week and room and board, which meant sleeping on the floor a lot of the time," she says. "But we all gained so much from our experience. . . . We learned we were capable of a lot more than we thought."

Delano farmworker Eliseo Medina, who was also twenty-one when he went on the boycott, thought he'd be back in Delano in a couple of weeks. It turned out he would stay in Chicago for more than two years. The first thing he did when he got to Chicago was look up names of grocery stores in the Yellow Pages, call them, and ask whoever answered the phone to stop selling grapes. Homesick, he called Cesar and asked if he could come home, but Cesar told him to stick it out longer.

■ *Eliseo Medina, a farmworker with an eighth-grade education, became a top organizer in California and Florida and directed the grape boycott in Chicago for three years.*

Medina had emigrated from Mexico when he was ten, and had an eighth-grade education. But he led the boycott in one of the biggest cities in the country, rubbing shoulders with powerful politicians and speaking at huge labor conventions. Medina found Chicago's supermarkets tough to crack, however, and he resorted to organizing sit-ins in the middle of stores. "We just totally disrupted all of these stores and we were a pain. I mean, we were terrible. We made life miserable for the stores and so they finally figured out we were more trouble than the grapes were worth. . . . In 1970, we got every single chain store in Chicago to stop selling grapes."

Some of the Chicano and Mexican farmworkers who volunteered for the boycott were considerably older than Eliseo Medina and Jessica Govea, and they had more children than young couples like Al and Elena Rojas. Maria and Hijinio Rangel, farmworkers who joined the union during the pilgrimage, agreed in 1968 to move to Detroit, Michigan, with their eight small children. Incredibly, the eldest child, Luis, who was fifteen, was put in charge of a boycott campaign in Ohio. "Cesar told me he needed me to go to Detroit," says Hijinio, who is in his late sixties now. "We didn't speak English too well, so I knew it was going to be tough. But I figured we had nothing to lose."

Detroit was a strong union town, so Rangel had little trouble attracting public support. The real challenge began when he met with the owner of a terminal that shipped grapes as far away as Canada and tried to convince him to stop. The man told him that he was wasting his breath: Two other boycott leaders had already gotten frustrated and gone home, and so would he. But Rangel rounded up all the union people he could find and led them in candlelight vigils at night in front of the terminal. "Then all the reporters came out, and the TV cameras came out. Nine days later, he signed a contract with me that he would stop taking in grapes," says Rangel, still proud of his accomplishment. "We shut down a whole region."

His son Luis organized an eighteen-mile march from Michigan into Ohio to publicize the boycott. The procession won support for the California farmworkers movement, but it also helped focus attention in the Midwest on strikes by the nascent—and UFW-inspired—Farm Labor Organizing Committee, which had started its own campaign against poor working conditions on Ohio vegetable farms.

Assisting the farmworkers on the urban boycott were young volunteers of all ethnic backgrounds. Chicanos were particularly loyal supporters, and they adopted the UFW eagle as their own, whether they were picketing stores carrying grapes or marching on campuses for Chicano studies. Keeping an eye on the boycotters were "red squads," undercover police officers assigned by various police departments. In San Diego, Chicano police officers were assigned to infiltrate UFW support groups, as well as the Brown Berets, a group of young Latinos who had formed their own, milder version of the Black Panthers.

Carlos and Linda LeGerrette, Chicanos who led a boycott group in San Diego, remember kicking out one supporter who they discovered was a police officer. They also remember Manny Lopez, code-named "Nacho," a twenty-two-year-old officer whose primary assignment was to infiltrate the Brown Berets and keep an eye on the

UFW in San Diego. Lopez's notes, and reports from a string of street-level informants, were passed on to FBI agents monitoring the same groups.

Lopez, who today is a private investigator, rose to the rank of field marshal of the San Diego Berets and often provided security for the UFW's Southern California boycott activities. In 1970, "Nacho" was one of the security guards escorting Chavez

■ *Brown Berets, young urban Latinos who patterned themselves after the Black Panthers, sometimes provided security for Cesar Chavez in the early 1970s.*

through the Coachella Valley during a tour Cesar took with Rodolfo "Corky" Gonzales, leader of the Colorado-based Crusade for Justice. Lopez came to admire Cesar's nonviolent stance, and he began to distrust the FBI. "I won't say that everyone in the FBI was racist and acted like J. Edgar Hoover, but . . . ," says Lopez with a pointed pause. "The FBI had a lot of money to spread around . . . and the informants knew that. They would tell the agents whatever it was they wanted to hear. I know of one informant, a pathological liar, who fed the FBI all sorts of reports about the farmworkers being subversives and extremists and being connected to Communists and the Black Panthers."

During the last week of August 1970, Lopez led the San Diego Berets to East Los Angeles for the historic Chicano Moratorium march against the Vietnam War. Ten thousand people marched peacefully against the war and protested the high number of Chicano casualties. The event ended in tragedy when *Los Angeles Times* columnist Ruben Salazar was killed by a police tear-gas canister. As word spread of Salazar's death, rioting broke out. Two protesters died, and dozens were arrested in several days of street fights between young Chicanos and police. Lopez, ironically, was among the many arrested; he was beaten as well. "While we were marching we would give the police our Brown Beret salute," Lopez remembers. "The officer who beat me looked at me afterward and said: 'That's for giving me that salute!'"

Lopez, a decade later, was invited to attend a party honoring a Chicano candidate for San Diego's city council. The party was at the home of Carlos and Linda LeGerrette. The pair recognized Lopez, who revealed his background to them.

BY THE TIME the first grape crop of 1970 was ripe on the vine, *la huelga* had been a cover story in *Time* magazine, and boycott supporters were in every corner of the country, from the radical chic circles of New York, where the wealthy hosted union benefits, to junior college campuses in rural Michigan. Sympathizers devoured eyewitness accounts of the strike in books like Peter Matthiessen's *Sal Si Puedes* and John Gregory Dunne's *Delano*.

The union added to the public pressure by demanding to know what chemicals were used in the fields. Jerry Cohen was told that the only information he'd be getting from California farm officials was an injunction declaring that pesticides were "trade secrets." An ensuing wave of negative publicity hurt growers even more.

On July 4, 1969, growers filed a lawsuit against the union, claiming they had suffered $25 million in losses because of the boycott. The boycott leaders had managed to stop sales of California table grapes in Detroit, Chicago, New York, Boston, Philadelphia, and Montreal and Toronto—all major distribution points. Ten growers from Coachella, which produced 15 percent of all grapes, called the union to negotiate in mid-1969. Talks stumbled. But by the following season, the industry had been brought to its knees. A special committee from the National Conference of Catholic Bishops convinced the Coachella growers to resume negotiations, and in April 1970, Lionel Steinberg—owner of three of Coachella's biggest vineyards—signed a contract. Steinberg, a Democrat who was heavily criticized by fellow growers for his action, remembers the union contract had unexpected benefits: "The immediate response from the other growers was dismay," he says. "But to my pleasant surprise . . . we found that six or eight of the major chain stores in Canada began calling us wanting our grapes and our brand because we had the union [black eagle] bug. So we had an immediate advantage over our competitors of one or two dollars a box."

Next to agree was Bruno Dispoto, one of the growers most violently opposed to the strike, followed by Hollis Roberts, whose corporate ranches totaled 46,000 acres of fruits and nuts. Both signed union contracts, and the bishops' committee began to nudge other growers in the same direction.

The big break came on the night of July 25, 1970. Jerry Cohen's phone rang around nine o'clock. It was Johnny Giumarra Jr., grandson of Joseph, calling from his tenth high school reunion in Bakersfield—the same class as Marshall Ganz. He and his father were leaving town, Giumarra said, and they wanted to meet—tonight. "I realized something important was going on, that this was the biggest grape grower in the state, twelve thousand acres of grapes," Cohen says, "And I was really excited, but I didn't want Johnny Jr. knowing it."

Cesar was on the road—there was trouble in the Salinas Valley—so Jerry left an urgent message with Helen. It was after one in the morning when Chavez and Cohen

■ *The first United Farm Workers grape contracts were signed in July 1970 at the union's Forty Acres headquarters. Pictured are: Cesar Chavez, John Giumarra Sr., John Giumarra Jr., Bill Kircher, Jerry Cohen, Larry Itliong, and Monsignor George Higgins of the National Catholic Conference.*

checked into room 44 of the Stardust Motel, where they waited until the Giumarras, senior and junior, arrived at 4:00 A.M. They hashed over the major details of the union proposal, and then Cohen hit them with a condition. This was the moment the union had been waiting for; it wanted to get contracts with twenty-eight other grape growers that had also been struck.

"We wanted them to round up all the growers. They said they could do it. So the next day, at the St. Mary's [Catholic] School, there they all were. We told them we wanted them to come to the Forty Acres and sign the contract. And to their credit, some of them decided if they were going to sign those contracts, they might as well be in a cooperative, celebratory mood."

By late the following night, negotiations were at last complete: The grape pickers would have a hiring hall and an immediate base pay increase from $1.65 to $1.80 an hour. Piece-rate bonuses were increased, and the growers agreed to set up joint grower-worker committees to regulate pesticide use. The employers would pay a dime an hour into the Robert F. Kennedy Health and Welfare Plan.

On July 29, growers who were exhausted by the boycott pulled into the parking lot of the Forty Acres, along with a large crowd of ecstatic farmworkers. They all filed into the union's Reuther Hall for the signings. Giumarra appeared cheerful and ready to make the best of it. He held up his hands in a gesture of surrender, then put pen to paper, signing one of the contracts stacked on the table. Everything had happened so quickly; many of the boycotters who had worked so hard for this day were still scattered in cities thousands of miles away.

The mood in the room was conciliatory. John Giumarra Jr., the company's young lawyer, beamed and offered an olive branch, along with a word of caution: "If it works well here, if this experiment in social justice as they've called it, or this revolution in agriculture—however you want to characterize it—if it works here, it can work elsewhere. But if it doesn't work here, it won't work anywhere."

■ *Richard Chavez displaying the first Coachella Valley grapes to bear the UFW's black eagle union bug.*

Chavez, wearing an immaculate, white high-collared Filipino dress shirt, spoke in a manner more spiritual than gloating. "Without the help of those millions upon millions of people who believe as we do that nonviolence is the way to struggle . . . I'm sure that we wouldn't be here today. The strikers and the people involved in this struggle sacrificed a lot, sacrificed all of their worldly possessions," he said. "Ninety-five percent of the strikers lost their homes and their cars. But I think in losing those worldly possessions they found themselves."

The archenemies shook hands all around, and the room thundered with applause and shouts of joy. Despite their happiness, Chavez and the rest of the union leadership knew there was other trouble. The Teamsters union was moving in again on Mexican farmworkers in the Salinas Valley. As Jerry Cohen put it, "We left poor Richard [Chavez] to administer that whole set of contracts, and a brand-new hiring hall by himself. . . . It was like a boa constrictor swallowing a big pig or something. We didn't even have time to digest the meal because we were off to Salinas."

■ FOLLOWING PAGE: *In 1973 growers refused to renew UFW contracts, signing instead with the Teamsters union, against most workers' wishes. Picket lines were attacked by sheriff's deputies in the San Joaquin Valley.*

> The opposition was fighting very hard and they would go to any lengths to discourage us from organizing, even as far as killing people. . . . So we just had to make a decision. Do we quit now or do we continue? We continued because that's what the people wanted.

—RICHARD CHAVEZ,
on the summer of 1973, when two UFW members were killed while on strike

CHAPTER SIX

BLOOD IN THE FIELDS

SABINO LOPEZ and his exhausted crew were picking the last heads of cauliflower in a field in California's coastal Salinas Valley in July 1970 when the foreman whistled and ordered them to listen up: "Hey, someone who's very important is coming and wants to talk to you. After you're done, I'd like you to get on the bus and give him five minutes." The men stomped the river-valley soil—dark as coffee grounds—off their boots and sheathed the knives they used to cut each vegetable from its tough nest of roots and leaves.

When the crew boarded the bus, a Chicano in a suit and tie climbed on, greeting them with an unctuous smile. Addressing the workers in Spanish, he explained how he'd been brought up from Los Angeles to tell them about the Western Conference of Teamsters and its plan to represent them in collective bargaining. The longer the Chicano spoke, the more anxious and irritated the men became, remembers Lopez. The twenty-year-old Mexican, who'd only been in the country four years, knew his crewmates were ardent *Chavistas*—and he feared the worst when the Teamster finally asked if there were any questions.

After a few seconds of stony silence, one of the workers exploded: "Who sent you? The company sent you here, right?" Others joined in. "Don't you think they should let the farmworkers union come talk to us? Isn't a union something you should choose?"

Caught off guard, the Teamster said he couldn't really disagree with them, and quickly disembarked. But the confrontation on the bus was just the beginning. The entire Salinas Valley—the "Salad Bowl of the World"—was soon to be engulfed in one of the biggest farm-labor disputes in America since the 1930s. In August and September of 1970,

■ *Cesar Chavez spent the last half of 1970 in the Salinas Valley, where farmworkers revolted when growers tried to sign contracts with the Teamsters union to block the UFW.*

thousands of immigrant workers and UFWOC sympathizers packed rallies and walked off their jobs. The conservative farm town of Salinas ended up split down the middle—Latinos versus whites—in a struggle that would strengthen the emerging farmworkers movement and embolden Chicanos to start chipping away at the Anglo lock on local politics.

It was a crisis that the Salinas Valley growers had invited upon themselves. In an extraordinary attempt to thwart the spread of Cesar Chavez's influence, almost every major grower in the valley that summer had secretly signed contracts that allowed the Teamsters union to represent their farmworkers; overnight, there were contracts that covered thousands of farmworkers. But the Teamsters and the growers never told the farmworkers that new contracts were being drawn up, and they excluded them from the clandestine negotiations. The terms of the contracts, they agreed, would be decided upon only after the contracts were signed. If the workers didn't like it—and didn't want to pay Teamsters dues—they would be fired. The Teamsters claimed the new campaign was a sincere effort to bring farmworkers into their union.

The Teamsters and a couple of growers had tried this ploy before in the San Joaquin Valley; this time, however, the collusion was on an enormous scale, affecting as many as eleven thousand workers from Santa Maria up to Salinas. And it was clearly designed to block the inevitable moment when Chavez would arrive on the coast, demanding that growers negotiate with the farmworkers union.

By July 1970, almost against his will, Chavez was pulled into a tense, high-stakes chess game with Salinas's growers. Ironically, it was Chavez, often labeled a radical, who tried hardest to control the workers' revolt and channel it into a settlement. For farmworkers, the summer of 1970 was pivotal in other ways as well. It marked the start of a years-long, relentless campaign of attempts to destroy the union—attempts that were orchestrated, in part, by political operatives in positions as high as the Nixon White House.

CHAVEZ NEVER INTENDED to lead another significant confrontation so soon after the Delano grape strike. The farmworkers union had been quietly organizing the vegetable and strawberry pickers who worked the Salinas Valley, a campaign that Cesar felt needed time to mature. He had only informed a few ranchers that their workers wanted to be represented by UFWOC. In the last week of July, Chavez was in the middle of negotiating the Delano grape contracts when he received an urgent call from supporters in Salinas. Rumors were flying that a backroom contract had been struck by the Teamsters and Salinas's growers, who were the world's largest producers of lettuce, broccoli, cauliflower, carrots, celery, strawberries, and artichokes. The sweetheart deals would rob workers of a chance to choose their own union.

"The grape boycott scared the heck out of the farmers, all of us," corporate lettuce-grower Daryl Arnold recalls, explaining why the Salinas Valley Grower-Shipper Association tried to use the Teamsters as a shield against Chavez. "[The growers] felt that the Teamsters were the best organization to represent the farmworkers in the area. . . . They had been dealing in their sheds and packinghouses with the Teamsters union, and they thought if they could sign a contact with [them] it would forestall Cesar trying to come in and take over the industry."

But the growers underestimated the resolve of their field hands as well as Chavez's ability to seize on the crisis and turn it to his advantage. Before the fight in Salinas was over, some of the world's top agribusiness chiefs—including those from the United Fruit company—would be calling on Chavez personally and appealing to him not to boycott their products.

Chavez was caught off guard by the allegations of the Teamsters' treachery, but he knew the farmworkers union had to counterattack immediately or risk losing the base of support it had cultivated in the vegetable industry of California's central coast and the Imperial Valley on the Mexican border. The union's executive board prepared a telegram to send to the Grower-Shipper Association that same night, and Chavez drove to Salinas to speak to supporters at a hastily called rally.

When Cesar arrived, cries of *"Viva la causa!"* and *"Viva la raza!"* thundered through the cavernous hall of a teen center on the Mexican side of town. Sabino Lopez remembers the thrill of seeing Chavez in person for the first time, noticing with surprise how campesino he looked under the brim of a straw Filipino hat. By then, Cesar was hated and admired in equal measure, and to assure his safety, he usually traveled with a security team and a German shepherd he named Boycott. But this time Cesar was in such a rush, he had no time to gather his security guards. Jacques Levy was the only available driver, and the pair raced to get to the rally on time.

A local reporter, Eric Brazil, covered Chavez's appearance. A native of Salinas and the son of a prominent Monterey County Superior Court judge and former district attorney, Brazil had grown up with many of the town's young ranchers, and his parents and their parents often got together for cocktails. Brazil's knowledge of the industry—and his reporter's instincts—told him this would be one of the biggest stories he'd ever cover for the *Salinas Californian.*

After a rousing speech in Spanish, Chavez addressed reporters in English: The telegram to the Grower-Shipper Association demanded immediate negotiations with the farmworkers union, declaring that it represented the vast majority of field workers in the valley. "A prompt reply will avoid the bitter conflict experienced in the Delano grape strike," Chavez warned.

Cesar then returned to Delano to put the final touches on the grape contracts, still hoping there was time to avoid a clash in Salinas. But on the eve of the contract ceremony on July 29, he found out that the rumors of the Teamster takeover were true. Thirty Salinas growers had already signed the agreements, and more than 175 other vegetable ranchers were waiting in the wings. "They can't get away with this. It's preposterous," Chavez responded furiously. "They're going to have a big fight on their hands. . . . They're not going to sign up our people." The valley was risking "an all-out war between the Mexicans and the Filipinos . . . and the Teamsters and the bosses," Chavez said. Looking back on the origins of the Salinas rebellion, Eric Brazil agrees: "The rage of the workers was just palpable. They weren't going to take this. They had really been stabbed in the back."

Leaving Richard Chavez behind to administer the grape contracts, Cesar and the rest of the union leaders rushed to set up UFWOC's temporary headquarters inside the Salinas office of the Mexican-American Political Association. Salinas, the birthplace of John Steinbeck, was the heart of the valley and its industrial center, from which produce trucks and trains rumbled out as the morning fog wafted in from the Monterey Bay.

Local *Chavistas* had already started a campaign to dramatize their dissent: Hundreds of workers and their children began a four-day pilgrimage toward Salinas, trudging along county roads and main highways. Like the San Joaquin, the Salinas Valley was linked by a chain of small cities, from King City in the south to Watsonville in the north, with expansive fields in between. The pilgrims would march into Salinas that Sunday, determined to show they would not let growers select a union for them.

Chavez feared that the war for Salinas would be even more insidious than the one in Delano. He was afraid the battle could slide into ugly confrontations between the Mexican workers and the formidable bloc of white growers and Teamsters. Some of Salinas's local ranch families had prospered using cheap bracero labor, and they made it clear that they were not about to give in to the "crazies" from Delano.

Union lawyer Jerry Cohen respected some of the agribusinessmen for their pride and their hard work. But by 1970, he had become convinced that some ranchers thought it beneath them to deal face-to-face with Mexican farmhands. Not only did they refuse to look Chavez in the eye, they considered Huerta and Cesar inscrutable, and tried to funnel bargaining positions through white attorneys. "Some of them didn't like the notion of dealing equally with people of a different race," Cohen says bluntly. And even if they realized it was becoming socially unacceptable to say such things publicly, Cohen believes many ranchers were dogged by one single question: "Are we going to have Mexicans telling us how to run our damn operations?"

Other companies in the valley were run by locals, but owned by faceless conglomerates that had turned vegetable and berry farming into global ventures. The biggest was InterHarvest, a lettuce and vegetable firm owned by United Fruit, the infamous "Mama United" that controlled plantations—and dictatorships—in Central America. Another lettuce firm, Fresh Pict, was owned by Purex, which made its fortune on bleach and other cleaning products. A third Salinas Valley company with wealthy corporate connections was Pic N Pac, a strawberry company owned by Boston-based gourmet food producer S.S. Pierce. At peak harvests, Pic N Pac employed up to two thousand berry pickers at a time, sending entire Mexican families out to harvest enormous yields of fruit.

Court challenges to the Teamsters-grower pact could take months to settle, and by that time the harvest would be played out and the Teamsters entrenched in the fields. The union's best weapons, Cesar believed, were the threat of a boycott and a general strike. Still, these were weapons of last resort. More cautious than growers realized, Chavez hoped to avoid a strike with an offer: He challenged Governor Ronald Reagan to call for secret-ballot elections for the Salinas Valley farmworkers. Reagan had only recently, and glibly, called the Delano grape contracts "tragic" because they were signed without elections. At a rally in Watsonville, Chavez called Reagan's bluff. "Of late he seems very concerned about farmworker rights," Chavez announced to

■ *In the summer and fall of 1970, the Salinas Valley was filled with strikers waving flags and exhorting other workers to leave the fields.*

farmworkers and the press. "Now he has a swell opportunity to demonstrate his interest." Reagan didn't respond, and the growers made no move to retreat.

On Sunday, August 2, more than three thousand farmworkers responded to the Teamsters contracts by "voting with their feet," as Dolores Huerta called marches. The farmworkers streamed through the tree-lined streets of downtown Salinas and jammed onto Hartnell Community College's football field. Chanting *"huelga"* repeatedly, they carried a sea of of red-and-black UFWOC banners, waved American and Mexican flags, and held up pictures of the Virgin of Guadalupe and Martin Luther King Jr. Chicano students at Hartnell showed their support by hosting the rally over the objections of conservative townsfolk, and young Brown Berets stood on the rooftop of the college's library to provide security.

Chavez had walked with laborers from different towns at various times during their march into the city. Now he took to the stage, alternating between English and Spanish, accusing the growers and the Teamsters of intimidation as well as an arrogant "Pearl Harbor attack" on the farmworkers. His angry words underscored the racial divide the sweetheart contracts symbolized. "It's tragic that these men have not yet come to understand that we are in a new age, a new era," he told the restless audience. "No longer can a couple of white men sit together and write the destinies of all the Chicanos and Filipino workers in this valley." Urging cheering farmworkers to refuse to sign Teamsters cards, he instructed them to form ranch committees to report to

UFWOC's headquarters the following week. He again warned corporate growers that the farmworkers union was prepared to launch a boycott on their products, including United Fruit's high-profile Chiquita bananas and Purex's Dutch cleanser.

Priests offered mass and passed out Communion wafers, and the crowd eagerly voted to be ready to strike if necessary. "There was so much anger among people, dating back to the history of the braceros," Sabino Lopez says. "The farmworkers movement gave us a chance to force people to know we existed, that we had decided it was time for better conditions and respect. Back then, we were regarded as strange people who nobody really wanted to see around."

Lopez, the son of a bracero, abandoned school in Mexico at sixteen to join his father on the migrant trail to help support the family. He remembers that he hated living in Salinas at the time—not because the work was so hard, but because he felt the townfolk were racist toward Mexican workers.

Lopez first heard of Cesar Chavez from a Chicano accountant who examined his pay stubs, shook his head, and told the young farmworker he couldn't believe he worked so much and never was paid overtime. "Someone's coming to town soon who might put an end to that," the tax man said. "Nobody had to even come talk to us, or to educate us," Lopez says, remembering the volatile summer of 1970. "All we had to do was hear about [the movement] and I think just about everyone in my crew was for it."

At the time of UFWOC's big Sunday rally, the Teamsters still had only a fraction of the valley's workers registered as their members. Over the next week, they fanned out to sign up others belatedly and were met with widespread resistance.

Meanwhile, laborers from ranches UFWOC had never heard of were pouring into the little union office on Salinas's east side. Tomato workers, vegetable harvesters, lettuce cutters, strawberry pickers, all arrived in eager clusters, asking UFWOC to organize them. The most militant were the lettuce cutters—the *lechugeros*—migrant workers who had a reputation for being some of the toughest and most independent of all the campesinos. The *lechugeros* usually migrated from the Imperial Valley in the winter to the Salinas Valley in the summer. Their base pay was not much over minimum wage, but they could more than double their earnings with piece-rate bonuses they earned from stooping, cutting, and boxing lettuce as fast as they could—excruciating, backbreaking work. Some of the *lechugeros* supplemented their income by weeding and thinning crops earlier in the season, using the dreaded *cortito*, the short-handled hoe.

Those first few weeks in Salinas, Chavez had to harness the workers' emotions and steer them into a strategy for winning contracts. "There was more work being done on more fronts in that period that at any time in the history of the union," he said, "but I wasn't feeling the pinch as much as at the beginning of the Delano strike because by this time we had developed a lot of organizers." The union dispatched organizers to other lettuce-growing areas to prepare for additional strikes should the dispute in Salinas escalate. Chavez flew to an AFL-CIO convention in Chicago to appeal to President George Meany for strike-fund assistance, and he negotiated a loan from the Catholic Order of Franciscan Brothers as well.

Chavez also directed the union to fire up its boycott machinery. Lettuce was not a luxury fruit, but a staple vegetable purchased for daily use: It would be far more problematic a boycott target than grapes. The most vulnerable item to start with, it was decided, would be United Fruit's Chiquita banana empire, which had a reputation for abusing its workers in Latin America. "I knew that if we ever started a public boycott against them, it would be very hard to turn off—too many disliked the company," Chavez would later say. "But there were other ways of applying pressure."

One of them was Leroy Chatfield's brainstorm: He worked the phone from the union's Los Angeles office, asking a friend who was a buyer for a supermarket chain to call United Fruit and announce that he was prepared to stop buying bananas should UFWOC declare a boycott. Another union friend in New York who had connections in business circles contacted two top executives of United Brands, owners of United Fruit, to urge them to investigate what was going on in Salinas. (The executives' college-age children had boycotted grapes and were upset when they heard their fathers' company was resisting signing with Chavez.) UFWOC was determined to pressure the conglomerates to negotiate their way out of the Teamsters' pacts.

Amid the crisis in Salinas, Chavez had to rush back to Delano: His daughter, Eloise, was getting married. But even as he walked Eloise down the aisle, Chavez could see Jerry Cohen motioning to him that he had an urgent message. After one dance with Eloise and the cutting of the cake, Cesar was off again—this time to a secret meeting with Teamsters in a secluded coffee shop south of Salinas.

Chavez had no love for the Teamsters, who had betrayed him once again, but he was always willing to negotiate. Under pressure from the Teamster hierarchy, which was starting to have doubts about the Salinas deal, local Teamsters agreed to hold a series of meetings with UFWOC mediated by the U.S. Catholic Bishops' Committee on Farm Labor.

By this time, however, growers were forcibly removing *Chavistas* from Salinas labor camps, and Fresh Pict had fired more than one hundred lettuce workers for refusing to sign Teamster cards. The workers voted to strike, and called on hundreds of coworkers to abandon Fresh Pict fields scattered throughout the valley. The next day, hundreds of Pic N Pac strawberry workers also refused to go to work, triggering a second strike.

As expected, Fresh Pict turned to friends in the local courts, immediately obtaining a restraining order forbidding all picketing, yet another injunction that didn't

■ *UFW organizer Marshall Ganz taking a vote with farmworkers in Salinas Valley in 1970.*

allow the union a chance to present its side in court. Chavez alerted the press and met reporters at the Fresh Pict office to expose what he considered an unconstitutional injunction. Tearing a page from Jerry Cohen's legal pad, he wrote, "I am here to be served the order." He pressed the paper up to a glass window, and when he was sure someone had seen him, sat down on some steps to wait. While journalists gathered around Chavez, Fresh Pict president Howard Leach looked for someone inside the building to go outside and serve him the order. Amused, reporters scribbled away in their notebooks, and cameras captured the bizarre scene, turning Chavez's nonviolent confrontation into front-page headlines.

By the next day, Chavez had begun a fast of thanksgiving and hope, and was preparing union members for mass arrests when good news put a halt to any further action. The Bishops' Committee had succeeded in persuading the Teamsters to sign a "no raid" pact, an agreement to stop poaching on UFWOC organizing campaigns, and—in a confidential side agreement—promise to use their "best efforts" to break the Salinas contracts they had signed behind closed doors.

In exchange, Chavez declared a moratorium on strikes for ten days. But, he warned the Salinas farmers, workers were willing to shut down the industry if growers didn't start negotiating with UFWOC. "They've got the money," he said of the companies. "We've got the people."

As A YOUNG REPORTER and native son of Salinas, Eric Brazil was in a unique position to understand why growers in his hometown were terrified of Chavez, yet so clumsy in their dealings with him and immigrant workers. "It sounds absurd now," Brazil says, but when he was growing up, residents of Salinas didn't even regard field hands as employees. "We thought of the shed workers—the Okies—who were the union people who packed the lettuce, the carrots, who packed the celery," he recalls. "The field was something you didn't even think about, it was so low down."

For years, Brazil says, the growers had enjoyed total control over field workers, 80 percent of whom were braceros. A grower, he adds, once sang the praises of the foreign workers by comparing them to a well-functioning irrigation system controlled with a spigot: "When you want the water, you push down. And when you don't want it, you just turn it off."

In 1964, local ranchers were thrown into crisis when the last of some ten thousand braceros employed annually left. "They had everybody in here trying to pick strawberries," Brazil remembers. "They brought in Indians from Sioux Falls. They brought in people from Guam. They had everybody trying to pick these damn strawberries." Strawberry production actually had to be drastically cut back because growers couldn't find workers who could pick fast enough. In time, ranchers began to adjust, once again hiring mostly Mexicans—ex-braceros who had obtained green cards and brought their wives and children to the United States. "In whole patches of the Salinas Valley, people from different towns in Mexico just settled in," says Brazil.

The phenomenon opened up the valley to legal workers whose expectations were higher and who were more likely to respond to unionization drives. And it was only a year later, in 1965, Brazil recalls, that he first interviewed Cesar Chavez at a friend's house in the Monterey Bay village of Pacific Grove, not far from Salinas. "He was one of the two people who I really misjudged, although I liked him. One was Ronald Reagan and the other was Cesar. I thought Reagan was crazy and he was going nowhere. . . . I thought [Cesar] probably didn't stand a chance of being a real leader."

AFTER CHAVEZ declared a moratorium on strikes, United Fruit's vice president and chief negotiator Will Lauer flew into Salinas. Chatfield's strategy had worked; word of the imminent strike and boycott had gone way over local executives' heads and reached the East Coast executives. All indications were that United Fruit was willing to work behind the scenes to try to prevent a boycott. At internal strategy meetings, organizers discussed how to take full advantage of this key moment. "Come in tougher than hell!" Chavez told Dolores Huerta and other negotiators. The plan was to force United Fruit, which farmed twenty-two thousand acres in California and Arizona, to persuade other growers to break their Teamster contracts as well.

The next high-ranking executive to fly in was William Tincher, chairman of the board of Purex, Fresh Pict's parent company. "We picketed his office in Los Angeles," Cesar recalled. "He took it very personally. He thought what we were doing was underhanded, illegal, and immoral. But what he had done about recognizing the Teamsters against the will of the Fresh Pict workers never crossed his mind."

In the meantime, Cesar kept fasting. "There was a lot of hatred building up," he said. "It wasn't clear in my mind what I would do in case violence broke out on our side." He ended the fast after six days, admitting it had only weakened him. He retreated to San Juan Bautista, a small mission town a half hour from Salinas, to recuperate with the Franciscan Brothers and Helen and his children, while staying in touch with the union with frequent telephone calls.

Chavez's retreat into seclusion may have seemed odd to outsiders, given the heat of the crisis. But it gave Cesar distance and an aura of power that lured Will Lauer and John Fox, the CEO of United Fruit, over to San Juan Bautista for a secret meeting. Jacques Levy, a journalist who was working on a biography of Chavez, happened upon the adversaries sitting under chandeliers in the Franciscans' library. "The talk is quiet, friendly: three cultured men meeting in a formal library," Levy wrote.

The executives insinuated that Dolores Huerta was unreasonable, but Cesar, while gracious, let them know he didn't intend to retreat or offer them any easier a deal. "You know, I think we've got you guys," he told the executives. "If we don't get this thing settled, the boycott goes on."

A few days into the moratorium, UFWOC discovered that the International Brotherhood's betrayal of farmworkers was even worse than originally thought. Teamsters had agreed to a mere two-and-a-half-cent increase in piece-rate bonuses spread over five years. The lettuce workers exploded when they heard the news, and they grew more militant. Trying to prevent another premature showdown, Dolores Huerta leaped onto a chair at a union meeting to beg angry workers to give union leaders more time to negotiate a resolution. As workers chanted *"Huelga! Huelga!"* and held up handwritten strike signs, Huerta read a statement from Chavez: "The only thing we have to lose by waiting is our chains," she read, straining to calm the frustrated laborers. Finally, after a heated debate, a majority voted to postpone the strike another four days.

The workers retreated to labor camps, where Mexican women in town helped them sew new banners in preparation for a possible strike. Relieved to have more time, Huerta and Cohen rushed back to InterHarvest negotiations, still hoping to reach an agreement that would break the impasse. Now it was Lauer's turn to talk tough: he threatened to break off the talks if UFWOC couldn't prove it had a majority of workers on its side. Monsignor George Higgins of the Bishops' Committee was given the onerous task of verifying the legitimacy of all 856 union cards that UFWOC had collected from the company's workers. Thumbing through personnel files in a back office, it took fourteen tedious hours for him to finish matching the cards. But the priest finally emerged triumphantly, certifying that 95 percent of the company's field laborers supported Chavez. Lauer agreed to sit down and negotiate in earnest.

Meanwhile, the Grower-Shipper Association and the Teamsters dropped a new bombshell. At a press conference, director Herb Fleming announced that the vast majority of growers insisted on keeping their contracts, and the Teamsters union was obliged by law to honor them. The news shattered Chavez's attempts to avoid a strike.

On August 23, at another mass rally at Hartnell College, workers thundered their pledge to strike nonviolently. Dolores Huerta read a statement from Cesar, who was still in seclusion. "Everything we have done, we have done in good faith. Our good faith has been received with a slap in the face of the farmworkers." The statement also quoted Henry David Thoreau: "Where injustice prevails, no honest man can be rich." By dawn the next morning, the world's biggest vegetable industry was paralyzed.

Jerry Cohen remembers that red union banners with their Aztec eagles flew next to fields for more than a hundred miles in every direction, from South Monterey County up through Santa Clara and Santa Cruz Counties. "It looked like a revolution," he says. "And some of these right-wing growers thought it was. They had bumper stickers that said, 'Reds, Lettuce Alone.'"

On the second day the growers obtained injunctions that prohibited picketing on more than thirty ranches, arguing that the Teamsters had jurisdiction over the fields. More injunctions were slapped on the union with each new day. On top of that, growers hired guards armed with shotguns to patrol their property, and Teamsters sent in thugs wielding baseball bats to frighten off the *Chavistas*. One of the most infamous of these goons, Jerry Cohen recalls, was Ted "Speedy" Gonsalves, who wore black-and-white pinstriped suits and drove an armored limousine. (Gonsalves, secretary-treasurer of a Modesto local, was later suspended and indicted for taking bribes, using $25,000 of his local's money to put up other thugs in hotels, and for running liberal bar tabs.) The imported thugs menaced pickets, pounded on the walls of rooms where UFWOC negotiators were meeting, and knocked over coffee cups and cursed at UFWOC members whom they encountered in restaurants. The Seafarers' Union in San Francisco sent some burly members to Salinas to provide another layer of security for UFWOC organizers.

Jerry Cohen still winces when he recalls walking onto a ranch to check on the safety of workers who were staging a sit-in to protest the injunctions. He was stopped by grower Al Hansen, who ordered him off the property. Cohen refused. Turning his back, he heard Hansen shout: "Get 'em, boys!" An enormous Teamster named Jimmy Plemmens grabbed the two-hundred-pound Cohen and lifted him up by his jacket.

Cohen yelled to Jacques Levy, "Hey, Jacques, take a picture of this!"

"I was thinking, Well, we'll screw them," Cohen says. "We knew whatever happened locally, we were going to turn around and use it all over the country. . . . I mean, Hollywood could not have provided a better cast of goons than some of these Teamsters."

But in a split second, a fist cracked Cohen in the jaw, and he sank to the ground. Someone grabbed Levy's camera, smashed it, and drop-kicked it into some bushes, punching Levy as well. Cohen stumbled back to the entrance of the ranch. The last thing he remembers before losing consciousness was a sheriff's deputy leaning over him, complaining that there were too many pickets outside the ranch. "That shows you what the law enforcement around here was like in 1970," concludes Cohen, who ended up in the hospital for a week with a concussion.

With nowhere else to turn, Chavez fought back by declaring a boycott of all non-UFWOC lettuce, including produce grown by the biggest company in the valley, Bud Antle. Robert Antle had a nine-year-old Teamsters contract covering some farmworkers and still owed the truckers'

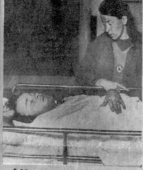

Valley Produce Strike

Court Bans Pickets As Tension Grows

Against conflicting claims, increasing tension and with no immediate resolution in sight, a strike of Salinas Valley agriculture entered its second day.

Yesterday's strike — called by the United Farm Workers Organizing Committee (UFWOC) — virtually shut down operations and a report from the Federal - State Marketing News Service indicated things looked even tighter today.

The marketing service said only three firms were working full speed ahead — Bud Antle, Sears Packing Co. and Jim Mapes, all of them firms with Teamster contracts of long standing.

Grower spokesman Rich

Vreeland, however, said 12 firms had some crews in the field today and "the number seems to be growing."

Chavez' farm workers have been striking to win for their union contracts signed last month between growers and Western Conference of Teamsters. Last Friday growers announced they and Teamsters have decided to hold fast to the contracts.

Major new developments in a tangled scene this morning were:

— A temporary restraining order issued by the Monterey County Superior Court at mid-morning to 22 of the struck firms. The restraining order prohibits picketing as a viola-

tion of the state's Jurisdictional Strike Act.

Separate but similar restraining orders were issued late yesterday and today to the Garin Co., Eckel Produce, Mann Packing Co. and Pic 'N Pac.

— The first indication of violence in the now more than three-week-old dispute.

At mid - morning UFWOC reported that Jerry Cohen, union attorney, and one other person were beaten up at the Hansen Ranch on East Alisal Street and Bardin Road.

Cohen, who was reportedly there to talk with workers, was taken to General Hospital. Attendants there said both men were still being examined just before noon and it was un-

known whether they would be hospitalized.

Monterey County Sheriff's office verified the incident but had no comment other than that an investigation was being conducted. UFWOC called a press conference for 12:30 o'clock, presumably to present its statement.

— Talks between Interharvest and UFWOC were reported at mutual impasse, but with both parties still leaving the door open to further sessions.

(Talks between UFWOC and Freshpict were suspended Friday in a disagreement over whether to include foremen.)

William Lauer, vice-president of United Brands and a

STRIKE   Page 2, Cols. 1, 2

Mideast Talks Opened

UNITED NATIONS (UPI)—United Nations mediator Gunnar V. Jarring opened Middle East peace talks today with the Israelis and the Arabs but there was no hint of compromise from either side.

The Arabs insisted on complete Israeli withdrawal from the territories occupied in the 1967 war, and the Israelis hinted that they still think that face-to-face talks—which the Arabs have always refused—are the best way to end the 22 years of strife in the Middle East.

The parties to the talks are Israel, Egypt and Jordan. Jarring saw each representative separately in his closely-guarded 38th floor office at U.N. headquarters starting with Israeli Ambassador Yosef Tekoah at 9:30 a.m. edt. Jordan's ambassador to Washington, Abdul Hamid Sharaf, followed at 11 a.m., and Egyptian ambassador Mohammed Has-

Attorney Hospitalized

United Farm Workers Organizing Committee nurse Peggy McGivern stands over union attorney Jerry Cohen at General Hospital of Monterey County this morning. Earlier today Cohen was kicked in the head, he said, during a fight at the Al Hansen ranch on Alisal Road.

Assaults By Reds Repulsed

PHNOM PENH, Cambodia (UPI) — Cambodian commanders today rushed three battalions of reinforcements into the Prek Tameak area nine miles north of here following a seven-hour battle with Communist troops. Other Cambodian units recaptured the village of Sdong Chey without bloodshed.

Field commanders told UPI photographer Kent Potter that a Communist regiment of five to six battalions took part in a series of attacks in the Prek Tameak area starting shortly before midnight.

The commanders told Potter all the assaults were repulsed with at least eight Cambodians killed and 22 wounded. Communist casualties were unknown but army spokesmen said "numerous" bodies were left on the battlefield. Potter said he saw no bodies.

If five or more Com-

■ *UFW attorney Jerry Cohen was seriously injured when Teamsters attacked him in a field in Salinas during a 1970 UFW strike.*

union part of a $1 million loan. He was outraged and threatened to sue Chavez for using his ranch as a wedge to pressure the rest of the industry.

Meanwhile, acts of violence were escalating. A ranch foreman drove a bulldozer into strikers' cars, a picket accused of throwing rocks was shot in the foot, and several others were hospitalized after being attacked by strikebreakers wielding chains. Police, not content just to issue citations, started brandishing guns and shoving pickets roughly to enforce injunctions. A distraught Chavez summoned a group of UFWOC boycotters to Salinas and assigned them to picket lines as captains. "The people don't

Blood in the **F**ields   •   **171**

understand how dangerous this is," he said. "They're sitting ducks to any stupid cop who pulls out his gun. I expect you to stop violence, and I hold you responsible."

At the height of the strike, three *Chavistas* were arrested for shooting a Teamster organizer in Santa Maria, where a simultaneous strike was raging. Chavez immediately accepted blame. Longtime press aide Marc Grossman remembers that reporters were expecting Cesar to defend his union members. Instead, Chavez said the shooting was inexcusable. Later, Cesar admonished members in a meeting, telling them sternly, "If we mean nonviolence, we have to say, 'Damn it, we mean it, and it's not going to happen.'"

In the midst of chaos, however, there was a stunning victory. Dolores Huerta, whom Brazil remembers dashing from rally to rally in a cocoa-and-white poncho, had driven a hard bargain with InterHarvest, finally winning a generous contract. The union's two-year contract far outstripped that of the Teamsters, giving InterHarvest workers a base-wage raise from $1.75 an hour to $2.10 and later to $2.15. (Minimum wage was $1.60 an hour in 1970.) The piece rates for each carton of lettuce packed by crews would go up by eleven cents, compared to the Teamsters' two and a half cents. The union also eliminated the use of DDT and other dangerous pesticides—a good two years before the federal government would do the same.

Two Salinas-based executives quit InterHarvest over the contract, and growers' wives showed their displeasure by lining up at an InterHarvest cooler for several mornings for *huelgas* of their own, waving American flags and attempting to block the company's produce trucks. Racial animosity deepened, with Anglo truckers and Teamsters driving around town waving American flags, attempting to claim them as a counterpoint to the Aztec eagle. "The social bones of Salinas stood out," recalls Eric Brazil.

Within weeks, several other growers, including Purex's Fresh Pict and Pic N Pac, recognized the groundswell of support for Chavez's union and rescinded their Teamsters contracts: They then signed with UFWOC. Community pressure against those contracts was so strong that Purex chief executive William Tincher later offered a disclaimer in a trade magazine: "Purex had been called communistic because we signed with Mr. Chavez. . . . Purex is not communistic. We pride ourselves on our contribution to the capitalistic life in America."

But after about a month, union members still on strike were hobbled by continuing court injunctions against picketing, and they were fast running out of five-dollar-a-week strike payments. Although they preferred to strike, some workers were forced to return to the fields. By November, the harvest was waning, and many migrants had either left for Mexico or were moving on to other jobs. Tension grew, however, as Chavez's boycott on all non-UFWOC lettuce companies continued. On November 4, unknown assailants planted dynamite that blew the door and windows out of a

UFWOC office in Hollister. A month later, a judge jailed Chavez indefinitely, until he complied with an order to stop boycotting Bud Antle lettuce.

In his Salinas jail cell, Chavez made the most of his solitude. He was placed in a small cell by himself, near the drunk tank, where he spent hours reading mail and books. Respecting his growing affinity for vegetarian meals, the union sent him soybean enchiladas. Mexican and Chicano supporters erected a shrine across the street, where they sang *corridos* and prayed.

The serenity was shattered, however, when Ethel Kennedy came calling the evening of December 6. The widow was greeted by more than two thousand Chavez supporters, with whom she celebrated mass before visiting Chavez. As Kennedy began walking toward the jail, she was surrounded by a hostile mob that had also converged on the scene. Several hundred John Birch Society members and antiunion demonstrators whipped themselves into a frenzy, shouting "Reds go home!"

The mob surged around Kennedy and her bodyguard, Olympic decathlon champion Rafer Johnson, and a phalanx of other security guards and union leaders. Brazil saw Jimmy Plemmens, Jerry Cohen's assailant, waving an American flag wildly and making obscene gestures. Another man tried to grab Kennedy's hair. "As soon as she showed up, Jesus Christ, it was incredible," remembers Brazil. "You could hear this roaring, 'Ethel go home! Ethel go home!' I couldn't believe what I was hearing. I mean, it was dangerous—it felt like someone could be killed."

Moments after Kennedy entered the jail, a Chicana teenager rushed up the steps of the building to wave a black eagle flag. She was knocked down by antiunion demonstrators, and police, deputies, and

■ *When Ethel Kennedy visited Chavez in jail in Salinas in December 1970, an antiunion mob jeered her and tried to attack her physically.*

Brown Berets had to hold back both sides to prevent a riot. Kennedy left through a back door, telling her husband's old friend Paul Schrade: "You really throw some weird parties here in California." It was "an incredible travesty," she said years later, that Chavez had been put behind bars.

On Christmas Eve, 1970, Chavez was released pending the outcome of an appeal. Four months later, the California Supreme Court unanimously ruled that the Antle injunction against the lettuce boycott was unconstitutional. (More than two years after the strike, the same court ruled that the Salinas growers had used the Teamsters in 1970 "as a shield" against the wishes of a majority of their workers.) The court also overturned injunctions that had forbidden most UFWOC picketing. Those high-court decisions failed to stop an even more violent—and fatal—battle that would take place in California's fields in 1973.

BY 1971, Chavez and the farmworkers union had won key contracts, against all odds and without a law giving farmworkers the right to organize. Union members' lives were improving dramatically, so much so that a generation of farmworkers was for the first time able to settle down, buy modest homes, and save money. They could even make sure their children finished high school instead of dropping out to work in the fields. Vegetable workers like Sabino Lopez, who was able to abandon the migrant trail and stay in Salinas, were among those who benefited most from the union. "Everyone wants to be acknowledged and respected as a decent person, even when you are covered with dirt," says Lopez, who got a job with InterHarvest. "I remember how satisfying it was to leave work on Friday, when we would get paid, and go to the bank and cash my check. The tellers used to look down on us, but when they saw our checks their expressions started changing. . . . We were making a lot more than people who were cleaner than us."

Chavez continued the boycott on non-union lettuce through the early part of 1971, but agreed to a moratorium when the Bishops' Committee again mediated a truce between UFWOC and the Teamsters. Vegetable growers once again sat down with Chavez to negotiate, but by the end of the year the talks had stalled and the lettuce boycott resumed.

During that time, the union was becoming an official member of the AFL-CIO, with voting rights in the national organization. In exchange, however, the new United Farm Workers of America—the UFW—had to forfeit the $10,000 monthly subsidy that the AFL-CIO had been giving it and assume full responsibility for its finances.

It was around this same time that Chavez startled supporters by moving the union's headquarters from Delano. Chavez had found an abandoned tuberculosis sanatorium fifty miles south, on the slopes of the Tehachapi Mountains separating the San Joaquin Valley from the Mojave Desert and the Los Angeles basin. The building was a collection of wood-frame cottages and meeting halls sheltered by trees and rolling hills. (Helen Chavez, coincidentally, had been placed in the sanatorium briefly as a child and was initially reluctant to move back there.) The union called the new headquarters La Paz, short for Nuestra Señora de la Paz, or Our Lady of Peace.

Union members were split over the move, which many thought foolish because it seemed to distance the union from the fields. But those who supported the relocation thought it would give Chavez and other administrators a place to retreat, plan strategy, and avoid the constant stream of workers filing through Delano with complaints better handled by field offices. The union's Forty Acres office remained open, with the credit union, hiring hall, and medical clinic intact. But the top administrators and their families moved to La Paz. "There were people who thought he was removing himself from the workers," Paul Chavez says of his father's decision. Part of the rationale for the move, Paul says, was his father's desire to preserve the union's national profile and to keep it from becoming too closely identified with one place or one part of the farmworker population. He also acknowledges, however, that his father was eager to know there was a place where he could retreat from the media and rejuvenate himself during the all-too-brief pauses between organizing campaigns and crises.

Moving to La Paz was not out of character for Chavez, who relished the communal atmosphere of the self-contained California missions, where he sometimes retreated with staff or family to rest. As early as 1965, at the beginning of the grape strike, Chavez told union organizer Eugene Nelson, "What I would really like is to be alone somewhere—in Mexico, or in the mountains—and have time to read all the classics that there are in English and Spanish." Chavez managed to steal time almost every night to read, or, later in life, to practice yoga or cultivate pesticide-free vegetables in his private garden. Christmas and Easter were very important holidays at La Paz, and the compound would come alive with parties and baseball games as the Chavez clan gathered.

More often than not, though, Chavez was on the move, traveling between field offices and around the country promoting the lettuce boycott and, later, another grape boycott. Although he sometimes took the opportunity to indulge his passion for art and photography by visiting museums, he didn't always enjoy the traveling, recalls UFW attorney Frank Denison. Chavez used to stay with Denison's wealthy brother, a New York City financier who managed the union's money and arranged fund-raisers in the early 1970s. One socialite was so taken with Chavez that she confided to an aide that her generous donation might be acknowledged with a private rendezvous. Denison remembers that Cesar cringed and wouldn't even joke about the proposition.

Chavez didn't like it when farmworkers fawned over him either, friends say. He seemed most at ease with family and close aides, like Mike Ybarra, an ex-Marine who was a Chavez bodyguard. While they drove from appointment to appointment, Ybarra saw a side of Chavez that few others did. They would listen to old jazz tunes and amuse themselves on drives back to La Paz by replaying their own version of a scene from *Casablanca*. Chavez would lean back and say, "Play it, Sam." Ybarra would respond, "Oh, come on, boss, I don't know it." They'd go back and forth, and then Ybarra would pop in a tape of "Chattanooga Choo-Choo."

"Every time we'd get to La Paz," Ybarra remembers, "I'd wake him up, 'Cesar, you're home.' 'Oh, La Paz!' We were happy . . . He never said, 'Go get some rest.' He said, 'Brother, see you at eight.' "

As it turned out, La Paz could not fully insulate Chavez from trouble. In late July 1971, Jerry Cohen had a disturbing phone conversation with agents from the federal Bureau of Alcohol, Tobacco, and Firearms. "So I met them," Cohen recalls, "and they had a picture of a guy who an informant said had been hired to shoot Cesar! I took it seriously because they were worried enough to give us the information. They were worried enough they didn't want to get caught holding the bag." The person who reported the plot was Larry Shears, a Bakersfield mechanic who was also a police informant. He told ATF that a drug dealer had been given $25,000 to hire a hit man to kill Chavez. Shears said the drug dealer, in turn, offered him $5,000 to burn a union office containing important records.

Armed with these details, ATF agents went undercover to secretly record conversations with the drug dealer and set him up for a bust. In the tapes, recorded on August 16, 1971, the drug dealer repeated much of Shears's story, adding that some Delano grape growers were behind the plot, but had canceled the hit because they no longer felt pressured by Chavez. If true, the growers' change of heart had taken place four days after the alleged "hit man," Buddy Gene Prochnau, was arrested while driving toward Salinas, where Chavez was speaking at a rally. Prochnau was arrested, however, in connection with a contract murder that had taken place in Visalia. He was later convicted for the Visalia murder and sentenced to life in prison.

The ATF dropped the investigation into the alleged plot to kill Chavez. But the agency did arrest the drug dealer on narcotics charges and found a total of more than $29,000 in cash in his house. All these mysterious connections were enough to convince the AFL-CIO to demand that the state attorney general's office investigate. A year later, the state concluded there had been no conspiracy against Chavez. The Civil Rights Division of the Justice Department had also investigated. But it too ended its probe without charges.

Jacques Levy suggests there may have been pressure from Washington to end the probe. Robert Mardian, an assistant attorney general in the Justice Department, had written to the ATF in 1971, requesting that he be kept informed of "all steps of the investigation" into the Chavez case. Mardian was later convicted of conspiracy to obstruct justice in the Watergate scandal; his family also owned a vineyard in Arizona that had signed a contract with the UFW—an agreement the family said drove it to bankruptcy.

The claims made by Shears were the most serious potential threat to Chavez and the farmworkers yet, Cohen recalls. Cesar was advised to leave La Paz and go into hiding for most of August of 1971. "For a long time I didn't believe there was a plot, but Watergate made me feel otherwise," Chavez said. "There's a certain amount of jealousy about your country. You don't want to think the worst of it until you're shocked into it. Then you have to realize it's true."

Cohen and other union leaders wondered if they were being too cautious. But in retrospect, "we did have something to worry about," remembers former UFW press aide Marc Grossman. "There were people out to get Cesar."

One of those people was Jerome Joseph Ducote, a veteran John Bircher and former Santa Clara County sheriff's deputy. Ducote confessed to orchestrating a series of break-ins of UFW offices between 1966 and 1970.

The theft of union documents, Ducote later told state investigators, was part of a clandestine effort to gain inside knowledge of suspected subversives like the UFW and peace activists. The effort was sponsored, he said, by growers and conservative businessmen. Ducote said that Kenneth Wilhelm, then the Santa Clara County Farm Bureau secretary, owned the post-office box to which he was supposed to mail his findings. Ducote also shared his pilfered information, he said, with federal agents and the anti-Communist Western Research Foundation in Oakland—a group that had close ties to the John Birch Society and was used as a source by the FBI.

Ducote was an authentic spy, says a former agent for the California Department of Justice who investigated Ducote in the early 1970s and cooperated with the FBI. After breaking into the union's office in Delano, Ducote took many of Cesar's personal notes, minutes of meetings, and letters between Chavez, Ross, Saul Alinsky, and SNCC activists—irreplaceable items that were never recovered.

The UFW first knew this man as "Fred Schwartz," the name Ducote used when he called to set up a meeting with Jerry Cohen in March 1973. Ducote told Cohen he was a lawyer whose client possessed confidential union documents stolen from Delano, and that for $35,000, the union could regain the materials. Cohen declined, and instead called the FBI. Agents instructed Cohen to arrange another meeting with "Schwartz," and the pair planned to meet at the lounge of the Ramada Inn in Bakersfield on April 4.

Schwartz had already filled an ashtray with Camel cigarettes by the time Cohen arrived, and was working on a bottle of sherry. He gave Cohen a glimpse of the documents, but didn't allow a close inspection. Cohen again declined the offer. The FBI photographed Schwartz as he left the meeting, and he was soon identified as Ducote. The FBI told Cohen the agency would look further into the matter. At one point, agents met with Ducote and his lawyer, but didn't interview the self-confessed political burglar. FBI records show that after agents met with Ducote, Justice Department officials asked them to halt the investigation.

"I was pulled from the case and told to shut up," said a former state investigator, who asked not to be named. "[Ducote] never said a peep again as well." Ducote eventually was sent to San Quentin for six months for credit-card fraud and theft in an unrelated case. As for the fate of the missing UFW papers, Ducote, if responsible, took that secret to his grave: He died of a heart attack in 1987.

IN THE EARLY 1970s, the union faced a surprisingly different challenge from a new quarter: administering the provisions of its hard-won Delano grape contracts. While Cesar and the other UFW leaders were in Salinas, Richard Chavez was saddled with the thankless job of supervising the registration of some nine thousand workers at twenty-nine ranches and teaching them about seniority, grievances, pensions, and medical benefits. The union itself had barely started learning how to put these concepts into practice. "Those were very awful days for everybody, I guess especially for me," Richard Chavez recalls. "We knew how to go out there and raise all kinds of hell and all that, but administration? We didn't have that kind of experience."

One of the worst tasks was running the new hiring hall, a contract provision growers fought most fiercely. Workers were supposed to join the union, sign up at the hall, and receive a dispatch to a union company based on seniority. Many of the ranchers still wanted the freedom to rely on labor contractors, who would do the dirty business of disciplining workers or firing them at will if they objected to working conditions. Chris Hartmire enlisted a group of earnest college students to help Richard register workers, but in their rush, they didn't follow directions. They abbreviated the names of ranches, causing great confusion because there were five different Zaninoviches, two Caratans, and a number of other names that began with the same letters or had the same initials. Crews had to be registered over and over again; duplicates had to be sorted out and properly filed. The union hadn't yet set up a filing system, so thousands of cards were at first stacked haphazardly around the office. "There was more confusion to add to the confusion at the hiring hall," Richard says. "There were thousands of people waiting to get dispatched at the same time."

To cope with huge lines snaking around the hiring hall, the union finally instructed workers to report to work and pick up their dispatch cards later after the system was smoothed out. Ranchers grumbled, with some justification, about the hiring hall, complaining that the union failed to send out enough workers when the grapes were ripe. They complained, moreover, that workers sent to the fields were too slow, and they objected bitterly to the union having authority to dismiss workers for refusing to pay dues. But the hiring hall, difficult as it was to implement, gave farmworkers the first legal tools to resist discrimination and arbitrary firings.

The traditional concept of seniority was controversial, as well. Not everyone in the union wanted a system based on it, recalls Al Rojas, who became UFW field office director in Poplar, a valley town in the shadow of the Sequoia National Forest. "Of course, there was a need for it," he recalls. "Cesar was strong on that."

But it was a challenge to introduce seniority into vineyards, which were worked by a mixture of local residents and migrants who followed a harvest trail from Coachella in the spring, up through the valley in the summer, into Stockton to harvest asparagus in the fall, and then back to Coachella for grapevine pruning in the winter. "A lot of the workers, the seasonal *mexicanos,* didn't have a concept of the union," Rojas says. Workers were not accustomed to democracy on the job, and disputes often arose over new seniority rules, dues, and grievance procedures. The local Chicano and Mexican immigrants were the core of support, but eventually other migrants became strong union loyalists, too, including Arabs from Yemen who had established a community in the San Joaquin Valley. "Arabs would come in with suits and ties on. 'I want to buy union for one year,'" Rojas remembers, slapping his hand on a table in imitation of their struggles with English.

Growers still trying to skirt the hiring hall would turn to labor contractors—Arabs, Chicanos, and whites—who would send workers over to a ranch without dispatches. Organizers would then have to go to the fields and object. The union also had to put up with continuing assaults by racists: In the summer of 1972, Al Rojas's office in Poplar was shot up by a mob of young whites.

Inexperienced field directors sometimes faltered under the weight of complaints. One meeting on grievances was attended by corporate grower Hollis Roberts, a large rancher who referred to Mexicans as "my boys." "Cesah Shavez, ah'm glad to meet you," Rojas remembers Roberts saying in his Okie drawl. "He hugs him, Cesar disappears into his suit, all you see is Cesar's black head. We go into a meeting, and it was pure disaster." The union field directors fumbled while presenting the grievances at Roberts' farms, and the businessman lost patience and flew off before anything was settled. Chavez was livid. He told the directors they had to sharpen their management skills, and he later brought in AFL-CIO trainers to help professionalize the staff.

Field director Ray Huerta, a native of East Los Angeles who'd picked almonds and apricots as a child, remembers feeling like a "one-man show" in the desert agribusiness valley of Coachella in 1971, where he would run the service center by day and visit labor camps at night. His biggest challenge was educating workers and growers about the new grape contracts. "The contracts were all in English, so we had to interpret them in Spanish, every little paragraph. You can imagine the tremendous amount of work you had to put into it. But as a result, you had a very, very strong constituency. I mean, they knew how to defend the contract." There were so many contracts at once that it was hard to keep up. "And there was super, super resistance from the growers," he says. "They weren't used to [organizers] coming into the fields." Even though union representatives were supposed to have access to workers, supervisors would chase them off.

Coachella rancher Lionel Steinberg, who was one of the growers most faithful to the union contracts, remembers the times differently. He found some union members unduly abrasive. "You had to work with the ranch committee," he says, "hard-core, arrogant people who were not what you would call rank-and-file workers. They started out rank and file; they gradually became union pros who were antigrower, and had more of the attitude of the hierarchy of the United Farm Workers."

But in the short time since the union had emerged, growers and unionists agreed, the farm-labor world had changed: There were water jugs, breaks, and bathrooms where none had been before, and workers were given more respect and opportunities for advancement. A Lionel Steinberg ranch in Coachella, Doug Adair remembers, was the first place where women, at the union's insistence, got higher-paying jobs like stacking boxes on trucks and pruning vines. "In different places, the union was really doing its job beautifully. The workers were becoming masters of their own lives."

IN THE SPRING and summer of 1973, as the union was learning to do its job better, the Teamsters union once again violated its pact with Chavez and stormed the vineyards of Coachella and the San Joaquin Valley. Events leading up to the UFW's bloody clash with the Teamsters and the growers began thousands of miles away, inside the Nixon White House.

When he ran for reelection in 1972, Nixon courted Teamster president Frank Fitzsimmons and won his only major union endorsement. Fitzsimmons associates say he thrived on the attention he got from Nixon, boasting about his visits and showing off golf balls embossed with the presidential seal. But there were other gifts: The Justice Department dropped its prosecution of Fitzsimmons's son, Richard, who faced fraud charges; and former Teamster president Jimmy Hoffa received a presidential pardon.

In the early 1970s, White House Counsel Charles Colson issued audacious memos to the Labor and Justice Departments and the National Labor Relations Board: Any federal intervention in the struggle between the Teamsters and UFW, he warned, should take the side of the Teamsters. "Only if you can find some way to work against the Chavez union should you take any action," he declared in a May 1971 federal memo obtained in 1973 by the alternative *Real Paper* in Boston. A second Colson memo in 1972 reiterated those instructions: "We will be criticized if this thing gets out of hand and there is violence, but we must stick to our position. The Teamsters union is now organizing in the area and will probably sign up most of the grape growers this coming spring and they will need our support against the UFW."

The memos cemented the White House's close ties to the Teamsters at a time when Nixon's campaign was receiving thousands of dollars in contributions from the union. (Colson became an attorney for the Teamsters after he left the White House under the cloud of the Watergate scandal. He was later convicted for his part in the conspiracy that began with a break-in at the offices of the Democratic Party and ended with Nixon's resignation.) In December 1972, Colson is reported to have arranged for Fitzsimmons to speak at the American Farm Bureau Federation's annual convention in Los Angeles. The UFW, Fitzsimmons told the receptive ranchers, was a "revolutionary movement that is perpetrating a fraud on the American public."

Meanwhile, in the spring of 1972, federal labor regulators apparently obliged

■ *A Teamster stomping on an effigy of Cesar Chavez in the Coachella Valley in the spring of 1973.*

White House wishes to wage war against the UFW by filing a lawsuit to stop secondary boycott activities. The UFW was boycotting nine wineries technically under the purview of the NLRB because a handful of employees worked in commercial sheds. The agency used this as an excuse to take legal action. Chavez said of Peter Nash, the Republican NLRB general counsel: "He was stretching the meaning of the federal law very far to get us." In response, the union picketed Republican headquarters across the country. When the UFW made plans to send tens of thousands of farmworkers to demonstrate outside the Republican National Convention scheduled for San Diego that summer, the GOP moved its meeting elsewhere. The NLRB dropped the lawsuit, and the union agreed to stop boycotting the wineries, but the episode had badly drained the union's coffers and energy. Nor would it be the last such challenge.

Not long after Fitzsimmons's appearance at the Farm Bureau meeting, one of the smaller growers in the Coachella area invited Chavez over for drinks, then confided to him that the Teamsters planned to move in on the grape industry as soon as the UFW contracts started expiring on April 15, 1973. The Teamsters, who controlled trucking, "have got us cold," the grower told him.

Ray Huerta also remembers the ominous signs of another Teamster takeover attempt that spring, long before the contracts expired. "The workers came and said, 'Ray, there's people coming out and signing up Filipinos in the fields,'" he recalls. Many of the Filipinos were living in labor camps at the time and were virtual captives. They had nowhere else to go and feared that if they didn't do as the foremen commanded, they would be evicted.

Huerta soon discovered that the Teamsters had opened an office in the Coachella Valley town of Indio. He realized, too, that the office was staffed by ranch foremen and labor contractors who had lost power under UFW contracts. With Coachella only an hour from Mexico, the contractors knew they could easily bring in strikebreakers should UFW workers resist.

After the grower's warning, Chavez quickly called for early contract negotiations with the ranchers to try to address their concerns. He offered to have a third party review the hiring hall, and to accept recommendations for running it jointly with employers. But the talks collapsed when a bloc of ranchers walked out ten days before the contracts were to expire. Nine hours after the UFW's Coachella contracts expired, most of the valley's growers signed quick agreements with the Teamsters—new contracts that eliminated the hiring hall and all the pesticide protections Chavez had insisted upon. Moreover, the contracts did away with grievance procedures, and left pay ten cents behind the $2.50 an hour that workers earned at two ranches that shunned the Teamsters and stayed with the UFW.

This new Teamster double cross unleashed six months of turbulence that rolled from Coachella up to the San Joaquin Valley, following the path of the grape harvest. On April 16, more than one thousand workers struck in Coachella to protest the Teamsters' invasion. UFW activist Clementina Carmona Olloqui remembers that when she was an eighteen-year-old Coachella farmworker, she drove her car up to a field that morning and blew her horn three times to give her coworkers the signal to strike. "Everybody came out," says Carmona Olloqui, a nurse whose eyes still shine with tears when she talks about that summer. "It was like being in a war. They arrested farmworkers; they hit them with sticks. Everywhere you looked there were Teamsters. If the truckers saw that you had eagles on your car, they would stop you and break your windshield."

After being pushed out of stores and knocked down for wearing a union button, Carmona Olloqui says it took her years to feel that Coachella was her home again. The intimidation did not stop even after she left farmwork: When, at Cesar's urging, she enrolled in a community college to study nursing, a grower's son tried to humiliate her, calling her "an ass who belongs in the fields."

The threat of violence intensified when caravans of Teamster enforcers—paid nearly seventy dollars each a day—poured into Coachella to barricade the vineyards and threaten UFW pickets. With their tattoos, heavy biker boots, and enormous beer bellies glistening in the hot sun, the Teamsters seemed oblivious to the bad publicity they generated outside California's agricultural valleys.

Local judges, concerned with the potential for violence, churned out more than a dozen injunctions limiting UFW picketing or completely banning it. Few restrictions were imposed on Teamsters who arrived on picket lines, sometimes carrying concealed weapons. At one of the ranches where a judge had forbidden all picketing, UFW volunteers and strikers converged in protest. They knelt in prayer, enduring Teamsters' threats, until police dragged them away by the hundreds. The arrests filled local jails, and it was clear there would be no end to the detentions if the injunctions were not relaxed. Riverside County judge Fred Metheny, curious about the dispute, decided to visit the vineyards to check on conditions. After a Teamster confronted the judge with a stick, Metheny indignantly expunged the UFW arrests and rewrote the restraining orders to allow picketing sixty feet from fields.

Ray Huerta remembers that, even faced with the venom and weapons of the Teamsters, Chavez was adamant about remaining nonviolent. "We had standing orders, 'You do not do violence. You do, you're out,'" Huerta says. "It was hard because these guys would come and get in your face."

"Cesar understood the importance of images and the importance of getting favorable press," Jerry Cohen says. "The growers were so arrogant on their own turf, it took them a while to catch up. And the Teamsters never seemed to care."

One night Chavez was sleeping at Huerta's house in Coachella, on the edge of the desert. It was so stiflingly hot that some of the security people who traveled with Chavez were sleeping outside in the front yard. Suddenly from the darkness, shots crackled all around them, and the men came rushing inside to check on Cesar. "There was someone way out there taking shots at us. Luckily, no one was hurt," says Huerta, who will never forgive the Teamsters.

But others were injured seriously. While sheriff's deputies accused UFW militants of blocking roads to prevent strikebreakers from leaving neighborhoods, and of hitting cars with picket signs, two Teamsters were arrested on attempted murder charges after they used an ice pick to repeatedly stab a man they mistook for a striker. Six others were arrested for standing on a truck and hurling huge rocks at a car Chavez was riding in. In a melee that Riverside sheriff's deputies blamed on Teamsters, a squad of nearly two hundred thugs armed with tire irons and clubs attacked four hundred UFW pickets in late June, sending four to a hospital.

In one of the more brazen and pointless assaults, a three-hundred-pound Teamster named Mike Falco attacked John Bank, a Catholic priest acting as the UFW's press officer. Bank was eating breakfast with a *Wall Street Journal* reporter at a Coachella restaurant when Falco came over and said, "I may go to jail for this, but it's worth it." He smashed the priest's nose with his fist, breaking three bones. The unrepentant Falco was charged with battery.

In late June, members of the Western Conference of Teamsters held a confidential meeting to plan a strategy to "escalate the violence" and run the UFW out of Coachella more quickly, according to FBI documents. "The Teamsters, according to the plan, would attack the UFW pickets in bursts of violence to last thirty to sixty seconds," an informant told the FBI. "Principal objects of attack would be UFW leaders. During the afternoon and evening, floating squads of Teamsters would rove the southern valley area performing 'vineyard patrol, camp visits, and house visits' both on UFW and Teamster members."

The strike was shaping up as the most vicious yet, and Chavez knew that he had to make regular appearances on the picket line to keep the peace. The contracts, he realized, were probably lost, but he had faith that the majority of workers would never slide back to the days when growers had total control. But he had little time to dwell on the future. As the strike unfolded, he was split in a dozen different directions—appearing on the picket lines, pleading with the AFL-CIO for a $1.6 million strike fund, and pursuing lawsuits against the Teamsters and growers for conspiring against the workers.

Chavez also traveled to Washington, D.C., to push for federal investigations into the violence and civil rights violations. Although some congressmen asked the FBI to investigate, acting director William Ruckelshaus declined: "We have no authority to conduct such an investigation," Ruckelshaus said in a July 5, 1973, response to Senator Floyd Haskell of Colorado. The agency did cooperate with a halfhearted civil rights probe by the U.S. Justice Department that yielded no action. Even so, the FBI remained committed to its surveillance of UFW picket lines and activities around the country.

As the turbulent weeks passed, some strikers rushed fields and began shouting obscenities at strikebreakers. When this happened, Chavez berated picket captains for not doing a better job at controlling tempers. He turned down an offer from the Seafarers' Union to send more guards to join the UFW on the picket lines. And while television cameras rolled, he stood before a group of snarling Teamsters, calmly but defiantly telling them the UFW had no intention of leaving the fields. He was thriving on keeping his cool while thugs who towered over him spat invective: "You stink, you smell, you're a lousy bunch of Commies!" Chavez told a reporter in Coachella, "This may be a blessing in disguise. This might be the ultimate confrontation. If we win, they'll leave us alone. Systems die a slow death, and the farmworker feudal system will take a long time to die."

But when the strike moved north to the San Joaquin, sheriff's deputies in Kern, Tulare, and Fresno Counties became the union's enemy. Unlike deputies in Riverside, who Chavez believed had treated the UFW with relative fairness, confrontations in the

San Joaquin Valley with police and growers' hired guards were violent—and fatal—and even Chavez began to lose faith in the power of the strike.

As THE FRUIT RIPENED north of Coachella, the UFW's contracts tumbled one by one. Gallo unilaterally ended a six-year union contract by signing with the Teamsters in July. The Delano and Bakersfield contracts that the union had fought so hard to win in 1970 were set to expire July 29. With growers already consulting with the Teamsters, the union started picketing, passing out leaflets, and staging marches, often in defiance of the dozens of injunctions issued against assemblies next to the fields.

About a week before the contracts expired, seventeen-year-old Marta Rodriguez was picketing on the edge of the Giumarra ranch with a group of other union loyalists when police helicopters swept low over the field, whipping up a storm of dirt in an attempt to scatter the pickets. Reeling and trying to cover their eyes, strikers shouted at police to stop, and deputies in helmets rushed in, swinging billy clubs and spraying cans of Mace. A deputy a foot and a half taller than the diminutive Rodriguez pinned her arms behind her and dragged her off into an orchard across the street.

"He screamed at me to shut up, and said lots of things I didn't understand," says the Mexican-born Rodriguez, who had only joined her father in California a year before the strike. "I was so thin then my wrists slid out of the handcuffs. He threw me down and put them on me again! And he put his knee on my waist so he could tighten them, almost breaking my bones."

Terrified and unable to understand the action swirling around her, Rodriguez screamed for help—a shocking image captured on film and in photographs. Frank Valenzuela, a former mayor of Hollister, California, and an organizer with the American Federation of State, County and Municipal Employees, stepped forward to calm her. But deputies lunged at Valenzuela, spraying him with Mace, clubbing

■ *Marta Rodriguez, seventeen, was handcuffed and restrained by Kern County sheriff's deputies after they broke up a picket line outside the Giumarra Ranch in 1973.*

■ *Frank Valenzuela, former mayor of the town of Hollister, was sprayed with Mace, beaten, and thrown to the ground by sheriff's deputies when he tried to calm a picketer being restrained by deputies.*

him in the stomach, and then handcuffing him as he lay facedown in the dirt, grimacing in pain.

A riot broke out as deputies pummeled strikers with batons, grabbing them by their faces to spray Mace directly in their eyes. Deputies locked one man in a chokehold so tight that he passed out. All in all, about 130 pickets were arrested, including Marta Rodriguez. A group of Teamsters who had jumped a UFW picket at the same field were not arrested.

The melee was just the beginning. The Delano growers had already made up their minds to throw their lot with the Teamsters, and negotiations at a Bakersfield hotel ended abruptly. An ugly pall of racial antagonism spread over the San Joaquin Valley, with Teamsters moving to Bakersfield from Coachella. Pickets complained that deputies called them "greasers" and roamed barrios late at night, rousting picket captains and other union organizers from bed and arresting them for violating anti-picketing injunctions.

The Teamster leadership also made little attempt to hide its bigotry. Einar Mohn, director of the Western Conference of Teamsters, said he didn't plan on letting any of the farmworkers he now had under contract attend membership meetings for at least two years. "I'm not sure how effective a union can be when it is composed of Mexican Americans and Mexican nationals with temporary visas," Mohn told a graduate student working on a report. "Maybe as agriculture becomes more sophisticated . . . as jobs become more attractive to whites, then we can build a union that can have structure and that can negotiate from strength and from member participation."

Publication of those remarks in the middle of the strike inflamed passions even more. By August 1, up to three thousand UFW pickets had massed on the sides of vineyards scattered throughout three counties. Chavez had resolved that the union's troops would face arrest rather than accept the injunctions against the strike. Paddy wagons in the valley were filled to capacity as farmworkers and supporters—including seventy priests and nuns—were carted off to jail, where many would remain for up to two weeks.

Chavez roamed the picket lines, spoke at candlelight rallies where workers would march around courthouses, and even visited jails to bolster spirits. He knew the only way to keep the union alive—much less win back contracts—was to phase out the strike and to start sending people back out on a new grape boycott. (Marta Rodriguez and her family would be assigned to Florida, where she was pleased to find sympathetic senior citizens who were living on their own union pension plans.)

On August 13, after thousands of detentions, the union finally had cause to celebrate. Fresno County decided to release all the pickets it was holding without requiring them to post bail. "It's like a resurrection," said Father Eugene Boyle of San Jose, who had been in jail since the beginning of the month. But the joy dissolved when the following morning the union learned that Nagi Daifullah, a twenty-four-year-old Arab picket captain, had been hospitalized and was in critical condition. Daifullah and other pickets had been at a party after a strike meeting and were standing on a street corner late at night outside a bar when a deputy ordered the group to disperse. A scuffle broke out, and Daifullah began running with a deputy in hot pursuit. When the deputy grabbed Daifullah, he clubbed him with a flashlight, and Daifullah fell to the ground. The deputy dragged him so roughly that his head bounced on the pavement.

While Arabs and Chicanos prayed outside the hospital, Daifullah lay unconscious. Finally, twenty-four years old and half a world away from his family back in Yemen, Daifullah slipped into death. Grief-stricken union members said blows from the deputy's flashlight killed the farmworker, but a coroner's jury ruled that Daifullah fractured his skull when he fell to the pavement. As union officials were desperately attempting to reach Daifullah's family in Yemen on August 15 to arrange for his funeral, a panicked striker called to report that a union member had been shot through the chest on a picket line near Arvin.

Juan de la Cruz was reaching for a container of water that his wife, Maximina, was offering him when someone in a speeding pickup truck fired five shots. A bullet pierced de la Cruz just below the heart. He collapsed into the arms of his horrified wife. The other strikers picked him up and rushed him to a clinic, where an ambulance was summoned to take him to a hospital. Within three hours, the sixty-year-old de la Cruz was dead. Devastated by the killings, Chavez immediately

■ *Deputies wrestling farmworkers to the ground during a wave of strikes in 1973 that resulted in more than 3,500 arrests that year.*

■ *Arab UFW member Nagi Daifullah's coffin was carried through the streets of Delano in 1973. He was chased by a police officer and died after being hit on the head with the officer's flashlight and falling.*

ended all picketing "until the federal law enforcement agencies guarantee our right to picket and see that our lives are safe and our civil rights not trampled on."

During Daifullah's funeral procession, the young Arab's coffin was carried past vineyards and through the streets of Delano for more than three miles, in a solemn march of more than seven thousand that culminated in a Catholic and Moslem ceremony at the Forty Acres. Juan de la Cruz's body was carried for five miles through Arvin and laid to rest at a cemetery where his grieving widow, her head swathed in a black lace mantilla, placed fistfuls of soil over her husband's coffin.

The movement, too, was fighting for its life: 90 percent of the UFW's grape contracts were gone, down from 150 to about 12, covering only 6,500 farmworkers.

"After people started to get killed, there was a visible change in Cesar," Luis Valdez remembers. "He felt personally responsible—I mean that in every sense of the word—personally responsible for the life and death of people in the union. And he didn't feel it was worth it." Chavez's brother Richard says, "He knew we were in very difficult times. Times were going to be very difficult from there on because the opposition was fighting very hard and they would go to any lengths to discourage us from organizing, even as far as killing people. . . . So we just had to make a decision. Do we quit now or do we continue? We continued because that's what the people wanted."

The strikes of 1973 had resulted in 3,500 arrests of farmworkers and their supporters, most of which were later thrown out of court. "We weren't all angels," Jerry Cohen explains. "There were some clod-throwing incidents, stuff like that. But I'd say over ninety-five percent of those arrests were on violations of unconstitutional orders, so they were cut loose." The strike left scars on families, however, and on a generation

■ *With Cesar Chavez and Joan Baez behind her, Maximina de la Cruz sprinkled
earth on the coffin of her husband, UFW member Juan de la Cruz. He was shot and
killed while both were on a picket line in 1973.*

of young Chicanos who witnessed police, deputies, and Teamsters batter and insult
Mexican farmworkers. Among them is Jose de la Cruz, son of the slain farmworker
and a strong union supporter who still lives in Arvin, not far from the cemetery where
his father was buried. He says the killing destroyed his mother's life: "She didn't last
much longer after that." He remains bitter about his father's killing, which went un-
punished. Several years after the incident, a jury acquitted the accused—who fired his
rifle from a moving car—because he claimed pickets were throwing rocks at him and
he feared for his life.

Unsure of their next step, Chavez and Cohen both knew it would be impossible to
win long-lasting changes to California's farm-labor landscape without having the law
on their side. No matter how successful boycotts might be, or how much the American public might support the cause, the UFW needed a special set of laws covering
farmworkers and agribusiness soon—or there would be no end to the violence and
the unfettered power of the ranchers.

■ FOLLOWING PAGE: *California governor Jerry Brown and Cesar Chavez met at
La Paz, the union's Tehachapi Mountains headquarters, to talk about the state law that
gave farmworkers the right to unionize.*

Whether it's Sacramento, Fresno, or Modesto, or the Imperial County, the farmworkers do not have the power the growers have. The growers have the lawyers, they have the allies, they belong to golf clubs, they talk to editors of newspapers. . . . This is power.
— Former California governor
JERRY BROWN

WITH THE LAW ON OUR SIDE

AFTER CHAVEZ CALLED an end to the grape strike in the summer of 1973, the UFW stepped up its nationwide boycott of California's non-union, lettuce, and the wines of vintners Ernest and Julio Gallo, who refused to negotiate with the union. By the winter of 1974, with the public still reeling from images of beaten and bloodied strikers, the boycott's impact was palpable: More than 17 million Americans had stopped buying grapes and many avoided Gallo wine. But even as the tide of public opinion flowed with the UFW, the dawn of the new year brought bad news from the East Coast. The *New York Times* shocked the union by reporting that critics had declared that the boycott was a losing battle. While the article said Chavez should be admired for *la causa,* a noble movement, it quoted cynics who predicted that the union had little hope of dislodging the wealthy Teamsters from vegetable fields and vineyards of growers who had welcomed them with open arms.

Fred Ross Jr., the son of Chavez's old mentor, read the article with a combination of anger and bewilderment. "It said the Teamsters were too rich and too powerful, and that this cause was over," he recalls. "That made a lot of farmworkers and organizers really mad," he says, shaking his head. He conceded that not everything had gone the union's way since Cesar had stopped the bloody grape strike to prevent further violence. Major growers like John Giumarra, for example, still refused to let their workers choose between the UFW and the Teamsters. So did Gallo, even though Chavez had offered to put up a million-dollar "performance bond" and end a nearly two-year-old boycott if Gallo's vineyard workers were to select the Teamsters in a secret ballot. Union membership remained low, and the bitter skirmishes in 1973 had also drained the

■ *The UFW marched to Modesto and the Gallo winery headquarters in 1975 to urge an election.*

union's reserves and used up the $1.6 million strike fund from the AFL-CIO. Worse, many UFW allies in politics and organized laber were quietly but relentlessly imploring Chavez to ease up on the worldwide boycotts of grapes and lettuce—goods that were often shipped, stacked, and sold by unions affiliated with the AFL-CIO.

But the *New York Times* story stung most because farmworkers, despite the contract wars, had made important gains. At La Paz, the union's school for members, the Huelga School, had won accreditation, and sixteen service centers were busy helping members with everything from rental agreements to citizenship classes. The UFW's energetic attorneys were drawing up drafts of an ideal farm-labor law, while organizers were supporting Arizona citrus pickers and Stockton tomato employees plagued by poor wages and working conditions. The UFW's Agbayani Village had opened its doors near the union's Forty Acres headquarters in Delano, and many of the old Filipino farmworkers who'd been thrown out of their longtime homes in labor camps at the start of the 1965 grape strike were enjoying a peaceful retirement there.

Fred Ross Jr. wrestled with what the union could do to vindicate itself after the *Times* article, and he came up with the idea of a march on Gallo. As Ross later explained, "We needed to do something really public and dramatic to show the world the farmworkers were far from dead." A march against Gallo wine appealed to UFW strategists who had chosen the super-rich Gallos as the union's official boycott target. The E & J Gallo Wine Company, based in Modesto at the northern edge of the San Joaquin Valley, was the nation's largest winemaker, with abundant political clout. It was a good target, with brand names that consumers could easily recognize. In San Francisco's heavily Latino Mission District, urban activists had already joined with UFW organizers for Operation Clean Sweep, a campaign that successfully persuaded many of the barrio's liquor-store owners and bartenders to stop selling the Gallo labels favored by street alcoholics—cheap, fortified wines like Thunderbird and Night Train Express.

Others within the union were leery of another mass parade of red-and-black banners. They predicted that, at most, two or three thousand people might actually take part—not enough to attract nationwide attention to counter an intense public relations campaign that Gallo was underwriting to show the public that all was well in its Teamster-run vineyards. But Ross persisted and won the backing of Chavez and Jerry Cohen. Cesar agreed that a march on Gallo's hilltop headquarters in Modesto would be a great way to rekindle the old union spirit. A long, peaceful march had worked marvelously in 1966, and it had been almost a decade since that legendary *perigrinación* through the center of the San Joaquin Valley had startled the world.

Cohen backed the march for another, more pragmatic reason: he believed it would force Jerry Brown, the new, young governor of California, to officially recognize the union and perhaps to back the union's proposed farm-labor law as well. Brown's victory in 1974 had signaled a power shift in Sacramento away from the pro-grower conservatism of Ronald Reagan, who in 1974 had helped kill a bill from Chicano assemblyman Richard Alatorre that called for election by secret ballot. A liberal with a background as a labor law specialist, Brown had long been an ally of the UFW. He'd come to know Chavez personally and had eagerly accepted the support of union volunteers in his gubernatorial bid. He even made Leroy Chatfield, Cesar's former assistant, an aide. But Brown had avoided any face-to-face meetings since he took office in January. The union's campaign for a farm-labor law was bedeviling the state's agribusiness giants, and the subject was apparently too sensitive for the new governor to address. In 1974, Brown had also won support for his recent campaign from rural Democrats who represented powerful growers in the state's farm belt.

The silence from the governor's office was frustrating for both Chavez and Cohen, who had dogged Brown for months in the hope of securing a meeting to propose their revolutionary law for farmworkers. Brown's neglect seemed odd for a man who had been thrilled when his father, former governor Pat Brown, introduced Chavez to him in 1968. As a former seminarian and student of Eastern cultures, Brown felt he had found a soul mate in Chavez, whose reading list included works on

Asian religions and philosophies. The younger Brown helped introduce the farmworkers movement to Hollywood's liberal glitterati: actors, singers, and film moguls who proved invaluable to future UFW causes.

Although the march on Gallo increased the pressure on Brown to renew his ties with the UFW, it was the growers who actually gave Brown his final push toward the union. As had happened at other key junctures in the short history of the UFW, farmers who were trying to outwit the union unintentionally provided Chavez with a badly needed media coup.

RATHER THAN STARTING the march in the Valley, the union chose San Francisco so it could attract more urban supporters. On February 22, several hundred people gathered on Union Square, bundled up against a Saturday morning chill. Ross Jr. was jubilant that the union was about to take on Gallo, but he knew this public test of wills was risky. If the march and the public pressure failed to force elections at Gallo ranches, the union would be humiliated, and its members further demoralized. And

■ *Juana Chavez joined her son on the Gallo march in 1975.*

Gallo was a formidable foe: Its founder's sons were notoriously ruthless businessmen and savvy enough to have withstood blistering attacks from social activists who claimed their cheap, high-alcohol wines were contributing to the acute problem of alcoholism in African American and Latino neighborhoods.

Before the protest began, however, Gallo had a surprise in store for the marchers. Three men unfurled a huge banner across the top floor of the St. Francis Hotel. In letters three feet tall, the sign read, "Gallo's 500 Union Farm Workers Best Paid in U.S. . . . Marching Wrong Way, Cesar?" Marchers scoffed at the banner: Gallo's Teamster contract did call for good hourly rates, but its medical and pension benefits, as they knew all too well, were only offered to full-time, year-round workers: a fraction of its total vineyard payroll. What's more, Gallo's workers had not been allowed to vote on the Teamster pact. That Gallo was nervous enough to display a banner was a good omen, and the marchers set off in exceptionally good mood.

The 110-mile procession made its way across San Francisco Bay, through the wild Sacramento River Delta country to Stockton, then south for another thirty miles to Modesto. Luis Valdez and Teatro Campesino provided entertainment for the marchers, which included an AFL-CIO contingent and political allies such as Dolores Huerta's longtime friend, San Francisco congressman Phil Burton. About 130 workers and union volunteers walked the whole week, while Ross played shepherd, troubleshooting along the route and herding stragglers who would drift in and out of the march. A southern contingent, tracing roughly the same route taken by Chavez and his original pilgrims nine years earlier, headed north from Delano. By the time marchers approached Modesto's Graceada Park, there were upwards of twenty thousand people, more than the 1966 *perigrinación* to Sacramento.

The singing and chanting of *"Chavez sí! Teamsters no!"* continued to the front of Gallo's corporate headquarters, where yet another banner awaited marchers—this one striking a bizarre, even surrealistic note: "73 more miles to go. Gallo asks UFW to support NLRA-type laws in Sacramento to guarantee farmworker rights." Cohen was astonished by this apparent capitulation. Today he believes Gallo's banners, which had attracted widespread media attention, were the final push needed to get Brown to call the various sides together for new talks on the farm-labor law. "It meant the growers were now telling Brown that there should be a law for farmworkers," Cohen recalls. "You've gotta remember now: There were growers, big growers, who had not been hit yet by us. Supermarkets were getting tired of us. There were the nervous rural Democrats [who represented farm districts]. And there were these county administrators all up and down the state who'd just seen thirty-five hundred farmworkers thrown in jail and didn't want a part of that. So all these forces are now ganging up on Brown. It was the strangest darn coalition you could imagine."

THE DANGER in such a coalition was that the growers obviously wanted a far different farm-labor law than that envisioned by the UFW. Chavez and Cohen wanted a farm-labor law that would go far beyond the NLRA by recognizing the special circumstances of the largely Spanish-speaking immigrants who moved from harvest to harvest. The law had to protect organizers, who were often threatened or beaten if they approached the ranches. In addition, it had to protect laborers, who were sometimes fired if they showed so much as a passing interest in collective bargaining. The law needed to set up a government agency to certify the results—a legal right that other unions had long enjoyed. Without such a watchdog, Chavez could call strike after strike, but the government and growers could legally claim there was no union activity.

The UFW wanted more: Cohen had been drafting a tough statute that would, for the first time, bind California growers to the results of secret farm elections. It would

Chavez and the UFW: A View from the California Farm Bureau

by Jack King

During the lettuce strike of 1979, California farmers criticized Chavez for starting a boycott of produce linked to SunHarvest, one of the companies in which union workers were on strike. In the following editorial, distributed to newspapers in California, Jack King of the California Farm Bureau Federation lambasts the UFW for its "pressure tactics."

Cesar Chavez has termed the current California farmworker strike a "dream come true." But it's a view which may not be as widely shared among rank and file UFW members as their leader would like to believe. There are signs of growing uneasiness among the members of the strike-idled union as the weeks turn into months with no sign of progress in the negotiations.

A federal observer to the talks reports there appears to be little interest on the part of the UFW to engage in serious negotiation. A regional director of the Agricultural Labor Relations Board admits he has no idea what the real motives of the union are. Some things are readily apparent: lettuce is rotting in the fields, more than four thousand farmworkers are out of work, and tensions are building.

Employer representatives have asked for federal mediation but the UFW refuses. Shunning the negotiations, Chavez instead is turning to his most familiar weapon—a national boycott. The target is Chiquita bananas marketed by United Brands, a parent company of SunHarvest, one of the struck companies. Through the use of the boycott, Chavez hopes to drum up public appeal, his strongest suit to compensate for his biggest weakness—a reasonable bargaining position.

When the California Agricultural Labor Relations Act became law in 1976, most observers felt the farm labor situation would stabilize and be marked by hard bargaining on both sides. With the capitulation of the Teamsters, the way seemed clear for the UFW to write impressive new labor contracts. Labor experts predicted it would be a time of maturation for the UFW during which they would develop their skills as seasoned contract negotiators.

But anyone who had negotiated an earlier contract with Chavez knew the transition would not come easy. Chavez doesn't negotiate; he demands. Guided by a master plan based on his ideology, Chavez pays little heed to the basis of labor relations compromise. And while the Teamsters are known for their business acumen and shrewd negotiating, the UFW is better known for its rigidity and pressure tactics. Whereas the Teamsters union knows that its future depends on the viability of the company it deals with, Chavez would seem to care less.

To the UFW leadership, the heart of the union is still the "cause." For Chavez it is far more gratifying to spread the gospel than to engage in prolonged negotiating sessions and mundane organizational matters.

■ *Chavez discussing strategy with farmworkers. Despite Chavez's efforts to prevent violence, ranchers charged he was uninterested in compromise.*

levy steep fines on farmers who interfered with organizing or voting, and would assure that farm elections take place no later than a week after workers petitioned for a vote, thus barring growers from stalling elections for months or even years. Chavez had also demanded that, unlike the NLRA, the California law preserve the farmworkers' right to secondary boycotts—setting up picket lines in front of grocery store chains that sold boycotted products. The boycott, he knew from experience, was the union's only truly effective tool against growers who refused to negotiate contracts.

Chavez considered the governor's help crucial in this new phase of the union's ongoing war with corporate farmers. California growers were avoiding direct confrontations in fields with the union. Instead, they were now hoping to write their own "labor protections" for farmworkers, seeking to adopt an NLRA-like law that would prohibit the boycotts and strikes that had won Chavez sympathy and boosted the farmworkers' power. Arizona had passed a law that outlawed boycotts and harvest-time strikes in 1971. The UFW had fought furiously against the Arizona measure, launching a recall drive against Governor Jack Williams after it passed and rallying around another well-publicized fast by Cesar.

Near the start of this fast, Cesar, surrounded by his aides and family in a Phoenix hotel, engaged in a short verbal debate with some of his Arizona supporters who doubted the UFW could prevail against established farmers in the decidedly conservative state. The exchange produced one of the movement's most enduring slogans: *Si se puede*. "It was May, and all these labor leaders and a state senator were gathered around Cesar, and they'd tell him repeatedly that the campaign against the Arizona law was futile and that he shouldn't fast," remembers Marc Gross-

■ *Cesar Chavez receiving the Martin Luther King Nonviolent Peace Award from Coretta Scott King in 1974 in Atlanta, Georgia.*

man, a longtime aide who traveled extensively with Chavez in the 1970s. "They would tell him in Spanish, *'No se puede, no se puede'* ['It can't be done'], and Cesar finally looked up at them and snapped, politely but firmly, *'Si! Si, se puede!'*"

This fast lasted three and a half weeks and was encouraged by a parade of visitors that included Coretta Scott King, widow of the Reverend Martin Luther King Jr. But when doctors monitoring Chavez noticed an erratic heartbeat, he was forced to abandon the fast. It ended in early June with a huge mass attended by members of Robert Kennedy's family and about one thousand farmworkers. The campaign had registered hundreds of thousands of new voters who, in subsequent elections, put into office several first-time Latino and Native American legislators, and Raul Castro, Arizona's first Mexican-American governor. But the UFW's bid to recall Governor Williams fell short, and the state's restrictive farm-labor law remained intact.

Hoping to replicate the industry's success in Arizona, California farmers had put a similar initiative, Proposition 22, on the ballot in 1972. That year, as secretary of state, Jerry Brown had thrown his weight behind the UFW by attacking growers and big business, whose petition gatherers had scoured phone books for "voters" and fraudulently attached their names on petitions that qualified this initiative for the ballot. Despite a half-million-dollar effort by farmers, voters rejected the proposition.

THE GALLO MARCH propelled Brown to action. "Brown was a great politician back then. He read his ticker daily, and he was sensitive to the immediacy of the moment," says Jerry Cohen. Within two days of the march, Cohen remembers, with a sly grin, the governor called. "That's when the real Kabuki dance started," Cohen says. "Brown started testing us, looking to see what he could get away with in this law that would satisfy the growers. Well, we really thought he needed us politically, and we were not going to accept a law that was unworkable."

What followed was a series of phone conversations and several unpublicized meetings attended by, among others, Chavez, Cohen, Brown, Chatfield, and the governor's agricultural secretary, Rose Bird. Once the group met at Brown's house in Laurel Canyon, just north of Hollywood, where Bird had laid out her initial version of a farm-labor law. To Chavez and Cohen, however, Bird's proposal too closely resembled the unsatisfactory NLRA. Cohen and Chavez had planned a unique version of "good cop/bad cop," with Cohen quietly negotiating while Cesar lambasted Brown and Bird in regular press briefings and at rallies, just to keep the heat on the embattled governor.

Brown appeared aloof, but he, too, was traveling throughout the state, testing the waters on behalf of the union's proposal. He privately quizzed business leaders and average citizens about a new law for farmworkers, and he went so far as to mail out eighteen thousand letters to "key people" in California, seeking their opinion. A series of climactic public hearings on the farm-labor crisis at the state Capitol in April and May of 1975 brought together the UFW, the farmers, and the Teamsters, who angrily roamed the halls, roughed up UFW supporters, and stood and shouted on top of lawmakers' desks.

The Teamsters' tactics made legislators even more eager for a break in the negotiations over the law. The hearings prompted another round of private meetings, on

■ *Teatro Campesino toured Mexico in 1974 while Chavez and the UFW began lobbying for a law to guarantee farmworker elections.*

May 3, at Brown's sparsely furnished apartment in Sacramento. There, all of the players were gathered for an all-night round of bargaining, away from the media. The Teamsters' lawyers were there, as were attorneys representing agribusiness. Chavez and Cohen had stuck to their plan, so with Cesar on the boycott circuit, Cohen represented the UFW. Also on hand were several lawmakers, including Assemblymen Howard Berman and Richard Alatorre, who had authored the farmworkers' bill. Once again Rose Bird laid out new proposals for the law, this time in neat, spiraled binders, only to have most of them spurned by various sides. During that tough session, Brown, who seemed determined to forge a peace treaty, deftly manipulated the seemingly gridlocked negotiations. He'd walk from room to room, passing around new proposals among the negotiators. At one point, Cohen and Bird argued over the NLRA's treatment of secondary boycott leafletting—Cohen said the national law permitted them; Bird argued that it actually forbade them. Brown called a time-out, withdrew to study the labor rules, and reemerged, saying, "Jerry's right. Let's move on."

The marathon talks ultimately yielded the state's unprecedented Agricultural Labor Relations Act, but not before rural legislators tried to change the law's wording to ban strikes during a harvest and allow Seventh-Day Adventists to decline union membership on religious grounds. The attempted sabotage outraged Cesar, who had just promised to end the Gallo boycott if the law was adopted before the legislature recessed for the summer, which would allow the fall grape harvest to proceed. The UFW prepared for another major march on Sacramento, but the law's opponents relented. After years of acrimony, boycotts, strikes, and deaths in the fields, California's farmworkers finally had a law guaranteeing them the right to defend themselves with a union.

On June 5, 1975, in an atmosphere of satisfaction and relief, Brown announced one of his first political achievements as governor at a press conference at the Capitol. "I think today marks a victory, not only for the legislature, not only for the farmworkers, but for all the people of California," Brown said. "In my mind [this law] restores great confidence to the ability of the legislative process and the goodwill of the people to work together even when emotions are running high and solutions seem very far and remote." In keeping with the union's strategy, Cesar stayed away from the ceremony, allowing the governor and the growers to have their day in the limelight.

Behind the rhetoric, Brown knew the law would be only a temporary truce in the struggle between the union and California farmers, but it was a worthy political victory, and not only for its altruism. "Certainly I liked the law because it got me off the hook," recalls Brown, looking back through the clear spyglass of history. "I didn't have to say, 'Oh, the grape boycott.' [I could say] 'No, we have a law, we take it to a vote, and we have a secret ballot election,' and that got it off my plate."

The new farm-labor law, which Brown signed with a flourish, was a great triumph for the union, but enforcing the law would prove to be like panning for gold: Chavez and the UFW would mine invaluable nuggets, but only after much frustration and hard work. The new law meant the UFW would have to devote hundreds of hours challenging grower objections to organizing and elections, a tactic that farmers mastered quickly and used to delay contract negotiations.

THE INK on the farm-labor law was barely dry, and already the controversies had begun. Growers were upset that Brown, despite his vow to appoint a balanced board not favoring worker interests or agribusiness, convinced enough legislators to vote in his first slate of appointments. The inaugural Agricultural Labor Relations Board included two old allies of Chavez: Leroy Chatfield and Roger Mahony, an auxiliary bishop in Fresno. Joining them were Joseph Ortega, a Latino activist from Los Angeles, and Joseph Grodin, a Democrat and a law professor with business ties to the Teamsters. (Grodin would later be named to the state's supreme court, joining Cruz Reynoso and Rose Bird as liberal voices on the bench. The trio would be recalled a decade later in a vicious campaign supported by growers.) The lone decidedly pro-farmer vote on the board was Richard Johnson, an executive with the Agricultural Council of California.

Despite the friendly majority, Chavez and Cohen were not convinced that progress would be easy. Field organizers were already reporting that Gallo supervisors were denying them access to vineyards where they could talk with workers. To publicize the violations of the new law and to recruit farmworkers directly from fields throughout California, Chavez once again turned to grassroots civil rights tactics. On July 1, 1975, he embarked on his least-noticed but toughest *perigrinación*, a "thousand-mile march" throughout California. Chavez began by touching a fence at the U.S.–Mexico border in San Ysidro, then strode with sixty supporters north to Sacramento and then south again, through the San Joaquin Valley to La Paz, just as the grape harvest was getting under way. All along the way the procession stopped almost daily to hold rallies and collect farm election petitions from laborers. At the end of each day of walking, Chavez would mark the spot where he had come to a stop, drawing an *X* into the ground with a stick or pushing a twig into the soil. The next morning, he'd lead the marchers out to the same spot to resume the journey. "In some places, there was no one around, and it could have been easy to cheat and skip a few miles," Marc Grossman remembers. "But Cesar would insist on finding that mark each day."

It was on this march, say Cesar's friends and aides, that the middle-aged Chicano warrior appeared to recover some of his monumental vigor, which had been sapped by many months spent on the road, sounding the boycott's message and thumping podiums in disgust over the lack of cooperation from politicians and growers in negotiations over the law. On the march, adhering strictly to his vegetarian diet kept him physically fit. More often than not, Cesar often ate simple lunches of watermelon wedges, while others devoured sandwiches and tacos. (Luis Valdez remembers many meetings when union members would feast on platters of carne asada—grilled steak—while Cesar dug into a whole watermelon or raw vegetables. Even during negotiations with growers, Chavez would not hesitate to unsettle the sessions by loudly juicing carrots in his blender.)

Although he struggled with flashes of the back trouble that had nagged him since the late 1960s, Chavez thrived during the march. "After a while we were so tough that we could run daily, or walk real fast," remembers Mike Ybarra, the ex-Marine who often worked as Cesar's driver, and who walked the entire route that summer with the tireless union president. One day, when the union parade left the San Joaquin Valley town of Merced, Cesar told Ybarra that they had to make up some ground: The union's big convention was just days away, and Fresno was still sixty miles to the south.

"Cesar said: 'I'm taking the point,' and he started walking real fast, and different television cameras would come by and his feet were not even touching the ground. He started running, and I'm going with him step by step," Ybarra remembers. As he sped up, Ybarra pointed to a distant telephone pole and said they'd stop there. "He says, 'Can't stop.' Then we get to a sign and I think he's going to stop. But he kept going and going, and oh my God, finally he stopped. We looked behind us and there was no line, everyone had scattered and was lying on the ground, people were hanging over the hoods of cars . . . but we did it!"

As it turned out, the July 1975 march would serve the union as physical and moral preparation for the difficult times that lay ahead: The acrimonious legal and political wrangling over the new farmworker law had just begun, and dissent over the union's future direction was springing up from all sides.

WITH a $2.8 million budget for its first year, the ALRB established field offices with arbitrators and paralegals throughout the state in anticipation of a flood of union-recognition elections. In only the first two months of the state's new farm-labor law, more than forty thousand farmworkers voted in union elections, many for the first time ever. But quick UFW elections at the InterHarvest farms in Salinas and El Centro soon sunk into the quagmire of company challenges. Cohen and his band of legal monkey wrenchers fanned out around the state, dispatching lawyers to read the law out loud in Spanish on farmworker buses, or to walk into ALRB offices with tape measures unannounced, telling startled bureaucrats that they were just measuring to see how many vocal farmworkers could be squeezed in the rooms for sit-ins.

Meanwhile, Teamsters advocates were being allowed easy access to the fields, while UFW organizers were barred, and sympathetic workers fired. In one case, UFW organizers had tried and failed to enter the Ballantine grape vineyards just off Highway 99, south of Fresno. Two ranch hands confronted the organizers, led by Juan Salazar, who reminded the foremen that the law allowed them access to field workers at specified times. "We know all that, and don't give a shit about the law," one of the foremen shouted at Salazar. "All we know is that you are trespassing on private property!"

"He then began pushing us out," Salazar said after the incident. "Other ranchers were around laughing and encouraging the supervisor to beat me up. [When] a sheriff came, he hugged one of the Ballantine supervisors and went around shaking the growers' hands. He told us to leave or be arrested."

In light of these flagrant abuses, the UFW filed more than a thousand complaints in just a couple of months against growers for threatening workers and scaring them away from the voting sites, blocking access to fields, and allowing Teamsters to continue bullying UFW organizers. In one of the more daring challenges to the growers' refusal to allow UFW organizers onto ranches, both Helen Chavez and Cesar's daughter, Linda Chavez Rodriguez, were arrested when they walked into Jack Pandol's fields to talk with workers. The pair were quickly released when police discovered who they were and tried, in vain, to avoid the negative publicity. The union publicly denounced Brown and the law's oversight board, including Chatfield and Mahony, for what it perceived as their inability to control the Teamsters. The Teamsters, in turn, charged that the UFW was using undocumented immigrants to "pad" the pro-UFW votes. All in all, eight of every ten elections were contested by one side or the other.

In Sacramento, the conflict overflowed into the ALRB's offices. In the fall of 1975, Teamsters stormed into the agency's offices, charging Chatfield and Mahony with pro-UFW bias. The protesters punched a hole through a wall, shoved Chatfield, hit him with picket signs, and rubbed dirt in his face. Before leaving, they grabbed Mahony and pinned Teamster buttons on him.

By the end of the ALRB's first fiscal year, farmworkers had voted in eighty elections, with the majority of workers voting for the UFW, and union membership would swell to more than twenty-thousand. One of the most significant votes was held not long after the law went into effect in the fall of 1975. Workers at the M. Caratan Ranch, a large vineyard and fruit operation outside of Delano, had pledged their support to the UFW by a three-to-one ratio. Ben Maddock, a longtime UFW staff member, was in charge of the organizing. The son of a San Joaquin Valley citrus farmer, Maddock had rebelled against his conservative upbringing in the early 1960s and drifted around the West for a few years before finding himself back in the valley, living with his sister in an apartment building not far from Delano. He soon met and befriended a neighbor, who turned out to be Gilbert Padilla. Before long, Padilla asked Maddock to come out to Delano and help the union for a while until he figured out his future. As Maddock recalled, "A month turned into twenty years."

The Caratan ranch election was his big test, Maddock remembers: It was the first state-supervised vote in Delano. The choice on the ballot was for the UFW or "no union." The farm, where a core of pro-UFW Arab immigrants worked, had earlier abandoned Chavez in favor of a dubious deal with the Teamsters. "We wanted to prove that in 1973, when the Teamsters had come in . . . it was phony," Maddock recalls. As the balloting continued inside a meeting room on the ranch, workers who had voted gathered, quietly at first, outside. With hands in their pockets, they whispered among themselves, wondering what would happen if they lost. Maddock could hardly stand the pain of seeing the UFW count fall, initially, fifteen votes behind. All he could think of was how he would explain the embarrassing loss to Chavez.

"As it went on, I knew we had made up the fifteen votes and we started going up. Then I couldn't wait to get outside; I mean, I just started yelling—the adrenaline was really running because I knew we'd won," Maddock remembers. The scene, a raucous mixture of joy and relief, was seen on news shows that night: There was laughter echoing through the farm, while Maddock hugged ranch hands, many of them Arab immigrants. A group of triumphant Yemenite workers celebrated with a traditional dance, turning and clapping their hands with arms held high overhead. (The celebration was probably the last happy time for UFW workers at the M. Caratan Ranch. Later, the grower pushed for a hotly disputed decertification vote, and ultimately won: The union was once again ejected from the ranch in the late 1970s.)

The Caratan vote was a prophetic loss for the Teamsters, who were already in trouble. By 1976, it had been two years since the union's friend, Richard Nixon, had left the White House after the Watergate scandal, and the Teamsters were starting to collapse from internal conflict. Former Teamster president Jimmy Hoffa hadn't been seen alive in two years, and several federal agencies were beginning to close in on the controversial union, which had been accused of violating an array of racketeering laws.

As the Teamsters began to retreat from representing field workers, pro-UFW votes became more frequent, and the number of laborers choosing Chavez's union grew. But the union soon noticed a clever shift in strategy on the part of growers and Cesar's growing list of political adversaries. Rather than fighting the ALRB directly, the grower-influenced state legislature simply tried to strangle it by leaving the agency with insufficient operating funds. After the mountain of elections, appeals, hearings, and unfair labor practices charges it had to investigate, the ALRB was broke. The board made an emergency plea for new money, but it was blocked by Republicans and rural Democrats who, still angry over Brown's appointments to the board, demanded changes to the law that would dilute its worker protections. A stalemate ensued, even though one influential farm, the Bud Antle Company of Salinas, threatened to sue the state if the legislature did not re-fund the ALRB. (Antle's friendly contracts with Teamsters were about to expire, and the company could not legally write a new deal without regulatory oversight.)

The deliberate legislative bottleneck frustrated Chavez. The ALRB's paralysis left hundreds of farm elections in limbo, and there were no regulators around to force growers to the bargaining table. Workers were also in jeopardy: During the agency's budgetary coma, from April to the end of May, the union estimated that some five hundred pro-union workers were fired from various farms.

Chavez again took his fight to the streets. Instead of a boycott, however, his supporters helped put an initiative on the November ballot to guarantee permanent funding for the ALRB and force a state-wide vote on any proposed changes to the farm-labor law, thus taking it out of the hands of legislators and the partisan budget-making process.

Chavez enlisted the group of volunteers and operatives that had helped defeat Proposition 22 in 1972. A platoon of assistants and farmworkers-turned–union activists—like Clementina Carmona Olloqui, who had come out of the bloody UFW strikes in Coachella—collected the needed signatures. Soon, more than 700,000 signatures were recorded, and the initiative got its official ballot name, Proposition 14.

■ *Cesar Chavez led a San Francisco rally in 1976 for Proposition 14, which he hoped would protect a new state farm-labor law with a constitutional amendment. The measure failed.*

The threat of the proposition pushed the legislature to re-fund the ALRB that summer, but by then the board was in disarray. Both the UFW and the growers were charging agency staff members with bias and incompetence; farmers were defying court orders as well as the ALRA by denying the UFW access to their fields; and two of the original five board members, Chatfield and Ortega, had left. Bishop Mahony's interest was waning, and he later resigned as board chairman. Under attack from both growers and the UFW, Leroy Chatfield had resigned, moving over to help Brown with his run for the Democratic Party's presidential nomination. He left in despair after spending too much time in what he calls "an impossible, no-win situation" of being harangued by growers, physically attacked by Teamsters, and verbally assaulted by his friends and former colleagues at the UFW. "Regardless of which or what decision I made, I would either be considered a sellout by the movement . . . or I would be attacked by the grower community as being unfair, biased, and not a proper board member," remembers Chatfield. "The fact that thousands of farmworkers voted for the first time made it worthwhile. But I did not want to be on the board. It was a very difficult experience for me," he adds, explaining that he only served because Brown had insisted.

THE ALRB'S near-death spurred the UFW back into political action. The summer signature drive became a fall voter-registration campaign to bolster support for Proposition 14. Chavez managed the campaign while Marshall Ganz oversaw the organizing efforts. The union called in whatever political favors it could, drawing

The Death of the Short-Handled Hoe
by Susan Ferriss and Ricardo Sandoval

In early 1968, California Rural Legal Assistance lawyer Maurice "Mo" Jourdane was shooting pool in a smoky cantina in Soledad, California, when a small band of farmworkers approached him, a couple of them walking with a rigid gait that spoke of constant pain. The men stopped to talk with Henry Cantu and Hector de la Rosa, Jourdane's billiards partners, who were outreach workers with CRLA. Cantu then translated a simple challenge from the workers to Jourdane: "If you really want to help the campesino, get rid of el cortito—the short-handled hoe."

El Cortito, "the short one," was a hoe that was only twenty-four inches long, forcing the farmworkers who used it to bend and stoop all day long—a position that often led to lifelong, debilitating back injuries. The pool-room meeting with a handful of its victims led Jourdane to try working in nearby fields for two days. Within weeks of experiencing firsthand the pain el cortito caused, he and other CRLA attorneys began a seven-year battle to outlaw the most insidious tool ever used by California agriculture.

For Cesar Chavez, who played a pivotal role in the long drama, there were few greater moments than when el cortito was finally banished from California's fields in 1975. In his youth, Chavez knew the hoe well, having used it to thin countless rows of lettuce and to weed sugar-beet fields along the Sacramento River. Later he would say he never

---

■ *Banned in 1975, the short-handled hoe was called* el cortito, *"the short one," or* el brazo del diablo, *"the devil's arm." Cesar Chavez fought to have the crippling tool banished from the fields.*

looked at a head of lettuce in a market without thinking of how laborers had suffered for it from seed to harvest. "[Growers] look at human beings as implements. But if they had any consideration for the torture that people go through, they would give up the short-handled hoe," Cesar said in 1969.

In the late 1960s and 1970s, *el cortito* was the most potent symbol of all that was wrong with farmwork in California: The tool was unnecessary, and farmers in most other states had long switched to longer hoes. Growers argued that without the control the short hoe offered, thinning and weeding would be mishandled, crop losses would mount, and some farmers would go bankrupt. As he prepared to take on California farmers, Jourdane quizzed many physicians—including Cesar Chavez's back specialist—who said that without a doubt, the hoe was responsible for the debilitating back pain experienced by many of their farmworker patients.

Jourdane was elated when he was finally able, in 1972, to tell Chavez that the first formal complaint had been submitted to the state's Division of Industrial Safety. The lead plaintiff was 46-year-old farmworker Sebastian Carmona. "I came from Texas in 1959 and had never seen a short-handled hoe," said Carmona during his legal battle. "It surprised me, but I thought I'd be able to handle it because it was smaller [than normal]. The first day, when I needed money for my family, I felt a tightness, but I was okay. The second, third, fourth days, it got worse and worse"—so much so that Carmona would feel a hot pain in his back each time he'd stand erect at the end of a row.

The DIS rejected the attorney's claim, but the state supreme court overturned that ruling, finding that the tool was a danger to laborers' health because it could only be used while stooping. A long-handled hoe, the court said, was just as useful. "Getting rid of the hoe felt as good as anything in my career," said Jourdane, now a retired state judge. "It was flat-out a symbol of oppression—a way to keep control of workers and make them live humbled, stooped-over lives."

The supreme court ruling failed to move state regulators, however, who in 1975 still had not written rules forbidding the hoe's use. A frustrated Jourdane called the UFW, which contacted the new governor, Jerry Brown. Under executive pressure from Brown, the bureaucrats finally wrote *el cortito* out of the fields.

But the tool refused to die. In 1985, as part of a drive to streamline government rules, state safety

■ *Jessie De La Cruz as a young farmworker. Years later she told one grower that, with the short-handled hoe, she had "measured his land inch by inch."*

officials amended the 1975 law banning the hoe, saying workers could be required to use the tool for five minutes each hour. The resulting howl from farmworkers forced new hearings that quickly killed the proposal. Former farmworker Jessie De La Cruz, who once told a grower that with *el cortito*, "I measured your land inch by inch," spoke at those hearings and invited state and federal officials to rise and do more than imagine what using a short-handled hoe was like. "I told them . . . just stand up and hold the tips of your shoes and walk up and down this room and see how many times you can do it."

enthusiastic support from presidential candidate Jimmy Carter and Hollywood celebrities. Twenty-five-dollar-a-plate fund-raisers drew big crowds, and more than a million dollars poured into the campaign treasury. In public, at least, Chavez was supremely confident his cause would again be sanctioned by sympathetic voters. "We're going to teach the growers a lesson they'll never forget once and for all," Chavez told a thousand cheering state labor leaders who'd gathered for a fall meeting in Sacramento. The campaign also brought Chavez and Brown closer politically, with Marshall Ganz joining Chatfield on Brown's presidential campaign, and Cesar himself nominating Brown at the Democratic Party convention in 1976.

Looking back to the fall of 1976, Proposition 14 campaign workers say they misjudged the mood of the public and underestimated the lengths that growers would go to defeat the initiative. On the unfamiliar turf of suburban Orange County, Clementina Carmona Olloqui was attacked by an angry student who hit her with a cup filled with urine when she spoke at a community college. Later, she and other UFW supporters were harassed by angry drivers who screamed at them when they set themselves up as human billboards on highway overpasses. Growers and oil companies spent lavishly on television and newspaper ads; campaign analysts said the anti–Proposition 14 forces invested nearly $2 million. The issue even split the state's Democrats, with rural legislators like Leo McCarthy speaking out against Chavez's initiative.

The No on 14 campaign—run by an influential public-relations firm, the Dolphin Group—hardly mentioned the fact that the initiative merely promised to add to the constitution the funding provisions and the rights of organizers to enter farm fields to talk with workers. These items had already been approved by the legislature, but the Dolphin Group branded the farm access provision a threat to property rights that had no place in the state constitution. Anti–Proposition 14 commercials featured concerned farmers and their wives complaining that their privacy and personal safety would be left unprotected. "They played the race card in that campaign," Jerry Cohen recalls, explaining that one ad showed a woman looking nervously out her window. "They were giving the public a message that, in reality, read: 'Do you want a Mexican on your property attacking your daughter?' The campaign was that crude."

Crude, but effective: Proposition 14 lost by a substantial margin. "It was a huge miscalculation," says Cohen, looking back on the UFW's optimism during the campaign. "We saw the limits of how much support there really was for us out there. . . . It was a different ball game, with a different bat, and we didn't know how to play that game so well." Cesar was crushed by the electoral rebuke. Although Chavez had started a fast several days before the vote, Ben Maddock remembers that Cesar was so heartsick that he ended it with uncharacteristic carelessness, eating too much and becoming ill the next day. "It was a most awful time. We were supposed to go to an event that night, but we ended up just driving back to La Paz," Maddock says. "We were not used to having an election and losing like that."

Flush with triumph, some growers stopped complying with the law or filed appeals of elections that delayed the entry of thousands of new members into the union. A bottleneck was rapidly building up, since the ALRB, still recovering slowly, had not settled the many legal disputes already on its books before the Proposition 14 vote.

AFTER THE DEFEAT of Proposition 14, friends and union workers noticed a change in Chavez. Although in public he remained optimistic about future political victories, his near-religious faith in the ballot box was fading. In private, former union leaders remember, Chavez seemed disillusioned with the initiative process. Rather than planning a new campaign for another initiative, union leaders began to funnel more money to political campaigns, hoping the contributions could reserve the UFW a place at the back-room tables where the real policy deals were being cut.

For his part, Chavez returned to mercilessly haranguing the legislature and the ALRB, while still trying to make a success out of the law he had fought so hard to create. He attended hearings and often spoke at schools and clubs about the law, accusing farm regulators of dragging their feet and failing to defend farmworkers' rights. "We've been asking the workers to be patient for a long, long time. We're not going to front for the board anymore," Chavez said at an ALRB meeting in early 1977. In an earlier meeting, he was even more critical: "Seventeen months after the farm-labor law went into effect, most farmworkers have yet to realize the promise and protection of this good law. Instead, for most, the law has been a cruel hoax."

Chavez was vexed by the mounting backlog of complaints against growers: illegally disrupting elections, barring organizers, firing thousands of pro-UFW workers, and failing to bargain after the union was certified. Regulators had failed to verify months-old elections covering thousands of workers at big farms like Gallo, the D'Arrigo vegetable operations in Salinas and El Centro, and the Egg City farm in Ventura County—delays that sometimes stretched out for years. These early disputes would later become the focal point of Cesar's claims that the government was thwarting union organizing and expansion. The bureaucratic foot-dragging was exemplified by the outcome of the union's fight against the Ukegawa Brothers Company tomato ranch near San Diego. In 1976, the union filed charges concerning retaliatory dismissals of thirty-four workers engaged in union organizing. Three years later, the ALRB finally ruled in the union's favor. In 1993— nearly fifteen years later—the ALRB was still trying to collect the awards from Ukegawa when the company finally claimed it was going broke and could no longer afford to pay.

The growers were still fighting the basic purpose of the ALRA. Like federal labor laws, the ALRA was established as a farm-labor *protection* statute—designed to bring peace to the farm fields and guarantee that field laborers could choose to have a union represent them in collective bargaining. "The law says it is the policy of the state of California to *encourage* collective bargaining," says Bill Camp, a former ALRB official who was assigned to educate growers on the nuances of the law—until pro-grower bureaucrats pressured him to leave in the mid-1980s. "If growers had all the money in the world to hire lawyers, it seemed only fair to level the playing field. We would go out and hand out fliers on the right to collective bargaining to workers, and growers were furious that these workers had the right to question their total control over the workplace."

Despite an ineffectual ALRB and resistance from growers, the union continued to show up at farms and participated in nineteen new elections in 1976, winning fifteen of them. The U.S. Supreme Court, meanwhile, refused to hear a grower's challenge to the state rules allowing UFW organizers access to their farms, in effect granting the union at least one of the guarantees it had sought with Proposition 14. And by the end of the year, the board announced its first substantial settlement on behalf of workers: Mapes Farms, with properties all over the San Joaquin Valley, was forced to pay nineteen farmworkers some thirty thousand dollars in back pay, then rehire several who had been fired for helping the union.

A few months later, in 1977, the UFW finally won enormous victory: After nearly a decade as roadhouse bouncers for the growers, the Teamsters officially gave up their claims to field workers. After several private meetings between Chavez and Teamster leaders, the two unions held a joint press conference to sign an armistice that let Teamsters keep the Bud Antle contract and organize packinghouse and farm transportation workers. The UFW was now free to organize field workers without having to fight the Teamsters as well.

It was the end of a bitter, violent era, yet everyone managed to exchange civil words. Teamster president Frank Fitzsimmons, ending months of internal struggle in his union over the cost of trying to organize farmworkers and compete with the UFW, thanked and praised Chavez: "Cesar's not the underdog, that's not the case as far as his organization is concerned." Chavez was equally diplomatic, expressing relief at ending the UFW's long and costly war with the Teamsters. "You know, it was one of the most difficult things to overcome," he told reporter Geraldo Rivera. "What do you win at the end, you know? And we couldn't put our resources in the direction of really building a union for workers, so it was a difficult experience. The Teamsters are a very, very formidable union."

With the Teamsters sidelined, the road looked smoother for the UFW. The union was still fighting growers to accept elections, but it had pushed away a nagging obstacle to building a firm, stable membership. Chavez, with his eyes toward the future, spelled out the union's new strength to thousands of followers who had gathered for a rally in Coachella, just before the spring grape harvest of 1977 was to begin. "All the money we've thrown into the fight with the Teamsters will now be used against the growers. All the legal talent that was tied up in litigation is now free for the main battle."

WHILE UNION LEADERSHIP appeared united after finally defeating the Teamsters, behind the scenes, fissures had developed and old friendships were strained. After the loss of Proposition 14, union staff members disagreed over a variety of thorny issues—from pay raises to political campaign contributions. Even issues like pay raises for staff members became a topic of passionate debate. Chavez had said publicly that the union was looking toward a future in which he would become less engaged in day-to-day operations. Some critics charged, however, that the UFW was instead becoming less democratic and that Chavez was consolidating his personal control over the union.

It was an old dilemma. In 1971, the struggle over the union's future moved Larry Itliong to quit the union he had helped start. "We in the top echelon of the organization make too many of the rules and we change the rules so very quickly that the workers themselves don't understand what the hell is going on," Itliong said after he had left La Paz. Other union members felt that Chavez was simply worried about the union's future in the wake of the Proposition 14 loss. He was obsessed with finding the right management technique for an organization that was evolving rapidly—perhaps too rapidly—into an unusual, and sometimes incompatible, blend of a traditional labor

---

■ *The UFW's executive board in 1977: Mack Lyons, Dolores Huerta, Richard Chavez, Eliseo Medina, Philip Vera Cruz, Gilbert Padilla, Marshall Ganz, Pete Velasco, and Cesar Chavez.*

union and a social movement. It was the drive to deal with these contradictions, some say, that pushed Chavez to study religious communal cultures like the Hutterites and to invite popular management consultants to lead training seminars at La Paz. Kenneth Blanchard, author of *The One-Minute Manager*, came out for some seminars, and Cosby Milne, another consultant, worked more than a year at La Paz without pay. According to Jerry Cohen, Chavez was in a quandary. "He didn't want the UFW to become a traditional labor union—he hated putting on a suit and going to glitzy AFL-CIO conventions—and he wanted to preserve the sense of community the movement had built since its inception."

During this time, Chavez became intrigued with the psychological techniques used in a drug-rehabilitation program known as Synanon. Its founder, Charles Dederich, was a dentist and acquaintance from Cesar's CSO days; he had offered food, clothing, and free dental care to the union in its lean early days. Synanon had received much media attention for its work with drug addicts, and Chavez, who had visited its facility in the Sierra Nevada, liked the group spirit and cooperativeness he found there. He was particularly impressed by "The Game," a self-awareness encounter session involving group pressure and "venting" to work out problems.

Chavez was drawn to The Game because he was urgently seeking a way to deal with the increasing disputes among his staff. Under intense pressure from other labor unions to distance himself from the left, Chavez had urged union members to keep their politics to themselves—not because he cared, he said, but to protect the UFW. Some longtime UFW organizers like Nick Jones had been asked to leave because Chavez considered them too openly leftist. Even before the UFW started playing The Game, arguments over strategy and politics had pushed away several union leaders. Itliong's resignation in 1973 was later followed by that of another Filipino on the leadership team, Andy Imutan. And in 1977 UFW board member Philip Vera Cruz also quit. One of the original Filipino strikers in Delano, Vera Cruz left, in part, because Cesar accepted an invitation to visit the Philippines and meet President Ferdinand Marcos. The trip drove a further wedge between Filipino farmworkers, who were divided over the revolution that was brewing back in their native land. Chavez was worried, former aides and union officers say, that the community of unified activists and farmworkers he had only started to assemble would unravel if he didn't intervene with a recipe for conciliation. Synanon's encounter sessions, he told colleagues, might do the job.

What Cesar didn't realize was that Dederich had become a dangerous man. By 1978 news accounts would accuse him of creating a cult and persecuting former Synanon members; that same year he was convicted of conspiracy to commit murder after some followers stuffed a rattlesnake in the mailbox of a former member. Even in 1977, when Chavez and the UFW became involved in The Game, some hints of Dederich's more unorthodox and unsavory leanings were emerging: He urged Chavez to simply purge his union of dissenters. When Chavez protested that he could never do that, and that the union had to vote on all such decisions, Dederich dismissed him as overly idealistic.

Dederich and his program still had a good reputation as rehabilitation specialists, however, when Chavez decided to use The Game to deal with tensions in the UFW. The Game was designed to clear the air and improve communications by allowing addicts to release emotions; as some union participants remember it, however, the sessions involved an uncomfortable amount of shouting and personalized attacks, usually directed at one person at a time. In early sessions of the game, Marc Grossman remembers a surprised Chavez took the brunt of the anger that erupted from his fellow staff members. Unexpected criticism of Cesar even came from his own family. In early sessions, Marc Grossman recalls, Paul Chavez vented frustrations about his father's long absences while the kids were growing up. Even though the words were sometimes harsh, Grossman says, Cesar weathered the criticism gracefully, figuring he had to take it if others were to be subjected to the same treatment.

Chavez's drift toward a more autocratic management style, former staff members say, was also inspired by Dederich. Instead of bringing the UFW leadership closer together and making it more efficient, as Cesar had hoped, the changes at La Paz caused greater divisions. Some union organizers, like Jim Drake, quit after The Game was brought to La Paz. Despite the upheaval it seemed to create, Cohen still believes that the experiment with Synanon techniques was not as great a problem as others have made it out to be, pointing out that few veteran union leaders actually left because of The Game itself. "It was just a little blip on the screen, [but] it was indicative of an internal problem in the union," he says.

Also controversial in the union was the knotty question of whether to have a volunteer staff or to pay organizers and directors a professional salary. Cohen supported the many union staff members who wanted regular salaries to support growing families, insisting that even a small but dependable salary would help the UFW keep some of the talented legal and organizing minds that Chavez had brought into the movement over the years. Chavez resisted, believing the union's great strength, even its soul, was the commitment of volunteers willing to live as precariously as the farmworkers. In 1978, the union laid off more than half of Cohen's legal team and announced the rest would be moved from Salinas to La Paz. This move resulted in the exodus of nearly all the UFW's highly experienced attorneys, who were replaced with young law students and volunteers. Jerry Cohen, although no longer general counsel once the legal staff was halved and relocated, stayed on to help Cesar negotiate contracts, as did a few other lawyers, including Tom Dalzell.

Loyal organizer Eliseo Medina quit in 1978, after a heated argument with Cesar over the changes in the legal department, a move the union said was necessary because of a declining workload. Medina recalls that the argument boiled down to Chavez challenging him to leave if he did not like the changes. "So I did," Medina says. Despite the abrupt and painful departure, Medina remembers that his years with Chavez were among the best of his life.

Internal difficulties were also disrupting day-to-day administration at La Paz, where the young volunteer staff had been struggling to learn how to run something as complex as one of the nation's first real farm laborers' unions. Federal auditors, tipped off by growers, investigated complaints that the union was not properly managing job-training grants, and the union was later ordered to correct some bookkeeping practices. (State campaign auditors several years later would find that "lax bookkeeping" had led to hundreds of thousands of dollars in contributions not being properly reported. The union issued an apology and was fined $25,000.) Doug Adair, who had moved to Coachella to work in unionized vineyards, remembers that he was so worried about lagging paperwork that he went up to La Paz for two weeks to write checks for medical claims that had been left unattended.

Despite growing internal problems, the union experienced a resurgence after the Teamsters abandoned farm organizing in 1977. It had gained power in both state and federal politics. And, in spite of the troubles and staff turnover, many key organizers stayed on, logging long hours in exchange for a few dollars a week and room and board. Dolores Huerta often bickered with Cesar over union strategy, but she remained committed through the difficult times. Fred Ross also stood solidly behind Chavez and continued to train a new team of union leaders in special classes started with money from the AFL-CIO.

The strong pro-union sentiment in the fields, and the absence of the Teamsters, led organizers to prepare for a block of new contracts with some of the state's largest vegetable operations—owned by international conglomerates like United Brands and the Purex Corporation—in the lettuce-rich expanses of the Imperial and Salinas Valleys. Contracts on those ranches were due to expire on January 1, 1979, so by mid-1978 a union task force was set up to figure out a strategy for the coming round of negotiations with the vegetable growers.

Chavez's goal was to negotiate with the whole industry at once, so that "wage competition" among growers would not create an unfair advantage for one farmer over another—an approach similar to what the United Auto Workers had been doing with the nation's big automakers. Another idea borrowed from the UAW and other industrial guilds was Cesar's demand that full-time UFW representatives be paid full-time salaries by growers, a proposal seen as an internal compromise with union leaders who were arguing for a professional staff. Further, the UFW would demand a greater contribution from growers to the union's underfunded medical plan, which had been established under old contracts before inflation had accelerated in the mid-1970s and skyrocketing premiums became a burden for the low-paid field laborers. Marshall Ganz investigated the fresh vegetable industry to better plan strategy and found that as a result of inflation and huge profit increases for farmers, lettuce workers were actually losing ground—making less per hour than the $3.50 they had earlier agreed to. Indignant, workers backed Cesar's tough proposals and caught growers by surprise. "The growers just freaked out. They had not expected those proposals. They didn't think the farmworkers were serious," Ganz recalls. "But we were serious, and the workers took strike votes and set up committees. Cesar Chavez was very serious about supporting these workers, and their solidarity was just outstanding."

Although they were girding for a contentious bargaining session with growers, the union had recently noticed a behind-the-scenes willingness by some farmers to talk about common problems. At one point, says Marc Grossman, Chavez sent a union lawyer to Sacramento to testify on behalf of corporate growers who were seeking changes to agricultural land-use laws. "He told me he never wanted to be friends with any grower, but he also knew that the one way workers were going to prosper was to have growers doing well," Grossman recalls. "The industry came to him for help [on the land-use law] and he saw it as a good chance to cooperate for the benefit of agriculture. That's why he felt so betrayed by the industry when it later tried to destroy the union."

In 1978, in the fertile farm belt of Southern California's Imperial Valley, reporter Joe Livernois awoke around daybreak on a mean winter's morning to a cacophony of angry voices trading epithets in a nearby lettuce field. The eager reporter, five months out of college and working for his hometown paper, stumbled out into the chill to investigate. Livernois got close enough to see two men—a ranch foreman and Marshall Ganz—standing opposite each other, their shoe tips hanging over the edge of an irrigation ditch, screaming at each other so loudly they were bending forward at the waist. "And then, just like that, they waved each other off, got their people into their respective cars, and drove off. Just as suddenly it was quiet again," recalls Livernois.

Livernois had just witnessed one of the first salvos of the great lettuce strike of 1979: a mass walkout that followed the lettuce harvest from El Centro to Arizona, up to Blythe, and then to the Salinas Valley. Most of the nearly five thousand workers who would come out of the fields in Imperial would still follow the lettuce crop that winter and spring, not as pickers, but rather as picketers. The protracted conflict tore apart communities along the way, even as it was unifying—for one last enormous

■ *Strikers threw back tear-gas canisters police had fired at them during the tumultuous Imperial Valley strike in 1979.*

fight against the growers—the old team of UFW leaders who by this time were privately at odds with each other.

It was the strike Cesar said he had always dreamed of: The incredible solidarity among the Mexican and Chicano workers made him proud. "I feel the way I have never felt before," Chavez told a crowd of strikers during a rally. "We have thirty years of struggle behind us, but I am spirited and encouraged. I feel I can fight another hundred years." Almost complete participation by lettuce crews buoyed the union and stunned the farmers who produced most of the country's domestic lettuce and vegetable supply. Ganz remembers it was the field-level workers themselves who ran the strike, making sure no union pickers lost heart and crossed back into the fields. "It was a coming of age for us as a union," Ganz remembers.

From the start of the campaign, the level of tension along the strike line was as high as the union's spirits. A cross was burned in one lettuce field, and Livernois discovered that the Ku Klux Klan was running a telephone job line to recruit strikebreakers. Klan leader Tom Metzger told Livernois that his "primary interest is the safety of Anglo, or white, workers." Growers recoiled at the presence of the Klan in their fields, but the union accused growers of refusing to acknowledge the UFW because they did not want to sit down and talk with a union led by Mexicans, Grossman recalls.

The Dolphin Group, Chavez's nemesis from Proposition 14, was also brought in to run a publicity campaign to counter the national attention that the UFW was once again winning. Farmers bought full-page ads in Spanish-language papers urging workers to question Cesar about the use of union dues, which growers said had been spirited away to secret bank accounts. In a show of solidarity, growers' mothers, wives, and children rode out to the fields in shielded buses to help harvest the crops. At the schools, confrontations over recruiting handbills that had been posted on classroom windows at Holtville High School led Mexican students to walk out in protest. Emotions flared when school administrators penalized the Mexican students for their walk-out, but granted the strikebreaking students excused absences for their days of picking lettuce. The union had to sue the school district to get the practice stopped.

The tension in the valley snapped violently on Saturday, February 10, when around noon, a group of UFW strikers entered a field owned by Mario Saikhon, one of the most vocal and antiunion growers in the valley. Rufino Contreras, a twenty-eight-year-old

■ *Cesar and Helen Chavez, Gilbert Padilla, and David Martinez joined other UFW members in a candlelight vigil for Rufino Contreras, a young lettuce worker who was shot to death during the Imperial Valley strike.*

Mexican national whose family had worked for years on the Saikhon ranch, led the group, which was intent on defying court orders, to confront some Filipino and Mexican strikebreakers. The group didn't get far, however, as shots—as many as fifteen rounds—were fired by ranch guards. A bullet caught Contreras below the eye, killing him instantly. Three ranch hands were arrested, released on less than eight-thousand-dollars bail, and ultimately cleared of the shooting for "lack of evidence." Contreras's death brought the whole valley to a standstill, with growers expressing sadness and halting farm operations for one day. Contreras had become another union martyr, and huge funeral marches were held in his honor. Cesar and Helen Chavez led the processions. Governor Brown attended the funeral mass.

Despite Cesar's unchanged commitment to nonviolence, many workers broke ranks and attacked strikebreakers who'd been trucked into the fields from Mexico and Central California. Union pickets also lashed out at Imperial Valley sheriff's deputies, lobbing back tear-gas canisters shot at them during what Cesar said were "police riots." Dozens of strikers were arrested, and dozens more people—including police—were injured. Looking back, Marc Grossman remembers that Cesar was disturbed by the violence, and by the tactics of some of the union's strike leaders. In public, however, Cesar would tell reporters, "Look at whose blood is being shed. Not a single grower has been harmed in our struggles."

The strike persisted in the Imperial Valley until the harvest season ended and the workers moved their pickets north. But scars remained. By the end of the year, the union had paid out more than a million dollars in strike expenses. Grower Carl Maggio sued the union and later collected more than $2 million for damage caused by union violence. Mario Saikhon would never fully recover. The UFW sued him through the ALRB for not rehiring two hundred union workers after the strike ended. In 1993, after serving a few months of a jail sentence for tax fraud, Saikhon died. Fourteen years after the strike started, the state forced his farm to repay its displaced union workers $2.5 million in back wages.

"I think I was a little skeptical at the time about the union's claim that it was disadvantaged in the valley, that the whole system was aligned against it," Livernois remembers. "But you know what? They were right." Livernois discovered that his newspaper's own police reporter was working as the media adviser to the sheriff's department during the strike. When Contreras's alleged killers were brought before Superior Court judge George Kirk, they were represented by the law partner of the judge's son. The grower community's attitude toward its workers was summed up by Dave Edwards, a foreman at one of the struck ranches: "It's pretty poor . . . when people who aren't even citizens can come over here and complain after we gave them work in the first place."

When the strike arrived with the harvest in Salinas, union workers at unaffected farms dipped into their own pockets to help those who had gone months without work. Strike committees coordinated by Jessica Govea shuttled striking workers up to San Francisco to work a few days with longshoremen to earn extra money, and the UFW engineered several work-slowdown plans so that those farmers could start applying pressure on their fellow growers.

The growers' front began to crack in August. First, Bud Antle renewed his contract with the Teamsters—for the same five dollars an hour in base pay that the UFW was demanding from other farmers. Then, as Labor Day approached, Chavez approved plans for two big marches—one coming south from San Francisco, and the other up from the Salinas Valley town of San Ardo—that would converge at Hartnell Community College in Salinas. More than twenty-five thousand people took part in the twin marches. Ganz was elated when workers would drop their tools and walk out of fields to join the passing parades. During a rally at the end of the march, farmworkers from the Salinas Valley's Meyer Tomato operation walked excitedly into the hall. Jerry Cohen announced that the company had agreed to all the union's demands— the five dollars an hour, the paid union representatives, and the improved medical plan. The crowd erupted in triumph.

A few days later, a faction of UFW workers walked off the job at West Coast farms in Watsonville in an unsanctioned wildcat strike that, while it caught union leaders off guard, actually sped the end of the strike. West Coast signed a deal within days, starting an avalanche of new contracts with most of the valley's other major vegetable firms. "These were great contracts," Ganz recalls, saying the strike had been won with the solidarity of the workers. "This was not done with a little core of volunteers out conducting vigils. . . . It was important to have workers win a contract through their own efforts. It was different than winning a contract with the work of some people on a distant boycott."

■ FOLLOWING PAGE: *The United Farm Workers were the first to demand protections for farmworkers against the dangers of pesticides, including drift from chemicals being sprayed from airplanes.*

In the old days, miners would carry birds with them to warn against poison gas. Hopefully, the birds would die before the miners. Farmworkers are society's canaries.

—CESAR CHAVEZ

CHAPTER EIGHT

THE POISONED EAGLE

AT THE BEGINNING of 1980, after nearly two decades of nonstop organizing, Cesar Chavez's drive to change farmworkers' lives had yielded an embarrassment of riches. With the hard-won victories in the lettuce fields of the Salinas and Imperial Valleys, membership in the UFW, according to a union count, surpassed forty-five thousand. The unionized *lechugeros* in California were probably paid at the highest rate of any American field laborers. With the new contracts and the increases in piece-rate pay, the fastest workers could earn up to twenty dollars an hour at the peak of the lettuce harvest, rivaling wages in nonfarm manufacturing. Moreover, the contracts included benefits and vacation packages previously unheard of for farm laborers. For the workers themselves, other crucial benefits of union membership were starting to materialize as well. The medical plan, built up by funds from growers who had signed the new contracts, was reaching more and more workers.

Farmworker elections sponsored by the ALRB were also becoming regular events, despite seemingly never-ending battles with growers in Sacramento and in state courts. On forty-six ranches in California, most of them stretching across the coastal vegetable empire near Salinas, farmworkers voted to join the union.

And despite repeated withering attacks from politicians allied with an industry bent on containing the union movement, it was clear that the UFW was not about to crumble and disappear as farm-labor movements had before. The strength of the UFW was undoubtedly one of the factors that pushed farmers into a decade of unprecedented reexamination of their business. "The UFW made us, as an industry, take notice that the potential out there did exist for labor unrest, and that we were going to have to work closer with these people and be fairer in our treatment toward them," said strawberry farmer Larry Galper of Watsonville, a decade after the farm-labor law was passed.

Although the changes that resulted were largely good for farmworkers, the single intent of most growers was to get the union off their farms. Corporate growers got smart, adopting a more professional, businesslike approach. With labor accounting for a third of their annual expense, growers found it necessary to improve their management skills to keep costs down. Many ranchers turned farm operations over to professional managers who drew up labor plans and dealt with the union. The California Farm Bureau, the state's largest agricultural trade association, invested heavily in the Farm Employers Labor Service, a program that made bilingual consultants available to growers and provided personnel manuals that advised growers on every aspect of dealing with laborers.

"There was no question that the farmers learned a lot from the United Farm Workers' effort," says former vegetable-company executive Daryl Arnold. "They found out how to deal with their workers and with the unions. They found out that they should have somebody—a worker, a true worker in the field—that could come in and give grievances to them, and [let them know] what their problems were to solve. It wasn't always dollars. It might be where the rest rooms were. It might be the working hours." Portable toilets mounted on trailers and company-supplied water jugs became a regular sight. On some larger, non-union farms, employer-subsidized medical plans and vacation benefits even became part of the wage package.

The Farm Bureau called the effort the Third Way—a happy medium between UFW-brand unionizing and the Teamster alternative. The state's other powerful farm group, the Western Growers Association, ran full-page ads in farm publications that urged growers to "Take Preventive Action!" by hiring outside consultants. Spanish classes and tapes with utilitarian titles like "Spanish for Farmers" were published to help ranch managers communicate with farmworkers, with such helpful phrases as: "Mañana empieza la pisca"—"The harvest begins tomorrow."

Not every change, of course, was beneficial to the workers. The industry developed an affinity for legal challenges to union elections that led to polite, but interminable, courtroom appeals. Growers also actively curried favor among seasonal workers—often with cash bonuses or promises of promotions—so they would call for quick union-decertification votes. In the early 1980s, *Chavistas* commonly complained bitterly that a new generation of farmworkers was more easily *"comprados con un taco"*—sold out to growers after a Saturday barbecue.

Some farmers believed the ALRA was a strategic gift for them because it blunted the threat of quick boycotts or strikes during contract talks. Bill Grami, the Western

Conference of Teamsters leader who squared off many times against Chavez, recalled a conversation with Salinas vegetable-company lawyer Andrew Church, in the wake of the signing of the ALRA: "I said, 'What about this new act in California?' He said: 'That's a joke, we can tie 'em up. They win an election [but] they'll never see a contract. It's a joke.'"

AS GROWERS were regrouping, the union's accomplishments made it tempting to move on to new challenges. Rumors even sprang up that Cesar was considering stepping down as UFW president. But these stories were quickly explained by Chris Hartmire. "We've been working more on a management team concept to share the leadership, to take more of the responsibility," Hartmire told a reporter in 1982. Underneath Hartmire's public position, however, was an intensifying debate that for several years had centered on questions about the future of the union: Was it to be a social movement led by Chavez? Or was Cesar just the chief executive of a traditional labor union? The answers were almost as varied as the number of people working at La Paz.

Some former union members say Chavez liked delegating authority and routinely created opportunities for inexperienced staff members who showed promise. He was especially patient with young staff members who came from farmworker families. Marc Grossman remembers that Cesar would sometimes bark out an order, but then back off to see how the task was carried out: It was his way of keeping an eye out for future union leaders. Others remember Cesar could also be a stubborn taskmaster, and there were few events that took place at La Paz or within the union without Chavez's direct oversight. He was tirelessly engaged. Al Rojas recalls that he'd fire off directives about almost every aspect of union business, sometimes in a curt manner that would unsettle people who didn't know him. In one instance Cesar personally "handled" one staff member's excessive phone use by climbing over a desk to pull out the phone cord; another time he ordered Rojas to take back a union credit card and a car from a young lawyer who had racked up huge business and travel expenses, and not to accept an explanation.

During this period, Cesar touched off a brushfire by selecting inexperienced staff members for the union's first training class for labor-contract negotiators, funded by the AFL-CIO and named after Fred Ross. "There were some who thought we should go out and get college-educated kids and get them into this program, but Cesar also wanted to give a chance to young [farmworker] staffers, who'd had fewer opportunities in life. He wanted to test them and see if they'd respond," Grossman remembers, adding that Miguel Contreras, who in the 1990s became the first Latino to head the powerful Los Angeles County Federation of Labor, was in the first graduating class of that training program. Though Chavez wanted the union to mature, he also wanted the union's future leaders to come from the same stock as its original farmworker members.

At times, Chavez's personality puzzled his coworkers: his strong desire to delegate was at war with an equally strong need to retain control. Unlike most trade unions, the UFW did not have local branches, or units, which would hold elections of union representatives. This disappointed UFW members who equated autonomy with union democracy. Rather than a simple rejection of autonomy, however, Chavez's position reflected a goal he had written into the union's first constitution twenty years earlier: The UFW would be a strong centralized union with no locals. In writing that first constitution, Cesar reasoned that a migrant workforce made local branches impractical, while a central union with many far-flung district offices would allow farmworkers on the migrant trail to vote, get help, or file grievances wherever their field work might take them.

■ *Cesar Chavez, who had been concerned about pesticides since the early 1960s, made their dangers a centerpiece of a new grape boycott in the 1980s. He spoke in New York City in 1986.*

The philosophical tug-of-war was played out at La Paz at a time when the union's public stature was rising. At the start of the 1980s, Chavez and the UFW were among the best-known forces in the Chicano rights movement, a status that infused the union with energy and self-confidence. Dolores Huerta had put her considerable political connections to work, transforming the union's campaign division into a valuable source of both volunteers and money. Since the defeat of Proposition 14 in 1976, the UFW had poured well over a million dollars, collected mostly from union dues, into the campaign treasuries of politicians who supported the farmworkers movement. Democrats were almost the exclusive beneficiaries of the union's largesse.

In the 1980s, Chavez's vision of the union's future came into conflict with the inclinations of his followers. Now that the union was financially stable thanks to the sweeping lettuce contracts won after the strike of 1979, Chavez hoped to establish a "Chicano lobby" to push the interests of Mexican Americans in Sacramento and Washington, D.C. That same lobbying team would also keep tabs on what Chavez called "welfare for growers"—the millions of dollars in government water, crop, and marketing subsidies.

The expanded role of the union was something Cesar had envisioned for years. "We must continue to keep our membership involved," Chavez told biographer Jacques Levy. "The only way is to continue struggling. It's just like plateaus: We get a union, then we want to struggle for something else. After contracts, we want to build more clinics and cooperatives, and we've got to resolve the whole question of mechanization. Then there's the whole question of political action. We have to participate in the governing of towns and school boards. We have to make our influence felt everywhere and anywhere."

But Cesar's vision had come into conflict with the goals of strong-willed union leaders who emerged from the victorious 1979 strike. Marshall Ganz remembers that Cesar tried to convince the strike's leadership team to scale back picketing and send farmworkers out on the boycott circuit. Chavez was unsettled because the success of the strike seemed to be bought at the cost of too much violence. Journalist Joe Livernois remembers that "the energy seemed to go out of the strike after Rufino Contreras was killed. Things changed after that for a lot of people." In the summer of 1979, just days before the industry's biggest ranchers started signing the new contracts with the UFW, union leaders met in private to discuss strategy. Chavez urged a return to the boycott, but Salinas-based ranch committee leaders would not let up on the strike, which they believed had galvanized farmworkers and was close to winning them new contracts. This behind-the-scenes defiance of the UFW leadership team caused further squabbles among union leaders. Some organizers pushed for a decentralized union, with more control at the local level, while executive officers insisted that union plans continue to emanate from La Paz.

As spontaneous as this struggle over union control seemed, it was really nothing new. As far back as the Delano grape-strike days, resistance to a strong central leadership was evident. Doug Adair remembers that in the 1960s, the free-thinking staff of *El Malcriado* created a list with their own names and gave each other demerits when someone would preface an order with "Cesar says . . ." When the union was young and individual roles were still evolving, those differences appeared harmless, even mildly amusing. But as the union matured, divisions over leadership, as well as goals, grew deeper and threatened the stability of the movement.

Adair traces some of the problems to the fame that enveloped Cesar after his initial fast in 1968. Before the fast Chavez "was our leader, but he was our brother. . . . But after the fast he had a role to play on the national stage. After 1968 it was very difficult, certainly for me, to disagree with him, because he had this tremendous moral stature and authority. And you couldn't go in and tell him anymore, 'Cesar, you're full of shit. That's a terrible idea.' . . . In a sense, [his fame] separated him from the other brothers and sisters. It put him on a pedestal . . . and according to him it was not of his choosing. It made it harder for him to really listen, which is what he was best at. And there was no other organizer in the union who was as good an organizer as Cesar. He was the best. He was tops."

Cesar's philosophy of nonviolence and fasting "was tremendously powerful for me," says Adair. "When I say things that seem to be critical of Cesar, it's . . . in no sense do I want to imply a lack of respect or a lack of love. We were brothers. And brothers need criticism from their brothers at times."

The hiring of full-time union representatives in Salinas and El Centro—people elected by workers at the different ranches and whose pay was provided by the growers—exacerbated the feuding. The idea, gleaned from autoworkers' contracts in Detroit, never had full support of the union leadership and staff members, especially those who were still collecting five dollars a week as pay. As the new decade began, many organizers left the union over the pay issue; others were put off by Cesar's insistence on using The Game to air personal differences. Some organizers dropped out because they thought the UFW was losing its edge as a social movement, while still others were distressed that it was not evolving into a traditional labor union with a professional management team.

The tumult was underscored by the departures in 1981 of Marshall Ganz and Jessica Govea, both highly regarded organizers who disagreed with Chavez on the direction of the union; Ganz, in particular, felt the UFW was not doing enough to support the union's grassroots field organizing. UFW attorney Jerry Cohen also left in 1981, on good terms with Chavez but unhappy with the controversy, which, he felt, was hobbling the union. Growing personal and philosophical differences also ended the longtime friendship between Chavez and Gilbert Padilla, who felt that he had to leave once the union leadership considered him "not a team player." The departures hurt and angered Chavez. *La causa* was all-important to Cesar; he admitted that he had given up his private life for the UFW. Although it was not expressed publicly, former union aides said that, to a degree, Chavez expected others to do the same. "There is no life apart from the union," Chavez once told a reporter. "It is totally fulfilling."

Most of the people who left the union talked respectfully afterward about Cesar and muted their criticism of the UFW in public. But privately, some gave vent to their dismay, charging that the union had become disorganized, with conflicting orders and shifting strategies. Former union supporters complained about an inconsistent stream of orders from La Paz—a centralized management that, they said, stifled local organizing.

Former activists also complained that the union was spending hundreds of thousands of dollars on politicians, while farm-level organizing drives went underfunded. Indeed, despite a long-held disdain for politicians, Cesar came to believe they could be used to shield the farm-labor law. "He so believed in the law that the union spent a lot of money getting into politics in order to protect it," Marc Grossman recalls. "When he saw that [new governor George] Deukmejian was coming into office, we were discussing how we could counter what was expected to be an all-out assault on the ALRA.

I suggested he get in with the new speaker of the assembly, Willie Brown." Chavez agreed, and the two would work closely for many years, even though Cesar had only recently opposed the San Francisco Democrat's bid for the speakership. Over the years Brown directed more than $700,000 in union contributions to his political allies.

The results of the UFW's energetic lobbying can be felt to this day: Despite almost yearly attempts to dilute the law by conservative legislators and farmers, the ALRA remains one of the few laws in California that has never been amended. (Even as late as 1996, politicians backed by growers were still trying, unsuccessfully, to cut government spending by eliminating the labor board or folding it into other state regulatory agencies.)

Viewed against the history of other movements and larger unions, the internal bickering could be considered a normal evolution. But this was the UFW, and this was Cesar Chavez, and the directors who left said that it was the union's special role in the Chicano rights movement, as well as Chavez's elevation to cultural icon, that made the behind-the-scenes turmoil so difficult to watch. "I think we are all responsible for what happened," says Marshall Ganz. "There was no space within the union to really be in opposition and [still] be in the union. But that's how we created it. It meant ultimately that if Cesar wanted things a certain way, you would sort of argue—you would disagree, but ultimately you respected his final authority. I think we wanted it that way. He was willing to take a certain burden that maybe some of the rest of us were not. He was willing to say 'the buck stops with me.' And maybe some of the rest of us didn't have the maturity or the courage to say 'no, the buck will stop with me.'

"We thought we were good people and we believed in what we were doing and that that would be enough to build a strong movement—a strong union that would serve farmworkers as we wanted to do," Ganz recalls. "I think one of the lessons is that it is not enough."

As the union's turnover and internal dissension undermined its ability to organize the growing farmworker population, critics seized on the UFW troubles. In 1979, Chavez had quietly endured a critical article in the *New York Times* portraying him as an autocrat who had fallen from grace. The reporting that followed was often characterized by glaring errors as well as its reliance on innuendo and anonymous sources. In 1984 the *Village Voice* ran a series on Chavez, who was described as a tyrant with a "mirthless grin." The story described a litany of internal battles and portrayed union leadership as a virtual cult borrowing from Synanon and worshiping Cesar. But the worst media attacks came from NBC News in 1979 and a column by muckraker Jack Anderson in 1980 that echoed the network's story. "It saddens me to have to report," Anderson wrote, "that the United Farm Workers (UFW) union, which lifted so many stoop laborers out of peonage and degradation, has become a violence-prone, tyrannical empire under the ironfisted rule of Cesar Chavez. . . ."

The article levied a series of charges against Cesar, and quoted former union members, including Tony Orendain, who claimed the UFW had strayed from its original mission. The column further charged that Cesar and his closest followers were raising large sums of money with the help of celebrities like Joan Baez, but not using it to help farmworkers and organizers. Moreover, Anderson wrote, Chavez had stood by as an "investigative reporter" was "badly beaten" at a union meeting in Arizona.

The allegations outraged Cesar, and he turned to famed San Francisco lawyer Melvin Belli for advice. Within a month, Belli had sent a letter to each of the newspapers that had published Anderson's column, refuting the major allegations. Reexamining the genesis of his report, Anderson found that his researchers had gotten their facts seriously wrong. Anderson sent another reporter to La Paz for a meeting with Cesar, and in a stunning mea culpa, wrote: "Flanked by three lawyers dressed in three-piece suits, a fuming Chavez sat in the hot California sun tearing clumps of grass from the ground as he spoke. He was angry. . . . But he was also persuasive."

The reporter and victim of the alleged assault, a friend of rival farm-labor organizers, told investigators from the union and Belli's office that he had not been beaten, and that Cesar probably never knew that he was in the audience. A priest, whom Anderson had credited with rescuing the reporter, also refuted the story. The union also showed Anderson's researchers stacks of financial documents that tracked more than a decade of fund-raising and revealed that the money had been used as strike benefits and seed money for organizing efforts not only in California, but in Arizona and Texas as well.

But what made Anderson's subsequent apology and retraction most extraordinary was his report that the union had gathered convincing evidence that the NBC News piece on the union had been shaped by Patty Newman, a paid consultant who, Anderson wrote, "collected money from a consortium of powerful growers in California and . . . had written a book assailing Chavez for a right-wing publishing house. In the book, she declares in large type: 'The assassination of Cesar Chavez is inevitable.' "

Anderson tried to contact Newman, but never got a response. In late 1979, Newman's research had tipped the UFW to the NBC report, and the union had fired off a letter complaining of her ties to growers and the nature of her work. The network responded, Grossman said, by softening its report, but it had still been enough to pique Anderson's interest.

P UBLIC EXONERATION wasn't enough, however, to keep the UFW's internal rift from widening, and it came to a head at the union's 1981 convention in Fresno. Salinas field representatives claimed that Chavez was avoiding their pleas for help in setting up a local credit union and for adequate support staff to handle the long new union rosters and a flood of claims for medical benefits. They also complained that the UFW board had too many college-educated members, who were unfamiliar with the difficulties of farmwork. In an attempt to have their own voice on the union's board of directors, the Salinas representatives nominated three candidates to run against officers who had been put on the ballot by the union's leadership. It was an unprecedented and miscalculated challenge: There was clearly not enough support for the Salinas candidates. When word started passing among delegates of an attempted coup d'état by the Salinas faction, Chavez responded by blaming the attempted rebellion on "external forces." (Rumors circulated on the convention floor that the dissidents had been put up to this test by Ganz—part of his plan to carve out a breakaway union based in Salinas. A saddened Ganz denied he had tried to usurp power.) The Salinas faction walked out of the convention, and its leaders were fired as union representatives.

Chavez's retaliation only worsened the problem. The seven field representatives, elected by their peers from Salinas-area farms, filed a lawsuit against the union for wrongful termination. In his defense, Cesar insisted that the representatives had actually been appointed by the union's leadership, and that the workers' vote was not the final word. In 1982, a judge sided with the Salinas contingent, saying the UFW had violated labor rules by firing the elected representatives, awarding them back pay and jobs at unionized farms. One of the fired representatives, Sabino Lopez, says he is not sure to this day why he was fired and refuses to discuss the incident, except to say that it was the darkest time of his life.

The public acrimony left the union vulnerable. Labor analysts began to suggest the union had abandoned its field-organizing efforts, and that by losing veteran leaders like Padilla, Ganz, and Govea, critical outreach work was being left to younger staff members with little experience. Sensing problems within the union, growers became even more recalcitrant. Many farmers skirted the union by hiring fewer laborers directly, instead turning to the network of unscrupulous labor contractors who filled fields with undocumented Mexican immigrants. The growers' legal strategy was increasingly successful, and they filed challenge after challenge against successful ranch-organizing drives and union elections. In 1982 the number of new elections at ranches fell to twenty-one, including several union-decertification attempts.

Such figures did not present the whole picture. The UFW was active in several farm regions of California in the early 1980s, leading spontaneous walkouts on some ranches and facing off against growers who flouted the law by refusing to allow the union onto their property to talk to workers. The union also succeeded in getting an affirmative ruling on a major back-pay case—against the Abatti Produce Company—from the U.S. Supreme Court.

Not all court decisions were so sweet, however. In one case a judge ordered the union to pay for damages caused by a strike, initiating appeals that would drag into the early 1990s. In another particularly bitter fight, the courts ruled that the union was overreaching in its control of workers by forcing them to pay into its Citizens Participation Day fund, which was seen as a partisan political fund-raising effort.

AFTER THE OVERTLY pro-grower George Deukmejian was inaugurated in 1982, he appointed David Stirling as ALRB general counsel. Stirling, an unsuccessful candidate for state attorney general, was on a mission to tilt the law away from what he saw as a "pro-union bias," and by 1983, the union started seeing a lot of its painstaking organizing work go to waste.

"There's no question that the standard we are applying to cases is much more stringent," said Stirling, shortly after taking office. "In the past, it is my belief that complaints would go forward that were simply not justified and should have been settled or not gone at all."

Union complaints that growers interfered with elections, fired pro-UFW workers, and refused to bargain for new contracts after elections went largely unheeded by the ALRB in 1983. Chavez complained bitterly that the labor board was preventing the union from gaining new members because of delays in investigations and certifications. Early into Deukmejian's first term as governor, the agency nearly imploded. While Chavez's organizers were trying to coalesce workers into voting blocs, ALRB lawyers and field representatives were quitting or were being fired as Stirling asserted his control.

Union critics look back on the 1980s and say Chavez didn't do enough to organize field workers, pointing out that Republican governors and presidents had not been able to stop the UFW's momentum in the past. But Dolores Huerta believes that the drastic change in politics in Sacramento and Washington, D.C., thwarted the union's traditional attempts to reach farmworkers. "People thought, well, Cesar should be running around the fields with a flag in his hand, right?" says Huerta, insistent that Chavez never abandoned field-level organizing. "People were unaware—because we got no press—that we didn't get the elections certified, and so people thought that Cesar wasn't organizing." Her claim of ongoing elections is backed by Stirling's own statistics, which show that, despite all the fighting over the farm-labor law and the reported turmoil within the UFW, there were still one hundred farmworker elections on California ranches in 1983. By the end of that year, the union still had 143 contracts in California as well as a number of pacts with growers in Florida and Texas.

While the number of workers covered by UFW contracts had fallen significantly from its 1980 peak, there was still enough strength to fund union operations, allowing

Cesar to focus on expanding the UFW base. Chavez was developing plans to work with the Mexican government on an unprecedented offer of medical benefits to union members' families south of the border. He was also setting up a new membership plan that would allow farmworkers to receive benefits even if they were not working under union contracts. Despite the problems, Chavez remained in high regard among farmworkers, who in 1983 told university researchers that they overwhelmingly viewed the UFW as their best avenue to get raises and to improve field conditions.

IN SEPTEMBER 1983, Rene Lopez, a nineteen-year-old worker, was relaxing with some friends in a labor camp on the edge of a dairy farm owned by Ralph Sikkema outside of Fresno. It was stifling hot, so Lopez and his friends were gathered in front of a labor-camp home, watching the fuzzy picture of a television that had been moved out of the roasting interior. The men were happy because they had just voted to join the UFW and were about to end several months of arguing with Sikkema over what workers said were unduly harsh work rules.

Lopez had led the effort to bring the UFW onto the ranch. Born in the Mexican state of Nuevo León, Lopez had moved with his family to the San Joaquin Valley in 1974, where the teenager quickly mastered English and attended Caruthers High. Soon after graduating, Lopez joined many of his friends and relatives at work in the area's abundant farm fields. He was, as campesinos put it, *muy listo*—bright and energetic. Because he spoke English well and was better educated than the other farmworkers at the Sikkema dairy, Lopez became the employees' informal spokesman. In one tense meeting, Sikkema had threatened to fire the whole workforce after Lopez and others protested his plan to arbitrarily cut salaries. Lopez then called on the UFW, whom he'd learned about in school, to organize the dairy's employees. Coincidentally, the union had been trying to organize workers on yet another Sikkema dairy near Delano, where the rancher had struck back with anger. He hired guards who routinely intimidated workers seen talking with UFW organizers at the Delano ranch. The guards pulled guns on workers and organizers, and in one incident fired shots in the air to break up an organizing meeting. The union demanded a court order disarming Sikkema's guards.

The UFW was still waiting for the injunction when Lopez managed to gather enough support from fellow workers to have the ALRB call a vote at Sikkema's Caruthers dairy. After the vote, while Lopez and his friends awaited the evening coolness, a car rolled slowly toward the camp, its occupants calling out to the men. Inside the car was a Sikkema guard, Donato Estrada, and the farmer's brother-in-law, Dietmar Ahsmann. Curious, some of the workers strode toward the car, with Lopez in the lead. A shot rang out; the bullet caught Lopez square in the forehead.

As with the three other union activists who had been murdered since 1973, Lopez's death weighed heavily on Cesar's soul. Lopez was the first UFW organizer to die while trying to use the beleaguered law that Chavez had so staunchly defended. He feared his union's nonviolent stance would be tested beyond endurance, and at Lopez's funeral asked reporter Henry Weinstein of the *Los Angeles Times,* "How long can I go to the well before it gets dry? We're asking for a lot of restraint."

Chavez blamed the shooting on Stirling, charging that his budget cuts were responsible for reducing the usual complement of ALRB monitors who would have been on hand to protect workers at the Sikkema dairy during their vote. The ALRB's own field officers said that prior to the budget cuts, four monitors would have been sent to an election that size. Ironically, the day after doctors declared Lopez dead, a Tulare County judge granted the union's wish to disarm Sikkema's hired guns.

"Rene is gone because he dared to hope and because he dared to live out his hopes," a somber Chavez said at Lopez's funeral, a quiet ceremony held outside of Fresno. Lopez's coffin, draped with a UFW flag, was surrounded by some thousand coworkers, friends, and family. "It was not possible to shut his eyes to situations of distress and of poverty, which cry out to God, or to keep silent in the face of injustice. How many more farmworkers must fall? How many more martyrs must there be before we can be free?"

Chavez challenged growers to sit down with the union and draw up a code of conduct for ALRB elections. Daryl Arnold, then head of the Western Growers Association, demurred, saying it was Cesar who was inciting the violence. The ALRB agreed to investigate the rancher's role in the shooting and to determine if there had been any unfair labor practices. Stirling's chief aide, Wayne Smith, told the press after Lopez's funeral: "There's no question in our minds that shooting somebody has a chilling effect on workers exercising their rights. The question is whether the grower had anything to do with it."

The ALRB's actions were "too little, too late," said board member Jerome Waldie, a Brown appointee and UFW supporter who, by 1983, was in a minority on the farm-labor board. It was unbelievable, Waldie said, that Stirling had waited until after a judge's order, and a week after Lopez's killing, to move to disarm Sikkema's guards. "It is one thing to be sympathetic to growers and to be antiworker in your policies, but to permit those biases to allow the ugly potential of more death and violence to continue in the fields of California is deplorable." Smith responded for his boss. He accused Waldie of using Lopez's death as a political wedge, a move "perhaps as large a tragedy" as the shooting.

In the end, it was determined that Estrada had shot Lopez in a spontaneous action. He spent six months in prison after a manslaughter conviction. The dairyman's brother-in-law was acquitted.

O N   T H E   O T H E R   S I D E of the Diablo Mountain Range from the Sikkema dairy, Pablo Romero was dealing with another kind of tragedy. Romero had made a remarkable journey from stoop labor to medical school in San Francisco, where he began to see the consequences of farm labor's harsh demands through the eyes of a physician. Working at the UFW's Salinas medical clinic, Romero saw hundreds of farmworkers needing treatment for malnutrition, bacterial infections, neck and lower back troubles, and worst of all, chemical exposure.

Coming face-to-face with the effects of pesticides changed Romero's life. He was paid five dollars a week for his work in the clinic, but he received the education of his professional life there. "One of the things that happens when you go to medical school is that you come out thinking you're *bad;* that you can do anything," recalls Romero, who now runs a thriving practice in Salinas and regularly treats many former and current UFW members. "But seeing what I saw when I came back, I told myself, 'You know what? I can do my best here.' I've never left."

Just days after finishing sixth grade in Mexico, Pablo was introduced to the American West's migrant trail by his father, Vicente. In 1964, father and son came to Salinas and decided it would be an ideal place for the rest of the family. Vincente and Pablo established themselves as hardworking *lechugeros,* and Pablo stayed with farmwork until he was nineteen and drafted by the U.S. Army. While in the service, he got word that his father and sister had become UFW activists, been arrested, and were helping negotiate farm contracts during the 1970 strikes.

After his discharge, Romero hit the fields again, but only for a short time. While in the service, Romero had resolved to go to college, even though he had never attended school in the U.S. and had difficulty reading English. He started out at a community college, taking double class loads, and eventually won a university scholarship. He earned a biology degree, and was ultimately accepted by the University of California at San Francisco medical school.

His medical boot camp in the UFW clinic convinced Romero that he wanted to practice in Salinas, and that desire got him a residency at Monterey County's public hospital. Not long after he arrived, hundreds of farmworkers had to be rushed to the hospital after two incidents in which they were accidentally sprayed with pesticides. "It was crazy, we saw all the obvious things: low heart rates, people salivating. Some were in shock, and others were vomiting." The incidents alarmed the community, and Romero quickly helped form a community task force that set new pesticide rules to minimize the risk of accidental exposure. "I believe those incidents helped change the mentality around here. I believe farmers realized that what they were using was not as innocuous as they had thought." Romero's instincts told him that science did not know enough about what happens to the human body after years of exposure

■ *Despite child labor laws, children like Alejandra Sanchez, eleven, photographed here with her baby sister, were still working in the fields in 1988. She was picking cucumbers.*

to levels of chemicals considered safe by the government.

In the mid-1980s, Monterey County would enact some of the toughest pesticide-use laws in the nation, prompting statewide revisions of chemical spraying guidelines. "The union helped make that possible," Romero says without hesitation. "The campesinos got to thinking that there were ways to collectively demand what should be theirs. Cesar was quite effective at teaching that. His work raised the consciousness of everyone, including the farmers."

Everyone would ultimately become acutely aware of pesticides. In the 1980s, according to government reports, American farmers were using some 2.6 million tons of chemical pesticides every year. And those same reports estimated that 300,000 people each year were suffering serious illnesses due to pesticide use. And if it wasn't the farmworkers who paid the price for working close to poisons for so long, it was their children. Startling accounts of cancer clusters—unusually high rates of disease, especially among the young—would divide small communities in the San Joaquin Valley and again pit the establishment against farmworker advocates throughout the decade.

Chavez was no stranger to the fight against pesticides. He and Dolores Huerta had negotiated away the use of pesticides such as DDT on grape and lettuce ranches well before the government stepped in with its own restrictions in the early 1970s. As early as the original Delano grape strikes, Chavez had been encouraging young farmworkers and activists to study medicine so they could return to the fields as doctors and nurses who might help laborers cope with the health threats that were a part of their daily lives. In 1969, Chavez had marched with a group of two hundred protesters in front of the federal Food and Drug Administration, demanding increased government surveillance of pesticide use on food crops.

Yet, despite his long record of environmental activism, Chavez was branded an opportunist, jumping into the antichemical bandwagon for his own gain.

FOR DECADES, the UFW had warned that the nation would continue to see images of human misery amid rich farmland as long as field laborers were not allowed to organize. Chavez was frustrated that successful elections at several Delano vineyards were subverted by growers who quickly filed challenges that dragged the vote counts into court. With Stirling now at the ALRB's helm, Cesar believed he had to find another way to pressure grape growers into returning to the bargaining table. Seeing an opportunity to organize around the issue of pesticide dangers, Chavez, in the summer of 1984, called on American consumers to once again stop eating grapes, this time because of health risks.

Richie Ross, a union activist who had become a high-powered political consultant, orchestrated a computer-generated direct-mail campaign that placed pleas from Chavez in mailboxes all over North America. The union warned of an impending health disaster and asked consumers to avoid California grapes until the industry agreed to stop using pesticides known or suspected to cause cancer in laboratory animals.

Marion Moses, once a UFW nurse who hailed from blue-collar West Virginia, became a doctor in the 1970s at Cesar's urging. Through her medical training, she also came to share Chavez's fear of pesticides. In 1983, she returned from residencies in internal and occupational medicine in Colorado and New York to work with Chavez, who at the time was still planning the new grape boycott.

Even before the new campaign started, Moses was plowing through data on various pesticides when she started hearing unsettling reports from farmworkers: Too many people, children mostly, were developing cancer in the small farm towns that hugged Highway 99 near Delano. The stories were coming out of McFarland, a town Cesar had gotten to know in 1962, when he walked from block to block, talking to farmworkers as they returned from nearby fields. The accounts were shocking: Thirteen childhood leukemia cases had been counted in a town with a population of 6,800 people, way above the number that medical researchers said was average for an American farm town. Those cancers, and Chavez's controversial boycott aimed at eliminating chemicals he believed were killing workers, would define the UFW in the 1980s. The devastation left in the wake of the illnesses, which clustered in communities dominated by farmworkers and former farmworkers, would occupy most of Moses's time for the rest of the decade.

The reports of the McFarland cancer cluster fueled the UFW's new grape boycott. Once again religious groups around the nation started recruiting boycott workers and spreading the anti-grape sermon. But while the UFW's campaign initially startled Delano grape growers, they quickly proved that they had learned how to use the media, and responded with a slick public relations campaign to counter the union's

The Spraying

by Helena Maria Viramontes

Exposure to pesticides is one of the gravest threats to farmworkers, injuring thousands a year and causing cancer and other illnesses. Writer Helena Maria Viramontes, who worked as a child in the fields and orchards of Southern California, recalls that the smell of pesticides was as familiar as the heat. In her recent novel Under the Feet of Jesus, *a coming-of-age tale of young migrants in California that is dedicated to Cesar Chavez, Viramontes gives readers a glimpse into that world. In the following scene, the author describes the terror of a pair of fifteen-year-old cousins who are working in a peach orchard when a crop-duster plane sprays the trees with pesticide.*

Alejo had not guessed the biplane was so close until its gray shadow crossed over him like a crucifix, and he ducked under its leaves. The biplane circled, banking steeply over the trees and then released the shower of white pesticide.

—What the...? They're spraying! But Alejo couldn't hear Gumecindo's response because the drone of the motor like the snapping of rubber bands drowned out his words.

—Run! Alejo screamed, struggling to get himself down the tree, get the fuck out of here!

Gumecindo dropped the sacks and ran, jumping over irrigation pumps, crunching the flesh of rotting peaches, running just ahead of the cross shadow.

Alejo slid through the bushy branches, the tangled twigs scratching his face, and he was ready to jump when he felt the mist. He shut his eyes tight to the mist of black afternoon. At first it was just a slight moisture until the poison rolled down his face in deep sticky streaks. The lingering smell was a scent of ocean salt and beached kelp until he inhaled again and could detect under the innocence the heavy chemical choke of poison. Air clogged in his lungs and he thought he was just holding his breath, until he tried exhaling but couldn't, which meant he couldn't breathe. He panicked when he realized he was choking, clamped his neck with one hand, feeling his Adam's apple against his palm, but still held on to a branch tightly with the other, afraid he would fall long and hard, like the insects did. He swallowed finally and the spit in his throat felt like balls of scratchy sand. Was this punishment for his thievery? He was sorry Lord, so sorry.

Alejo's head spun and he shut his stinging eyes tighter to regain balance. But a hole ripped in his stomach like a match to pepper, spreading into a deeper and bigger black hole that wanted to swallow him completely. He knew he would vomit. His clothes were dampened through, then the sheet of his skin absorbed the chemical and his whole body began to cramp from the shrinking pull of his skin squeezing against its bones.

He wheezed and almost fell, and if it wasn't for the fact that he was determined not to fall, he would have tumbled like the ripe peaches hitting the ground with a hard thud. His body swung forward and he caught himself by hitching on to a branch and he scratched his face against a mesh of leaves. As the rotary motor of the biplane approached again, he closed his eyes and imagined sinking into the tar pits.

He thought first of his feet sinking, sinking to his knee joints, swallowing his waist and torso, the pressure of tar squeezing his chest and crushing his ribs. Engulfing his skin up to his mouth, his nose, bubbled air. Black bubbles erasing him. Finally the eyes. Blackness. Thousands of bones, the bleached white marrow of bones. Splintered bone pieced together by wire to make a whole, surfaced bone. No lava stone. No story or family, bone. And when he awoke from the darkness of the tar, he was looking up into the canopy of peach trees, his forehead a swamp of purple blood and bruise and hair, and into the face of his cousin.

grape boycott. Consultants with the California Table Grape Commission demanded equal space for their views in newspapers that printed Chavez's opinion articles; and they followed union speakers everywhere, offering the farmers' view to reporters at UFW press conferences and rallies. The growers' new position seemed reasonable to many editors and consumers: "If food safety is the issue, then Chavez should say 'Ban the chemicals,' not 'Don't buy grapes,'" said the group's president, Bruce Obbink. "Chavez believes there is still enough residual support for himself from the old boycott days to do this. He chose grapes because that's where he was in the beginning. He's locked in a political battle, but he uses a chemical issue because it's a hot-button item."

Perhaps Obbink hadn't noticed that Chavez and the UFW had been pushing to protect workers from all dangerous chemicals. In 1985, Governor George Deukmejian vetoed a bill that would have forced growers to post signs warning that chemical spraying had just occurred. The governor said that law would have led to "a saturation of signs . . . which would diminish the effect of posting on those fields that were truly dangerous."

To supplement the media campaign, growers established a grape workers and growers coalition, which accused the union of interfering with what they said was an improving relationship between laborers and their bosses. The group also started a college scholarship fund for Delano-area children of farmworkers. What went unpublicized was that, once again, the Dolphin Group was quietly directing the show.

THE RHETORICAL BATTLE went back and forth, but grape sales, in reality, were unaffected. Cesar's message was muffled by the changing times. The UFW could no longer count on the broad spectrum of support it once had. American liberalism was changing, and antiwar students had grown up; even countercultural leader Jerry Rubin had become an investment banker. Labor unions of all sizes were reeling amid a decline in American manufacturing that once employed hundreds of thousands of unionized workers—people who a decade before had helped spread the boycott gospel around the country. The decade had started poorly for American labor, with President Reagan's dismantling of the air traffic controllers' union. By the end of the decade, only one-sixth of the nation's labor force would be represented by unions.

Worse yet, a deepening debt crisis was beginning to hobble agriculture, wiping out many small and medium-sized family growers and concentrating more political and economic clout in the hands of large-scale agribusinessmen and their bankers.

Meanwhile, other pesticide scares were being reported. In 1985, the nation was jolted by news that hundreds of people had become ill after eating California watermelons that had been illegally sprayed with the pesticide Aldicarb.

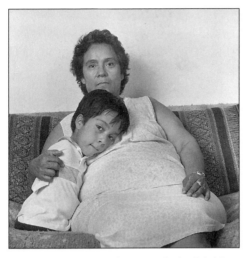

■ *Felipe Franco, photographed with his grandmother, was born with no limbs in the San Joaquin Valley. His farmworker mother, exposed to pesticides suspected of causing birth defects, won a settlement from a chemical company.*

In 1986, Moses got a call from a frantic union organizer who breathlessly said more than one hundred citrus workers had been burned by chemicals in a San Joaquin Valley orchard. It was one of the worst cases of chemical burnings ever reported on an American farm, and Moses remembers seeing workers with seared arms and hands, blistered by an untested combination of pesticides that had not been approved by agriculture regulators. She called Cesar, and the two quickly shut the farm down with a picket line of hundreds of workers. The protest worked; as the state launched an investigation, the farmers quit using the improvised pesticide formula.

In 1987, as more and more journalists were reporting on the cancer disaster in the San Joaquin Valley, new clusters were found in other towns, including Delano. Chavez decided to turn up the heat and enlisted volunteers from Hollywood to produce *The Wrath of Grapes,* a fifteen-minute video that was mailed to consumers, students, and elected officials throughout the country. It featured mothers and fathers from Delano and McFarland lamenting that public officials were doing nothing to investigate the hazards of pesticides for fear of the farm industry. Roughly fifty thousand copies of the short film were distributed as a fund-raising tool for the boycott effort. The video's images were heartwrenching: They linked decades of nonstop pesticide spraying to children who were born disfigured or who developed fatal childhood cancers.

But again, growers were equal to the challenge. The ersatz coalition of farmworkers and growers persuaded some McFarland mothers to criticize Chavez and to object to their own appearances in the UFW film. But it was not difficult to raise the ire of the mothers, who had already started grumbling that money raised by the film was not going to help their children. Some McFarland mothers sued the union over *The Wrath of Grapes.* Although they signed release forms that said their stories would be used as part of a boycott promotion, they insisted they were not told they would be used as fund-raisers. A judge agreed, and the union suspended distribution of the film.

Despite the controversy, Chavez did not abandon the McFarland issue. And today he is credited with keeping the media focused on the cancer clusters by arranging marches and supportive visits from celebrities and activists like Jesse Jackson and by loudly criticizing public health officials for not doing enough to solve the mystery.

Moses also fanned the flames. In 1988, she founded the Pesticide Education Center in San Francisco, and she continued to attack public health officials for not suspending the use of pesticides with well-established links to cancers. She warned that if nothing were done, the state would soon be overwhelmed with more cancers in other farm towns. "And then, in 1989, in Earlimart, indeed there was another cluster," Moses recalls. "This time the news came from the workers themselves. They had been coming to Cesar. I was asked to speak with these workers and seven of them had cancer. Six of them were children of farmworkers." Moses convinced state officials to once again expand their studies to include Earlimart.

Over the years, despite one inconclusive scientific study after another, McFarland mothers and union activists kept the faith. As late as the summer of 1996, scientific reviews of the cancer clusters kept coming up empty, and the federal Environmental Protection Agency was preparing to start its own investigation, spurred on by McFarland activists who petitioned the agency using provisions of the Superfund law.

Cancer's death march through McFarland continued into the 1990s, oblivious to the mothers, the UFW, and the politicians and government scientists. The number of childhood cancers in McFarland had risen to twenty. In other parts of the valley, cancer was being found at rates six times as high as those in comparable cities elsewhere.

WHILE McFARLAND became a symbol of the health hazards in the fields, a collection of coffin-shaped "caves" dug out of a hillside north of Salinas was emblematic of the inhuman living conditions that were on the rise among farmworkers. The inhabitants of these secret encampments were a new generation of undocumented farmworkers pushed north in ever greater numbers by Mexico's imploded oil economy. The workers became easy prey for growers or contractors trying to increase their income.

In the fall of 1985, Jose Ballin, a strawberry grower of modest means, drove onto his north Monterey County strawberry ranch into an ambush of television news cameras and reporters scurrying around his property. He saw grim-looking people in gray suits, along with a priest who looked very displeased. Off in the distance were Ballin's own workers, guiding strangers around the edges of his three-hundred-acre strawberry field. They were peering inside tractors filled with old blankets, and hovels fashioned from pieces of wooden pallets and cardboard. A toothbrush and a cheap transistor radio hung from a hook inside one of the huts, which was no bigger than a casket.

Some of the workers led reporters through a grove of eucalyptus trees, where they kicked aside leaves, and shined flashlights into dark holes that had been cut into the

■ *Farmworkers were reduced to digging coffin-sized holes in the ground to use as shelters.*

steep side of a hill. The caves were just big enough for a man to crawl into; some of them had been reinforced with corrugated tin and furnished with decaying mattresses.

When the reporters spotted Ballin, they flooded him with an avalanche of questions: What was he going to do about all the workers living in such wretched conditions on his farm? Why was he taking twenty-five cents for rent out of each worker's hourly pay, when it appeared they got only a hole in the ground in exchange? Why was their only water supply an irrigation pipe? Why were his workers forced to use the eucalyptus grove as a toilet? "I think everybody to leave tomorrow," Ballin said in broken English. "I don't tell them to live over there, and I don't charge them for water, electricity, or rent. . . . I pay $3.50 an hour, and those living here, I pay $3.25 because they get free rent."

A woman's voice interrupted: "This order forbids you from retaliating against the workers." It was Lydia Villarreal, a lawyer with the California Rural Legal Assistance, which had just filed a $3 million lawsuit against Ballin and had obtained an order barring him from evicting or firing the workers. Villarreal was sickened by the sight of the laborers' living conditions, and she blasted the industry for allowing such human rights violations to become what she called an international problem. The caves were a sobering dose of reality; some farmworkers were living and working in conditions as bad as the days when the Delano grape strike began, or when Edward R. Murrow interviewed migrants in shacks and under trees in the early 1960s for his documentary "Harvest of Shame." The conditions looked even worse than those recorded during the Great Depression.

Stories, photographs, and footage of *Rancho las Cuevas*—the "Ranch of the Caves"—were sent around the world, again reminding consumers that food was often produced under less-than-savory circumstances. Some of the news footage showed laborers like Jesus Carrillo crawling on all fours in and out of his tiny den. Inside, there was barely enough room for him to sit up. On one of the pallets that held back the dirt, he had tacked up small pictures of saints and letters from family in Mexico in an attempt to humanize an impossibly inhumane situation.

After the caves were exposed, social workers and community activists discovered more farmworkers—some fully employed—living in equally wretched conditions in

Monterey County and other parts of the state. Ballin would eventually be convicted of workplace safety violations and ordered to repay his tenant laborers almost $200,000 in back wages. What was missed during the initial media frenzy, however, was that Ballin himself was a victim of sorts: He was barely making a profit selling his berries to a corporate wholesaler, and his entire operation was based on exploiting illegal immigrants.

Ballin's ranch was similar to some of the strawberry farms worked by the area's sharecroppers, usually Chicanos or Mexicans, who grew berries on someone else's land to sell to large companies. The sharecroppers were largely unsophisticated about business, and they concentrated on the sowing, tending, and harvesting. The companies provided them access to corporate grocery store chains and food processors. In the 1980s, working conditions on these sharecropping operations were among the worst in the industry, according to UFW organizers.

Because Ballin's ranch was so isolated, CRLA community workers like Ricardo Villalpando spent many months trying to figure out what was going on there. Like Chavez in his early organizing days, the husky, chain-smoking Villalpando—whose own militancy was inspired by his days as a UFW lettuce picker in Salinas—would spend evening after evening quietly meeting with Ballin's workers. Some of them, like Jesus Carrillo, were up from Mexico for their first California harvests. Villalpando had to convince the laborers, almost one at a time, that they didn't have to tolerate the near-slavery just because others had done so. They had to constantly be reassured that they could stand up to Ballin, who had left no doubt that complaints would only bring unemployment and deportation. Villalpando also had to help the workers find new places to live, a daunting task in Monterey County where housing costs were among the highest in the country.

Salinas was by no means the only place where farmworkers lived in desperate conditions. Union organizers uncovered illegal housing in the Coachella desert, where laborers said they had no place but the

■ *Shacks of cardboard and wooden crates proliferated in the 1980s, when farmworkers found it harder to afford housing in California.*

fields in which to camp. Near an affluent area north of San Diego, undocumented Mexican workers, most of them Oaxacan Indians, had established their own community of ramshackle lean-tos and tin-roofed hovels in Devil's Canyon. This encampment became the communal home—complete with a governing council and religious ceremonies—for hundreds of families. The camp was finally demolished in 1995.

In Stockton, a tireless Mexican immigrant named Luis Magaña gained notoriety among government health officials for helping CRLA lawyers expose derelict camps along the San Joaquin River. Magaña pointed out dozens of campsites where Mexican immigrant laborers, passing through town to harvest the region's valuable asparagus crop, were living alfresco behind dense stands of reeds and tule grass, hidden from the view of nearby homes in a riverfront subdivision. The laborers had been pushed into these thickets because they could not afford even the cheapest motel rooms on Stockton's seedy south side. From the riverside camps, they would walk in the early morning darkness to freeway underpasses downtown, where they'd huddle, share cups of coffee, and wait for the labor contractors to sweep them out to fields. There, they would cut asparagus for $3.50 an hour.

"Some farmworkers now earn fair pay and have family medical plans, protection from dangerous pesticides, and paid holidays and vacations," Cesar wrote in a newspaper editorial in the summer of 1985. "Their children attend school and they make enough to live in decent homes instead of wretched camps. But this progress only highlights the miserable poverty too many other farmworkers still suffer in our midst: child labor; sexual harassment of women workers; pesticide poisoning; high infant mortality; short life expectancy; illegally low wages . . . living out in canyons and under trees."

MANY WORKERS were willing to accept low pay and subhuman housing, especially the ever-increasing numbers of undocumented aliens who were crossing the Mexican border. By 1984 Congress was debating new laws about illegal immigration. The Immigration Reform and Control Act, passed in 1986, finally made it a crime to hire undocumented workers. The law also created hundreds of thousands of newly legal workers by granting amnesty to illegal immigrants who had been living in the United States. Further, the powerful agribusiness lobby also convinced Congress to offer special amnesty to undocumented farmworkers who could prove they'd worked on farms for at least three months since 1985. Growers also won expanded rights to recruit foreign workers if there was a shortage of workers at the peak of harvests. California growers, in particular, opened their pocketbooks in the massive legalization drive that followed the law. They set up "Ag Help" centers throughout the state to help farmworkers get prized *micas*—"work permits"—and they hired consultants and lawyers to explain what the new law would mean to their ranches. The industry's response was fueled by fear that shutting the border would cut off the flow of laborers who helped keep costs low and would create shortages of workers at harvest.

The UFW vehemently opposed the new immigration law's provision for possible guest workers, which smacked, to many, of a revived bracero program. But Chavez and his union directors saw some benefits in the new law: It would bring thousands of workers out of the shadows—no longer afraid of deportation—and perhaps force the government to better enforce labor laws. Moreover, Cesar believed, legalization of workers could translate into new union recruits.

The question of undocumented workers had always been thorny for the UFW. Throughout its history the union had embraced anyone, regardless of residency, who wanted to join the organizing process. At the same time, the UFW also actively tried to stop the use of imported Mexican strikebreakers, a tactic growers turned to time and again to snuff out strikes and organizing drives. The union has sometimes been characterized as anti-immigrant because Chavez instructed union members to call the INS if they suspected undocumented workers had been brought into struck fields. But Chavez issued those orders in the heat of strikes. He had often appealed to labor officials who frequently turned their heads when growers used Mexican nationals as strikebreakers. On more than one occasion, Chavez had even gone to Mexico City to plead with Mexican labor leaders to persuade campesinos not to answer requests for strikebreakers.

■ *Migrant workers, impoverished by Mexico's economic crisis in the early 1980s, jammed into a car trunk hoping to cross the border.*

"We never worked out a solution with immigrants rights groups," Jerry Cohen says, remembering the issue of undocumented workers as particularly vexing for Cesar. "We were saying we wanted immigration laws enforced, and that meant dealing with the INS, which is like dealing with the devil. Cesar was never happy about that."

The government's refusal to stop growers from using *ilegales* as strikebreakers helped give rise to one of the union's most controversial episodes: the labor camp raids and beatings of undocumented Mexican laborers near the Arizona–Mexico border. In 1973, Cesar had sent his cousin Manuel to manage a contract dispute with citrus farmers near Yuma. The subsequent strike had been successful until growers started recruiting workers from Mexico. The union response was to set up what it called a "wet line"—a series of outposts manned by UFW supporters assigned to stop incoming workers from crossing the border in uninhabited desert areas and then crossing picket lines.

Some accounts of the Arizona strike say Manuel Chavez's people tried at first to talk strikebreakers into staying in Mexico, and then called the border patrol to report the illegal crossings. But when the INS did not respond to the UFW's pleas to halt the stream of undocumented workers crossing the border, some union members became vigilantes, beating workers as they crossed into the United States, according to police and FBI reports. No charges were filed against Manuel Chavez on either side of the border, but two dozen strikers were arrested for carrying weapons and for assaulting undocumented immigrants near the border and in orchards. Reports of the violence caused an uproar in Mexico and Arizona and civil rights groups condemned the union for its actions. Cesar stubbornly defended his cousin, however, insisting that the reports of violence in Arizona were exaggerated.

Still, the citrus campaign was a disaster. Even if the worst of the rumors of violence were not true, the campaign had yielded no organizing gains, no contracts, and in fact had alienated many farmworkers in Arizona. Earlier that year, Chavez had insisted on new language in the union's bylaws providing a "bill of rights" for farmworkers, no matter what their legal status. After the Arizona debacle, however, and increasing violence on other picket lines in California, Chavez shifted his goals and resolved to get a collective-bargaining law enacted that would make strikes less likely. In the meantime, Manuel Chavez stopped doing substantive work for the union.

As IMMIGRATION continued to hold center stage, in early 1988, Cesar was increasingly concerned with the deteriorating living conditions of farmworkers in California. The four years of the new grape boycott had been frustrating, and the industry kept reporting increased sales of the fruit. The UFW seemed no closer to getting growers to bargain over new contracts, let alone stop the use of pesticides. Union membership was also falling as more and more contracts disappeared, while growers increasingly returned to a tactic that had first been used in the late 1970s. The farms could void union contracts if they shut down their operations, then transferred ownership to a third party, usually a relative or a consultant. Many farms reemerged with new names. Firms like Abatti and SunHarvest used the ploy, leaving thousands of unionized workers, many of whom had worked on the farms for years, without jobs and even homeless. In one celebrated 1983 case, however, the California Supreme Court ruled that the owners of Rancho Sespe farms in Ventura County had changed its name solely to void a UFW contract, but the court decision did not prevent others from fashioning new methods of closing their doors to break union deals.

All of this—on top of the pesticide issue—was weighing heavily on Cesar in 1988, when he announced he would once again try to refocus the movement with a fast. Despite his age—he was sixty-one—he said he would not eat again until farmers agreed to talk about grape contracts and consider removing the pesticides from their vineyards. A medical team, which included Marion Moses; Dolores Huerta's physician son, Fidel; and Pablo Romero, would monitor Cesar's condition daily.

■ *Cesar Chavez fasted for thirty-six days in 1988 to protest pesticides. He broke the fast with Ethel Kennedy; his wife, Helen; his mother, Juana; and Jesse Jackson at his side.*

At first, the fast seemed to go well. Moses recalls that she rarely worried about Cesar's fasts because he knew what he had to do to protect his health. But within two weeks, it was clear that Cesar's health was rapidly deteriorating. Linda Chavez Rodriguez, Cesar's daughter and the wife of Arturo Rodriguez, remembers that as the fast continued, "It was scary. The grandkids were very upset and I know that Artie was very concerned and some of the union's board members were concerned. It took its toll on us, that fast," she says, recalling that the 1988 fast affected Helen Chavez more than anyone, even though she put up a stoic front to the waves of farmworkers that came to support Cesar. The fast even drew former union leaders like Ganz and Cohen to Delano, where they set aside whatever differences they had with the union to ask if they could do anything to help.

Fifteen days into the fast, a worried Arturo Rodriguez called Moses. Moses drove the same day to Delano to check on Chavez's condition, When she arrived, Cesar greeted her with a put-on: "What are you doing here?".

"That's what I want to know: What am I doing here?" Moses answered. "I said to him, 'Look. If you die, it's not my fault. I'm not the one starving to death!'" Moses then realized Chavez had wanted her there for a frank opinion about his condition. "We really need to find out how your blood is doing," Moses said. She asked Chavez, who was notoriously uncooperative about being stuck with a needle, how often he

■ *Chavez and Arturo Rodriguez, his son-in-law and union vice president, enjoy an outing in Canada with Mike Ybarra.*

would allow blood samples to be taken. "He answered, 'Mariana, as often as you want.' And I thought to myself, 'Oh boy, we're in for a long one.'"

The fast lasted thirty-six days. Chavez had lost thirty pounds and was advised to stop the fast or risk permanent damage, even death. So serious was Cesar's physical condition that former union leaders, including Jerry Cohen, Marshall Ganz, and Fred Ross Jr., had also driven out to Delano to check on his well-being. "It was very hard on him physically," remembers Chris Hartmire. "But worse than anything, the growers didn't call."

The end of the fast, on August 21, brought a crowd of eight thousand for a mass under a tent at the Forty Acres in Delano. Cesar's frail mother, Juana, then ninety-six years old, was there. Jesse Jackson and Ethel Kennedy and some of her children attended as well, along with new celebrity friends of the union, like actors Martin Sheen and Edward James Olmos, and liberal state assemblyman Tom Hayden.

Reading a statement from his father, who appeared at the ceremony for about an hour, Fernando Chavez announced the end of the fast. He also said a new one would start, to be shared in three-day periods, by union supporters: "Today, I pass on the fast for life to hundreds of concerned men and women throughout North America and the world who have offered to share the suffering." Jackson was the first to take a small wooden crucifix, Cesar's symbol of the fast, and begin the first three days of the new hunger strike.

Chavez had come close to incapacitating himself permanently, and he aggravated his condition while recovering at La Paz. Stepping out of his quarters, Cesar leaned against a rail that gave way. He tumbled eight feet to the ground below, fracturing a wrist and suffering bruises to his head and chest. It was months before he regained his strength, but he eagerly resumed his campaigns in 1989, traveling from town to town throughout the West, still trying to convince consumers to quit buying grapes until they were pesticide-free. "Health experts think the high rate of cancer in McFarland is from pesticides and nitrate-containing fertilizers leaching into the water system from surrounding fields," Chavez told students at Pacific Lutheran College in Washington State. He talked about the cancer clusters and deformities, all conditions he saw as signs that farmworkers were still treated like tools to be thrown away when they no longer served their masters. He told the students about the deaths of farmworkers who weren't much older than they were: Rene Lopez and Juan Chabolla, an undocumented worker in San Diego who, the union said, had died as the result of working in a freshly sprayed tomato field. The farmer put Chabolla in a car and drove him over the border to a clinic in Tijuana, an hour's drive away. He was dead on arrival.

Chavez said he'd been fighting for thirty years to control pesticides. And he told the audience that the federal Delaney Clause bans all food additives that are known to cause cancer, except one—farm pesticides. "Last year California's Republican governor, George Deukmejian, killed a modest study to find out why so many children are dying of cancer in McFarland. 'Fiscal integrity' was the reason he gave for his veto of the $125,000 program, which could have helped eighty-four other rural communities with drinking water problems," he said, telling the students there was nothing the union cared more about than "the lives and safety of our families."

"There is nothing we share more deeply in common with the consumers of North America than the safety of the food all of us rely upon."

■ FOLLOWING PAGE: *Richard built Cesar a pine coffin, just as he promised he would if Cesar died before him.*

Regardless of what the future holds for our union, regardless of what the future holds for farmworkers, our accomplishments cannot be undone. The consciousness and the pride that were raised by our union are alive and thriving inside millions of young Hispanics who will never work on a farm.

—CESAR CHAVEZ

CHAPTER NINE

A PINE COFFIN

WHETHER IT WAS the strain of the fast, or his evolving role as an elder statesman in the Chicano and labor-rights movement, as the 1990s began, Cesar was more reflective and measured in his public appearances. Growers had roundly pronounced the union ineffectual and Chavez a has-been, but people close to him confided that they knew he was biding his time, always patient, always circling, preparing for new battles.

The decade began with a confirmation of his accomplishments. On October 19, 1990, bathed in the sharp desert sunlight of Coachella, Chavez walked onto a stage in front of a new elementary school. The school, whose young pupils were almost all Latino, would be the first public building in California to bear his name. Chavez once told Al Rojas he would rest uneasy in death if monuments were named after him. But in the late 1960s, Rojas knew Chavez was forging history, and he used to joke with him about how California would one day be home to a statue of Cesar Chavez, like it or not. "I just want to stop those grapes, Al," Chavez told Rojas.

Twenty years later, Chavez's family finally persuaded Cesar to accept the invitation from Coachella to celebrate the opening of the new school. The union had faced some of its most violent, difficult days in this desert community, beginning with the grape strike of 1973. The name Cesar Chavez had been enough in some parts of Coachella to start a fight. Now he was a guest of honor. Clementina Carmona Olloqui, the young *Chavista* who had become a nurse at Cesar's urging, remembers what it was like to welcome him. It was a bittersweet vindication for Chavez and for the area's farmworkers.

"It was odd to see him up on the stage, next to people who in 1973, when we were marching and striking, wouldn't even let him sleep in any of the hotels around here," Carmona Olloqui says. "But now they love him, and they all say they marched with Cesar."

Chavez was well aware of the irony, but he accepted the gesture from city officials graciously, insisting that the ceremony carried a deeper meaning than just putting his name in front of the school: "I do not believe that the people meant to recognize me alone when they named this school. Rather, it is named in recognition of the farmworkers. Their work and sacrifice have built up this community."

Two months after the celebration in Coachella, Chavez was on the road again, immersed in another campaign against pesticides. This time he was helping California environmentalists tackle chemical companies and growers. In early 1991, he flew to Washington, D.C., where he delivered a keynote address to young organizers brought together by consumer-activist Ralph Nader. It was a few months after the November 1990 election, in which California voters voted down Proposition 128, the initiative to scale back the use of many chemicals, including several farm pesticides. Cesar found that campaign exhilarating—no matter that the proposition later lost—and he had driven around the state for more than a year, stumping with environmental activists and leading noisy, anti-pesticide rallies in front of the Capitol in Sacramento. Chavez was still committed to the political activities that he believed one had to use when the system refused to change. "Direct confrontation, making grassroots appeals—principal tactics of the environmental and other popular movements in the 1960s and '70s—are still viable alternatives in the 1990s," he told the Washington audience. "Americans who are truly interested in working for social change can increasingly look less and less to the political process for redress."

WITHIN WEEKS of that grueling campaign, personal tragedies would start taking their toll on Chavez. December 14, 1991, at the age of ninety-nine, Juana Estrada Chavez died in San Jose. In public, Chavez was stoic as he delivered a tender eulogy for his mother before three hundred family members and guests who had squeezed into Our Lady of Guadalupe Church, not far from the Sal Si Puedes home where Cesar's public life had begun in 1952. Cesar recounted his mother's long journey and described the impact she had left on everyone she met. "Juana Estrada Chavez does not need for any of us to speak well of her this day. The simple deeds of a lifetime speak far more eloquently than any words of ours about this remarkable woman and the legacy of hope and strength that she leaves behind."

Juana Chavez's death had a profound effect on Cesar. She had given her son a core set of personal beliefs—his deep Catholic faith, his commitment to the poor, his pride in Mexican folk culture and its wisdom, even his fascination with traditional medicine and spirituality. And Juana Chavez, despite her advancing years, had been at Cesar's side at important moments in the UFW's history. She sat next to him, quietly, when he broke his long 1968 and 1988 fasts. "We were the strikingest family in all of farm labor," Cesar told the audience during his mother's services, recounting the family's depression-era tour of California's farmlands. "Sometimes the contractor would cheat the workers by putting his knee under the sack so it would weigh less. Instead of getting credited for a hundred-pound sack, the worker would get marked down for only eighty pounds. All this would happen pretty fast and the victim's view was usually blocked. Well, Mama was pretty sharp. She saw the contractor cheating a worker who was in line ahead of her, and she called him on it. The contractor was furious. The entire Chavez family got fired."

IT TOOK TIME for the grieving Cesar to recover from his mother's death, but he forced himself to go back to work. Despite its limited success, the union was still pushing for a boycott of California grapes; George Deukmejian had passed the Republican Party's torch on to a new governor, Pete Wilson, who seemed equally content to let the ALRB hobble along with diminished funding. Chavez estimated that if the state would only force growers to bargain in good faith, the pro-union votes that had been recorded at many ranches during the 1980s could be turned into contracts, and the UFW's ranks would swell beyond ninety thousand members. Arturo Rodriguez remembers that it was in mid-1992 when his father-in-law began talking more about

■ *Cesar Chavez in March of 1990. He was immersed in UFW activities until his untimely death in 1993.*

rejuvenating the union's organizing efforts in the fields. His newfound spirit mobilized farmworkers and led to union walkouts from familiar vineyards in the Coachella and San Joaquin Valleys that year. This time the catalyst was the wretched living conditions of the workers housed in old labor camps. While they didn't win union contracts, the action yielded immediate fruit for the farmworkers: They got the first industry-wide wage hike for grape workers in eight years. Rodriguez muses that after that pay raise, growers proved that they had not changed that much in the twenty-seven years since the first Delano strikes. Still one old nemesis, the M. Caratan Ranch, struck back in early 1993 by firing 180 of its workers who had carried picket signs.

In 1992, Cesar marched again, this time, in a truly surprising setting. It was mid-summer when Chavez forgave two decades of animosity and agreed to allow members of Teamsters Local 890 to join him on a two-mile procession through Salinas. Teamster leaders wanted to show support for the UFW's drive to organize farmworkers. "It feels a little strange for everybody," said local Teamster chief Mike Johnson, whose union was under reformist leadership and was looking to make amends. "What farmworkers need is a union, and it doesn't really matter which one." With that unprecedented show of solidarity, and the pay increases for grape harvesters, Chavez had regained organizing momentum that the union hadn't seen in years.

Within weeks of the Salinas march, however, Chavez was overcome by personal grief again. Fred Ross, Cesar's mentor and his unwavering friend for forty years, died at eighty-two. Ross succumbed to cancer in a San Rafael hospital on Sunday, September 27. Fred Ross Jr. and his sister, Julia, had been singing folk songs to their father just days before he died. "He loved to sing," his son said afterward, "and the last song he ever heard was Friday when we played 'Hobo's Lullaby,' one of his favorites by Woody Guthrie, whom he had known in the Arvin labor camp in the 1930s."

Paying tribute to Ross at his funeral was one of the toughest tasks Chavez ever dealt with, Marc Grossman remembers. After writing a rough draft of a eulogy, Cesar wrestled with almost every word, looking for just the right sentences that would serve Ross the best. Just hours before the eulogy at San Francisco's Delancy Street Foundation, Chavez, slowed by a severe flu and laryngitis, was still amending his speech, crossing out whole sections and replacing them with more fitting anecdotes. Grossman recalls that just before he was to speak, Cesar hurriedly excised all the "I's" and "me's" that he was able to catch. The eulogy was for Ross, Chavez said to Grossman, with a hoarse voice, and no one else should serve as a point of reference during the day. Ross's ability to organize defied even death. For at least this one day, most of the original team that had followed Chavez and Ross into the farm fields was reunited. Gilbert Padilla was there, as was Eliseo Medina, and Cohen, Ganz, and Al Rojas. The past was forgotten for a day. "Cesar was talking about getting people back together. I think he wanted to do something [with us] again," Cohen remembered.

WITHIN SEVEN MONTHS, Cesar E. Chavez died of natural causes.

If it had been up to Chavez, the ceremony marking his final departure would have been as simple and straightforward as his entry into the world sixty-six years before. Over the years, Chavez had repeatedly spelled this out to family and close friends: His funeral should be a plain affair. No royal procession. No flamboyant send-off befitting a fallen hero or a civil rights legend. But when the day finally arrived, everyone knew it would be impossible to keep the event simple. For a man who had been called the "Father of all Chicanos," and who was the spiritual leader of America's farmworkers, it was inevitable that his funeral would take on the same epic proportions as his life.

The end had come sometime in the still hours of April 22, 1993, on yet another night of little sleep. It caught him unawares on familiar turf—in the city of San Luis—near his family's former homestead in an obscure corner of Arizona, where half a century before the Chavezes had been forced off their land and thrown onto the migrant trail.

The circle had closed: Just days before his death, Cesar had returned to his home state to battle on behalf of the UFW against the agricultural interests that had first cast a shadow over the Chavez family during the Great Depression. In 1939, the Chavez family lost its farm just outside Yuma, Arizona, to a local bank, which— ignoring both Librado Chavez's long history of paying off debts and his pleas for an extension—seized the property. It is ironic that the land upon which the Chavez family had built their homestead ultimately fell into the hands of the union's longtime nemesis, Bruce Church Incorporated, one of the world's largest lettuce producers. When he died, Chavez was in Yuma, not far from his family's former property, to defend the union in a lawsuit filed by the same Bruce Church.

■ *The unvarnished coffin, which was carried for miles in a final pilgrimage for Cesar Chavez, passing the fields in Delano where he began the UFW.*

In one form or another, the Bruce Church empire had dogged Chavez and the union at key moments throughout their history. The company had always been a tough foe. In 1970, Bruce Church was one of the companies struck by the UFW when growers and the Teamsters colluded to sign contracts covering workers in the rich fields of the Salinas Valley. Later, under supervision of the ALRB, workers at Bruce Church voted for UFW representation, but a contract was never signed. Bruce Church launched legal challenges to the vote, a fight that would last seventeen years. In 1984, the company sued the UFW, blaming it for lost supermarket sales after union pickets went up in front of supermarkets carrying its Red Coach– brand lettuce. The company chose to file its lawsuit in Arizona, where it has extensive landholdings. Unlike California, where courts upheld the right of labor groups to conduct so-called "secondary boycotts," Arizona supported laws forbidding the practice.

For nine years the Church lawsuit had dragged on the union like a ball and chain. The union was accustomed to legal battles; there were always one or two suits or countersuits in progress, it seemed. And throughout its history, the UFW had had to fight hard to score the occasional victory in court over the well-heeled legal teams representing growers or government agencies. But in 1992, the Bruce Church challenge loomed as the most serious threat to the union's survival. In the late 1980s, a federal judge had rejected claims for nearly $5.4 million in damages from Bruce Church because no secondary picketing had occurred in Arizona. But the company persisted, and in 1993, an Arizona judge was to hear a new Church lawsuit, this time for $3 million, which charged that the UFW had "interfered with business."

The UFW lost before when an Arizona Superior Court judge ordered the union to pay Bruce Church for its lost supermarket business. But the union had no intention of doing so. Friends recalled the fighting spirit that belied Cesar's slight stature and soft-spoken demeanor always emerged when it came to Bruce Church. Besides, total assets of the UFW at the time barely exceeded $2 million; the treasury had been drained by the dwindling number of farm contracts in the 1980s. There were only twenty-two thousand farmworkers officially on the union's dues list. The high-tech grape boycott that Cesar had launched in 1984 was floundering. Grape sales had remained steady since the boycott was launched; only in 1989 and 1990 was there a serious dip in sales. (While the union claimed credit for that sales drop, it was perhaps only partially responsible. Public awareness of the dangers of chemicals reached a new high in the late 1980s. Grapes from Chile were discovered to be tainted with cyanide, and a suspected carcinogen, Alar, was found being sprayed on Washington State apples to improve their look.)

With no reserves, the union strategy was to fight hard in Arizona: The cycle of court cases was extremely costly, but if the union lost in the local court, it would have to quickly file a federal appeal. In the federal courts, lawyers believed, the union stood a good chance of again showing that the Arizona judges had erred and a new trial was needed.

When Chavez traveled to Yuma for the first appeal of the lower court ruling, he knew he was in for what might turn out to be the toughest fight of his life: Lose the millions of dollars to Bruce Church, and the financially strapped union might sink out of sight for good. Determined not to let the same company prevail once again, he began one of his now-routine fasts to regain his moral strength. After two days on the witness stand, Chavez felt drained, even exhausted.

The day before he died, however, Chavez was in a particularly cheerful mood. Neither the Bruce Church fight, the continuing deterioration of field conditions for farmworkers that even his best efforts could not forestall, nor the public's slipping support for his cause seemed to bother him, on the surface. His high spirits, it seemed, had returned, and his friends were pleased. After his second day in court, Chavez and David Martinez, a longtime union activist and UFW board member, went on a tour of the working-class barrios of Yuma's south side. The drive sparked memories of Chavez's childhood years: some pleasant, like playing stickball with his brother Richard; others that still stung—the schools where teachers hit him for speaking Spanish.

Later, Cesar called Arturo Rodriguez to report on how the trial was going. Although his son-in-law thought nothing of it at the time—Cesar was always doing the work of two men, if not three—he recalls that Chavez had confessed to him, "I'm tired, Art. I'm really very tired."

Later that night, Chavez seemed relaxed, despite the gloomy outlook for the union and two days of often tedious questioning inside the envelope of cool manufactured air of the Yuma courthouse. He broke his fast with a simple vegetarian meal of rice and cabbage, and, retiring early in a friend's house where he was staying, took a book about Native American art to bed with him.

It was Martinez, a man who'd grown up in service of Chavez and the UFW, who found his mentor in the early morning hours, lifeless on the small bed, fully clothed, shoes off, the book open and still within reach of his cold hands. "It was like a bad dream. I had wakened him before when he'd read all night. I just thought he was asleep . . . ," recalls Martinez.

After calling for paramedics, whose arrival created a stir in the neighborhood, a badly shaken Martinez started a round of calls to family and friends. The news reached the press and touched off a series of emotional conversations as union supporters around the country scrambled to call one another. Sixty-six-year-old Dona Maria Hau, a retired farmworker who had put Chavez up that night, broke down when she found out what happened under her own roof. "I keep asking God why didn't I die, instead of Cesar. He was needed much more than me. He's the one who should be alive so he can keep on fighting the rich people."

As word spread among union field offices, farmworkers began to gather. In Salinas, vegetable workers, their clothes still muddy, met at the union office, where they cried and huddled together in prayer. Sabino Lopez got a phone call from a reporter who wanted a statement about Chavez. Lopez was long gone from the union, having moved on to help found a farmworker housing rights group, the Center for Community Advocacy. But the news ended his long absence from the union hall. He rushed over to confirm what he had heard, and then made plans to attend the funeral of the man who had changed his life.

Arturo Rodriguez was driving on a lonely stretch of road near Yuma and couldn't be reached before the news became public. He was heading toward San Luis to meet Cesar and was listening to the UFW's radio station, Radio Campesino, when he thought he heard his father-in-law was sick. He turned up the radio, worried, and began speeding toward the nearest telephone he could find. Then the crush of the news hit all at once; he spotted a phone, but at that instance the radio announcer said Chavez was dead. "I couldn't believe it—that he was no longer alive," he says. "That you wouldn't walk into his office and see him there. That we wouldn't be making those trips together. That he wouldn't be there to discuss strategy with. But I called my wife, Lu [Linda], and Helen, and it was true: He had passed away."

Dazed union leaders, pressed to comment on the future of the union before they had time to grieve, issued a statement containing Chavez's own words: "Regardless of what the future holds for our union, regardless of what the future holds for farmworkers, our accomplishments cannot be undone. The consciousness and the pride that were raised by our union are alive and thriving inside millions of young Hispanics who will never work on a farm."

Richard Chavez was among the first in the family to receive word of his brother's passing. He wept with Cesar's son Paul before they composed themselves and told Helen and the rest of the family. "I just couldn't believe it," Richard says. "I thought somebody had probably shot him or something like that . . . But he died a very peaceful death. When I saw him, he looked like he'd just . . . like you could move him and he would wake up.

"I kind of blanked out for a while," Richard continues. "Then Paul says, 'I want you to come with me to tell my mother.'

"Helen told us afterward that as soon as we called them all together, she knew. She said she knew exactly that something had happened to Cesar. She just felt it," Richard recalls.

Richard Chavez and Cesar had been the closest of brothers; as youngsters they hardly spent a day apart. But in the year before Cesar died, union work had become so demanding that Richard had rarely seen Cesar alone. In March 1993, however, Cesar celebrated his birthday at La Paz with his family and close union members. "He cooked for everybody," Richard says. "He liked to cook this Indian cuisine, from India, and he did a beautiful, beautiful spread. I think it was the best feast I've had in my life. He and Helen put it together. We had a lot of fun, and that particular night he was sitting, not at a head table, but a table toward the corner."

Cesar called Richard over to talk alone and made an unusual proposal.

"Richard, you want to go to Mexico?"

His brother was game. "Yeah! I always want to go to Mexico. Who's going?"

"I am," Cesar answered.

"Really? Who else is going?"

"Nobody, just you and I."

"Really?"

"Yeah, let's take off!"

The brothers spent a week together, traveling a bit, resting a lot. "We spent a whole week together, which we hadn't done in ages," says Richard. "Something told him that something was going to happen."

Dolores Huerta also believes Chavez had sensed he might be nearing the end. Over the years, she and Cesar—despite their close friendship—had often argued over union issues. On more than one occasion, Dolores would leave La Paz in a huff, vowing never to return. Within a day of the argument, however, she was usually on the phone, talking with Chavez about union business. Sometimes, Huerta laughingly recalls, Cesar would just fight with her about anything just for the sake of fighting, adding that he was *"peleonero,"* someone inclined to pick fights. It seemed unusual to her, Huerta says, that before he died, Chavez had been uncharacteristically tame: "For the last six weeks we hadn't had a single fight."

Huerta, as energetic as always, was still the firebrand Cesar had met and befriended some thirty-five years earlier. She had withdrawn some from regular union work by the 1990s, pouring energy into other social causes. She was also still on the mend from a clubbing she suffered at the hands of a San Francisco police officer in 1988, an attack that almost killed her, breaking several ribs and destroying her spleen. Huerta had joined protesters demonstrating against then-president George Bush. Pressed against a police barricade by the crowd, Huerta was trying to comply with an officer's order to move back when he suddenly drove his riot stick into her abdomen.

Controversy gripped the city for days afterward, as Cesar and civil rights activists demanded a thorough investigation. Initially, there were doubts that the officer had acted improperly; but the case broke wide open when a subsequent review of television news footage clearly showed the officer striking Huerta. In 1991, the city settled with Huerta out of court for $825,000, a sum that inspired her to joke that she would have to open a bank account. In all her years of volunteerism with the UFW, she had never so much as applied for a credit card. Huerta said she intended to use the money to help groups organizing women.

CHAVEZ'S WISH for a simple burial was symbolized in the smooth, unvarnished wooden casket crafted and planed by Richard. The pine box was Cesar's idea.

Max Benavidez is a Los Angeles essayist, critic, and poet who has written widely about Chicano art and culture. In this essay, he describes Cesar Chavez's impact on Chicano creative expression, which blossomed as a result of the farmworkers' moral and political struggle.

Cesar Chavez died with an art book in his hands. This final image of the great visionary is appropriate—and poetic—because he and his struggle for justice were intimately intertwined with creative expression by Chicano visual and performing artists.

■ *Chicano artists bloomed with the farmworkers movement. Ester Hernandez designed this poster, which dramatized the hazards of pesticides, in 1982.*

The veteran union organizer in him would have appreciated the death-bed image because Chavez intuitively understood the power of symbols. By the mid-1960s, the United Farm Workers flag, with the black eagle-like thunderbird dramatically set on a background of blood red, conveyed an instant sense of grassroots struggle and cultural pride. Later, he used a banner depicting the *Virgen de Guadalupe*, the quintessential guardian of the oppressed, when he and the farmworkers went on the famous Pilgrimage to Sacramento. Perhaps it is fitting that Chavez's own visage has already superseded that of Emiliano Zapata, the beloved peasant leader of the Mexican Revolution, as the paramount Chicano icon.

Chicano art was born in 1965 when Chavez gave budding theater director Luis Valdez permission to mount primitive *actos*—sketches—on the very picket lines of the Delano fields in California's Central Valley. In fact, says Valdez, "El Teatro Campesino emerged directly from *la huelga*"—the strike against the growers and agribusiness interests. Visual artists were inspired to join the cause, eventually including a Who's Who of the genre: Antonio Bernal, Malaquias Montoya, Rupert Garcia, Ester Hernandez, and members of Sacramento's Royal Chicano Air Force, a socially activist art group of Beat- and Dadaist-influenced painters, poets, and pranksters.

What Chicano art has ultimately become in all its globally hip and often subtle permutations—including the irreverent satire of the comic trio Culture Clash—owes a lot to Chavez and his labor movement. The post-modernist work of Gronk, the often harrowing meditations on existence by the late Carlos Almaraz (who was once an illustrator for the UFW newsletter), the dreamscape paintings of family life by Carmen Lopez Garza, the sardonic conceptual videos of Harry Gamboa Jr., or the big-budget plays and films of Luis Valdez, Chavez influenced them all. It is no exaggeration to say that Chavez was one of Chicano art's foremost progenitors.

When Chavez told Valdez to make something out of nothing, he was articulating the essence of early Chicano art. In the mid- to late-sixties, there were no grants, no art degrees; just a burning and sometimes indefinable need for artists to say something about themselves and their situation. That was what the work was about then. Now it is displayed in high-priced, fashionably stark galleries, not on decaying street-corner walls. The plays are now performed in

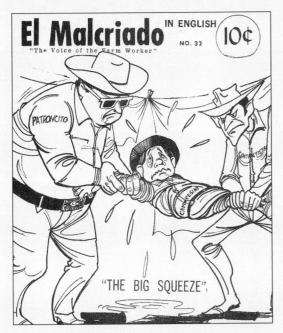

El Malcriado

IN ENGLISH

"The Voice of the Farm Worker"

NO. 32

10¢

PATRONCITO

CAMPESINO

CONTRATISTA

"THE BIG SQUEEZE"

■ *Farmworker being squeezed by both his boss* (el patroncito) *and the labor contractor* (el contratista). *Cartoon characters like El Patroncito, who appeared regularly in the UFW newspaper* El Malcriado, *often showed up as characters in plays by Teatro Campesino.*

state-of-the-art theaters, not on dusty roads. But the work's roots are far less glamorous. It comes from the smell of the earth, scrawled signs, and the naked will to create under any conditions.

As long as he was alive, Chavez served as a reminder that Chicano creative expression had a moral basis and was founded on the political struggle to overcome oppression. Chavez was an archetype, a bigger-than-life symbol of Chicano art's distinct origins.

Songwriter Lalo Guerrero, himself a legend in the Latino community, remembered traveling up and down the state in the early 1960s. "We went to places like Fresno and Stockton and we heard about the organizing that was going on. I became familiar with the farmworkers' situation. I was very impressed with what Cesar was doing. In 1964, I wrote a *corrido*, a folk ballad, called *'El Corrido de Delano.'* Chavez often remarked that the *corrido* helped him tremendously.

"I remember the last time I saw him," Guerrero recalled. "It was in 1992, at a tribute for me in Palm Desert. When he walked into the theater, everything stopped. Quiet. I was amazed to see how everyone

stepped aside so he could pass. I got goose bumps. He was revered. Most of the Chicano painters, writers, and actors were very close to the movement. They all found a need to join him."

One artist who joined him early was Barbara Carrasco. She painted the banners for Chavez's funeral, which took place in Delano.

"I was nineteen when I heard him speak at UCLA," she reminisced. "I volunteered right after the speech. Cesar told me at the time: 'We need artists very much.'" Carrasco started painting banners for the union and has been doing it for fifteen years.

In 1989, she was in New York City to unveil her computer-animated billboard, *Pesticides,* in Times Square. Chavez was also there to promote his campaign against the use of the pesticides in the fields. He held a press conference as Mayor David Dinkins raised the UFW flag over City Hall.

"Cesar was very interested in the billboard and wanted to know how the computer animation worked," Carrasco said. "He was surprised to learn that the animation would run for a whole month. The timing was perfect. It showed a farmworker picking grapes and being sprayed by a crop duster. Then the grapes are shown in a supermarket where a woman buys them and takes them home to her children. The children eat them and become ill. The animation ends with *calaveras*—skulls—the symbol of death, as the crop duster comes down on the farmworker again."

Chavez lived to witness Chicano art as it evolved from simple agitprop in the open fields to a computer-driven image among the big city lights of Times Square.

In death, Chavez has become a mythic figure. For the Chicano community, his posthumous image will resonate long into the next century as a symbol of strength and creativity. He will appear in poems, songs, and films. He will become a reflection of the organic nature of Chicano art and its powerful origins as a synthesis of the spirit, culture, and politics. The Chavez legacy will inspire artists for decades to come. As an art lover, he would have liked that.

■ *The irreverent comedy trio Culture Clash is one of the many groups of Latino actors with roots in Teatro Campesino, the farmworkers' theater.*

When Richard heard that Cesar was dead, one of the memories that haunted him was a promise he'd reluctantly made to his brother. "He always asked me to build his coffin when he died," Richard says. "And I used to kid him, I said, 'Cesar, come on, Cesar. I'll never outlive you. You don't smoke, you don't drink, you have a good diet, you do exercise, yoga, and I do just the opposite that you do. I'll never outlive you, so you better look for somebody else to build your coffin.' 'No,' he says, 'when I die I want you to build my coffin. I want you to make it out of pine.' He told me what kind of wood and everything."

The casket took Richard thirty-eight hours to build. He was interrupted by grieving and by friends and family who came to watch or talk to him. At last it was ready.

The pine coffin carried Cesar to an unremarkable burial site near the entrance of La Paz, just a few feet from where Boycott, Chavez's trusted German shepherd, was buried. The site was chosen after a small debate between family members and union leaders, who had suggested that Cesar be buried at the Forty Acres compound in Delano.

Mike Ybarra, Cesar's longtime driver and bodyguard, accepted the duty of dressing Chavez's body for the open-casket wake that would follow a long march through Delano. The casket would also carry simple ornaments that would serve as powerful reminders of Cesar's impact on a race of people: the carved eagle—the symbol of his movement—and the short-handled hoe, the insidious farm tool that had ruined many a farmworker's back and whose banishment from California's fields was among the UFW's proudest accomplishments.

Ybarra last saw Cesar alive when Chavez dropped by his home before he left for Yuma. Chavez ate some supper with Mike and his wife, Maria, complimenting Maria on her delicious hot salsa. The three talked in Spanish about Mike coming out of retirement and traveling again with Cesar. "He took care of himself so well," Ybarra says, still incredulous that Chavez didn't live longer. But the two had talked about dying before, and Ybarra was comforted to remember that Chavez had said he wanted to pass away in his sleep.

The plan was for the funeral procession to follow the Garzas Highway out of downtown Delano past People's store and bar, where so much union business had been accomplished over billiard tables and bottles of beer. Lucio Gonzalez and his wife, Butter, who were running People's now, listened as the steady, distant drumming grew louder and announced the parade's arrival in Delano's west-side barrio. Gonzalez vividly remembers sighting the head of the procession: The Aztec dancers in shimmering feathered headdresses rose slowly from the other side of the railroad grade that cut through town. The couple summoned their small grandchildren, telling them they would understand one day why Cesar Chavez was such a great man.

"Others would have done things violently," says Lucio. "The growers are lucky it was Cesar who led the movement."

LIKE SONS AND DAUGHTERS long absent from home, Cesar's *gente*—his people—found their way back to Delano for the funeral on April 29, 1993. Some forty thousand people came to pay their last respects and follow Cesar once again down the dusty, shoulderless roads of Kern County. The proud red-and-black flags with their thunderbird eagle were held high, along with hundreds of American and Mexican flags. There were banners adorned with the work of movement-bred artists, and a sea of black-and-white mourning flags the union had silk-screened overnight. Dotting the march like candles were thousands of white gladiolas—in Mexico, a symbol of mourning, and the favorite flower of Helen Chavez. The banners whipped back and forth in the mild spring wind, lending an odd rhythm to the songs—*corridos* and *cantos*—that for decades had leavened marches like this, raising them to the level of celebration. Cries of *"Viva Chavez!"* were punctuated by periods of heavy silence and the dry sounds of thousands of feet treading asphalt and sandy soil.

Scattered among the marchers were the celebrities who had come to Delano countless times before when Cesar needed a hand with one event or another, or when he would break a fast with a religious rally. Jesse Jackson and Edward James Olmos took turns as pallbearers, as did some of the children of Robert Kennedy. They'd come again to Delano with their mother, Ethel, who strode slowly near the front of the procession, holding Helen Chavez's hand. Luis Valdez walked alongside the coffin, greeting famous and not-so-famous fellow actors and old friends. Comedian Paul Rodriguez agreed that Chavez "was the closest thing we've had to a national hero," adding that "my father once told me, 'If that crazy Mexican doesn't get shot, he's going to do a lot of good for us.'"

The bulk of the crowd was made up of regular people, people like seventy-seven-year-old former farmworker Richard Lopez. A proud Lopez waved away the

■ *Ethel Kennedy comforted her friend, Helen Chavez, while both stood by Cesar's casket.*

POSTSCRIPT, 1993: REMEMBERING CESAR CHAVEZ

by Peter Matthiessen

Writer Peter Matthiessen, known for both his novels and his nonfiction, became friends with Cesar Chavez while covering the grape strike in Delano. His series of eyewitness articles on Chavez and the union became the basis for his 1969 book Sal Si Puedes: Cesar Chavez and the New American Revolution. In this excerpt from his 1993 eulogy, published in the New Yorker, Matthiessen pays tribute to the man behind el movimiento.

With the former scourge of California safely in his coffin, state flags were lowered to half-mast by order of the governor, and messages poured forth from the heads of church and state, including the Pope and the President of the United States. This last of the UFW marches was greater, even, than the 1975 march against the Gallo winery, which helped destroy the growers' cynical alliance with the Teamsters. "We have lost perhaps the greatest Californian of the twentieth century," the president of the California State Senate said, in public demotion of Cesar Chavez's sworn enemies Nixon and Reagan.

For most of his life, Cesar Estrada Chavez chose to live penniless and without property, devoting everything he had, including his frail health, to the UFW, the first effective farmworkers' union ever created in the United States. "Without a union, the people are always cheated, and they are so innocent," Cesar told me when we first met, in July 1968, where he lived with his wife, Helen, and a growing family. Chavez, five feet six, and a sufferer from recurrent back pain, seemed an unlikely David to go up against the four-billion-dollar Goliath of California agribusiness.

With the funeral march over, the highway empty, and all the banners put away, Cesar Chavez's friends and perhaps his foes are wondering what will become of the UFW. A well-trained new leadership (his son-in-law has been named to succeed him, and four of his eight children work for the union) may bring fresh energy and insight. But what the union will miss is Chavez's spiritual fire. A man so unswayed by money, a man who (despite many death threats) refused to let his bodyguards go armed, and who offered his entire life to the service of others, was not to be judged by the same standards of some self-serving labor leader or politician. Self-sacrifice lay at the heart of the devotion he inspired, and gave dignity and hope not only to the farmworkers but to every one of the Chicano people, who saw for themselves what one brave man, indifferent to his own health and welfare, could accomplish.

Anger was a part of Chavez, but so was a transparent love of humankind. The gentle mystic that his disciples wished to see inhabited the same small body as the relentless labor leader who concerned himself with the most minute operation of his union. Astonishingly—this seems to me his genius—the two Cesars were so complementary that without either, la causa could not have survived.

During the vigil at the open casket on the day before the funeral, an old man lifted a child up to show him the small, gray-haired man who lay inside. "I'm going to tell you about this man someday," he said.

■ *Cesar and Helen Chavez and family in front of their house in the small California town of Delano, where they moved in 1962 to start a farmworkers union.*

hands of fellow marchers held out to steady him after he had collapsed in an almond orchard during the funeral procession. In a sure voice, he told reporters that he and Chavez had long before struck a deal: "If he died first, I would walk in his funeral, and if I died first, he would walk in mine. I'm going to honor that pact."

Disaffected friends were also in the crowd, one-time followers who had been put off years before by UFW infighting or by Cesar's flirtation with Synanon. Many had returned to bury their political differences with Chavez and pay tribute to the man who'd done more for farmworkers than anyone. Eliseo Medina, his hair now streaked with gray, returned as a highly successful organizer with the Service Employees International Union. Medina and other ex-UFW activists were among the architects of the successful Justice for Janitors campaigns in Los Angeles and San Francisco—campaigns that mobilized immigrant workers often dismissed as too downtrodden to ever be organized. Employing the tactics he learned from Chavez, Medina had scored impressive gains for the underpaid, mostly Mexican workers who quietly cleaned the high-rise law offices and high-tech labs after professionals went home.

Al Rojas was at the funeral, too, swapping stories of the glory days in Delano and on the boycott with old union buddies. Rojas had left union work in the early 1980s, at the advice of his doctor, and settled into a new career as a mediator with the state Department of Industrial Relations. But the old activist streak that Cesar had encouraged was still there and had gotten him involved in a new campaign for democracy in Mexico. That movement would later take Rojas down to the Mexican state of Chiapas for meetings with the *Zapatista* rebels and one of their masked leaders, *Subcomandante* Marcos.

Former UFW organizer Marshall Ganz also returned, talking with his old friends about the exhilarating times when the union had brought the state's largest corporate farmers to their knees and won landmark contracts for thousands of vegetable harvesters. Jerry Cohen was there, pushing aside criticism of the union he believed had stopped thriving in the early 1980s, when it shifted its resources to politics and away from field organizing. Cohen, who still breaks down when he talks about Chavez's death, says his first reaction when he heard the news was a longing to return to the hotel room at the Stardust Motel, where John Giumarra Jr. had asked to meet Cohen and Cesar to tell them he was ready to sign a contract.

Former governor Jerry Brown, who paid tribute at the funeral, believes Chavez left a profound imprint on the pages of California history. "Cesar wasn't afraid of anybody. He was a real fighter. The farmers were scared to death of him," said the former governor. "I remember the first time I saw him. He walked into my father's house in Hancock Park and he was dressed the same way as when I saw him a month ago. He never lost his modesty and simplicity."

The Peasant and the Pauper

by Jose Antonio Burciaga

The late Jose Antonio Burciaga was an artist, writer, and resident fellow at Stanford University. One of the original members of the Latino comedy group Culture Clash, he is also the author of Drink Cultura *and other books on Chicano culture. In this essay, carried by Pacific News Service, he surprised some readers by comparing Chavez to the great Mexican comedian Cantinflas.*

As we celebrate the springtime ritual of el Cinco de Mayo, it would be well to recall Cesar Chavez and Cantinflas, two giants who died this month and who portrayed, spoke, and fought for the peasants and the paupers.

With their roots in two countries, they shared a common soil and culture. Behind their disguises as a pauper and a peasant, they were a couple of idealistic Don Quixotes. With wit and wisdom, one of them elegantly and effectively skirted life's trials. The other was involved in an obsessive struggle against the incredible poverty and trials of the people who pick the crops for the best-fed country in the world.

Chavez fasted frequently and often began his day at 2:30 A.M. He took no salary and never wore a coat and tie. Cantinflas wore a tie, but it was as raggedy as the rest of his famous outfit. Off camera he was a natty dresser. But at the same time, Cantinflas gave much of his fortune away to the poor.

Both touched the lives of millions, both offered hope: one through nonviolent organization, the other through humor. One fought death, the other laughed at it. They both belonged to us and they both left together: one with a laugh, the other with a quick smile.

They befuddled authority and the upper class with their ability to walk with kings and keep the common touch. They loved children: one as a clown, the other concerned for their health and welfare against poverty, disease, and pesticides.

Cantinflas had been there forever, making us laugh in Spanish-language movie houses throughout the Southwest and Mexico. His films remain as timeless as *Casablanca*.

Cesar Chavez was our first national Chicano hero. In launching a nonviolent struggle to better conditions for farmworkers, he inadvertently created a Chicano movement that begat a whole revolutionary artistic movement in *teatro*, art, and literature. Cantinflas was a major influence on Chicano theater.

Chavez turned down the leadership of the Chicano movement in 1970 because he had dedicated his life to farmworkers regardless of their ethnicity.

Chicanos became like Cantinflas. Taking on mainstream culture, the Vietnam War, and politics, they utilized elements from two cultures and two languages to forge and fuse new symbols and icons, a new aesthetic.

Unfortunately, when Cantinflas and Cesar Chavez left, our youth, who had reaped the benefits of those two people, knew very little, if anything, about them. Time, politics, memory, and ignored history had passed them by.

On the day Cesar Chavez died, Stanford University was visited by hundreds of prospective high school students who had been accepted by the university. One young woman wanted to know why Cantinflas was so important. Though hailed as the greatest comedian in the world by Charlie Chaplin, too few people in this country knew or understood Cantinflas's comedy beyond his visual pantomime. He was a master comic of Spanish double-talk. The three U.S. films he acted in never came close to capturing his genius.

When a young woman claimed Yuma, Arizona, as her home, a university administrator said, "Oh! You must be very sad today. Cesar Chavez was from there." The young woman was puzzled: "Who was he?"

The influence of sports and television over history was evident when a student who knew about Cesar Chavez still thought he was named for a boxer.

The students knew very little about Cantinflas, and today we hope they know something about the 1862 battle in which the French, at that time the most powerful military army in the world, fought a force of poor Mexican campesinos and indios. The Mexican forces were led by General Ignacio Zaragoza, who was born in Texas when Texas was part of Mexico. Had Mexico lost that battle, France would have concentrated on freeing the port of New Orleans to help the south in the Civil War. And U.S. history might have turned out differently. Is this taught in our history books?

Cesar Chavez, Cantinflas, and the Cinco de Mayo battle are living proof that the poor do not lack leadership, courage, vision, ideals, or a future.

Brown had reconnected with Cesar in the months before his death. The two men who had constructed California's landmark farm-labor law were once again talking about organizing, perhaps even working together on establishing new farmworker cooperatives. "I really think he was on the right track. He didn't get there, didn't get to the promised land. Neither did Martin Luther King. Nobody ever gets there."

Other longtime political allies also marched with Cesar's casket, including then U.S. trade representative Mickey Kantor, the former pro bono farmworkers' lawyer, who carried a letter from President Clinton to Helen Chavez, in which he wrote that Chavez was "an authentic hero to millions of people throughout the world." Cardinal Roger Mahony, who had spent countless hours as a mediator in the early days of the union, offered personal condolences from the Pope. Richie Ross, a Democratic Party political consultant in Sacramento who received his start in organizing with Chavez and the UFW, said, "He was the spring, the root, where it all started. . . . He helped an entire race of people find its place in America."

He would not find disagreement from the farmworkers marching in the procession. A seventy-year-old farmworker named Remigio Gutierrez said for him and for other farmworkers, Chavez was a giant. "For all the workers," he said, "Cesar was strong." Manuel Amaya, a sixty-one-year-old, said, "God has taken the strongest arm we have, but we will continue."

The procession through Delano was a magnet for farmworkers from all over the United States. Baldemar Velásquez, president of the Farm Labor Organizing Committee (FLOC) in Ohio, arrived with a small contingent from his union—an organization Chavez had inspired and counseled for years. Velásquez had emulated UFW tactics and legal strategies, modifying them for the soup and canned vegetable industry he was up against in the Midwest. Cesar had been there at each crucial step, Velásquez remembers. "We were working on ambitious plans to do more things together, but all that was put aside when he died."

Cesar was the keynote speaker at FLOC's first convention in 1979, and the UFW sent organizing help to FLOC during its own tough and bloody strike against farmers growing vegetables for the Campbell Soup Company. When FLOC called for a boycott of Campbell products, Cesar called on UFW volunteers nationwide to help. The bonds of the two organizations were strong, perhaps, because of the kinship between the charismatic Velásquez and Chavez. Velásquez was a fellow devotee of Gandhi and had learned the Indian leader's nonviolent philosophy from his wife's grandfather, a cleric who had traveled with Gandhi. As a college student and a volunteer in Cleveland with the Congress of Racial Equality, Velásquez had watched closely how the UFW managed nationwide boycotts and organized workers in an

Elegy on the Death of César Chávez
by Rudolfo Anaya

*R*udolfo Anaya has been called a father of Chicano literature. From his classic novel Bless Me, Ultima *to his southwest mysteries, his novels are among the richest chronicles of the Mexican-American experience. Following is an excerpt from his elegy on Cesar Chavez's death.*

I weep for Adonais—he is dead!
O, weep for Adonais! Though our tears
Thaw not the frost which binds so dear a head!
—Shelly

We have wept for César until our eyes are dry,
   Dry as the fields of California that
      He loved so well, and now lay fallow.
   Dry as the orchards of Yakima, where dark
      buds
      Hang on trees and do not blossom.
Dry as the Valle de Tejas, where people cross
   Their foreheads and pray for rain.

This earth he loved so well is dry and mourning
   For César has fallen, our morning star has
      fallen.

The messenger came with the sad news of his
death—
   O, kill the messenger and steal back the life
   Of this man who was a guide across the fields
      of toil,
   Kill the day and stop all time, stop la muerte
   Who has robbed us of our morning star, that
   Luminous light that greeted workers as they
   Gathered around the dawn campfires.

Let the morning light of Quetzalcoátl and
Christian saint
   Shine again. Let the wings of the Holy Ghost
      unfold
   And give back the spirit it took from us in
      sleep.

Across the land we heard las campañas
doblando:
   Ha muerto César, ha muerto César.

His name was a soft breeze to cool the
campesino's sweat
   A scourge on the oppressors of the poor.

Now he lies dead, and storms still rage around us.
   The dispossessed walk hopeless streets,
   Campesinos gather by roadside ditches to
      sleep,
   Shrouded by pesticides, unsure of tomorrow,
   Hounded by propositions that keep their
      children
   Uneducated in a land grown fat with greed.

Yes, the arrogant hounds of hate
Are loose upon this land again, and César
Weeps in the embrace of La Virgen de
   Guadalupe,
Still praying for his people.

     "Rise, mi gente, rise," he prays.
     "Rise, mi gente, rise. . . ."

industry once thought impenetrable. In 1983, Chavez joined Velásquez for the last leg of a 580-mile march from Ohio to Campbell Soup's Camden, New Jersey, headquarters, a *perigrinación* that helped draw Campbell Soup to the bargaining table. To this day, the UFW and FLOC remain close, working together in Texas and Florida.

"Cesar opened the world of possibility for us. He showed us that we could win in our own struggle," Velásquez says, crediting Chavez's farm-labor law as the inspiration for the Dunlop Commission in Ohio, an independent board headed by former U.S. labor secretary John Dunlop, that oversees FLOC's labor agreements with farmers. "When I started organizing farmworkers, I didn't know anything about the UFW. But then I saw what they were doing and I was elated—there was someone else succeeding at what I was trying to do. The fact that the UFW had succeeded and survived was an important step for us."

Cesar's death had brought everyone within the union's embrace, if just for the day, and if only to grieve. Middle-aged Chicanos arrived from Los Angeles and San Francisco, leaving behind the suits and ties they'd become accustomed to in their professional lives. Many had dusted off the brown berets they had worn on the

■ *Filipino friend paying a final tribute to Chavez at a shrine decorated with carnations shaped into a UFW eagle.*

streets years before when their passions were less inhibited. They brought their own sons and daughters, so they, too, might feel the electricity that had stirred the soul of a generation, inspiring Chicanos to pursue careers in medicine and law and education.

Farmworkers and former farmworkers brought their children to the procession; some had been born on the boycott or on the migrant trail. Many of the sons and daughters of the field had been pushed by their parents and, inspired by Chavez, and against all odds, had made it through college and graduate school.

They were all back now, to honor Chavez's memory and reflect on what had been. Among the farmworkers, Filemon López was singing in his second language—Spanish—and holding up the banner of his own fledgling farmworkers group, made up mostly of Mixtec Indians from Mexico's southern state of Oaxaca. It was Lopez's people—called "the cloud people" for the dramatic misty formations that grace the skies of Oaxaca—who were now occupying the lowest rungs of the farm-labor ladder. As the hours passed and the procession poured onto the Forty Acres, the funeral went beyond mourning. The gathering grew festive. People remembered the good times—and there were plenty of them, too—throughout the years they had devoted to their near-impossible mission. One by one, supporters paid tribute with speeches. Some were holding the memorial program the Chavez family had designed, bearing words Cesar had spoken nearly thirty years before at the beginning of the movement: "I am convinced that the truest act of courage, the strongest act of manliness, is to sacrifice ourselves for others in a totally nonviolent struggle for justice. . . . To be a man is to suffer for others. . . . God help us to be men." When the mass farewell was over, Cesar Chavez was laid to rest in a private ceremony, away from television cameras, surrounded only by his closest aides, Helen, and his children. It was the simple burial he'd wanted.

Paul Chavez tries to sum up his father's life and his legacy with a story. A reporter was interviewing Chavez, probing for the reasons below the surface that motivated him. Paul felt the journalist was mining for veins of dogma or political ideology, and so did Cesar. "My dad says, 'Let me see if I can explain it to you in its simplest sense.' He talked about going to the eighth grade and having to stop going to school to help put food on the table. He said, 'I don't ever want to have another parent have to make the decisions that my parents made. That's why I'm doing this, and why I'm going to keep on doing this. Very basic.'"

Pete Velasco, a Filipino immigrant who met Chavez during the 1965 grape strike, credited his Mexican friend with stepping up fearlessly and confronting giants while others were too afraid. "Cesar was a gift to the farmworkers, to all people, and to me," said Velasco, who died not long after Chavez. "He taught us how to walk in the jungle and not be afraid. He taught us to maintain dignity." With Cesar's passing, "the spirit within every one of us has become renewed, just like the spirit of 1965 has come back to life. And that was a beautiful legacy that we received from our brother Cesar Chavez."

The impact of Cesar's death—its effect on farmworker organizing and the drive to protect the rights of the working poor—was perhaps best summed up by Jose

Padilla, a young lawyer, who was just a child when Chavez joined the Delano strike in 1965. Padilla, director of the California Rural Legal Assistance, was awaiting the start of the funeral march when a farmworker reminded him of a Mexican *dicho:* " 'When a pine tree falls, it falls distributing millions of seeds that will one day create a stronger forest' . . . Cesar was our pine tree," Padilla added, "we are his forest."

In death, it seemed, Cesar belonged to everybody.

■ FOLLOWING PAGE: *Arturo Rodriguez, who is married to Cesar's daughter, Linda, became the new UFW president after Cesar's death.*

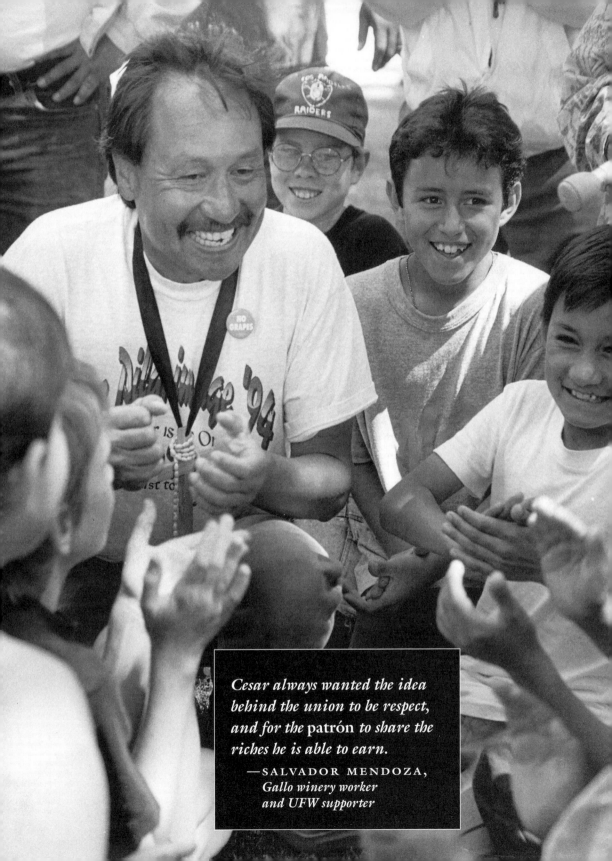

*Cesar always wanted the idea
behind the union to be respect,
and for the patrón to share the
riches he is able to earn.*

—SALVADOR MENDOZA,
*Gallo winery worker
and UFW supporter*

EPILOGUE

IN THE SUMMER of 1993, in the small California farm town of Livingston, Vicente Domínguez pushed a button on a videocassette recorder, filling a television screen with an image of his sister, Lucina. Wearing an old silk turban around her head, an embroidered blouse, and a traditional ankle-length skirt, she rested her spade on a hillside where young corn sprouted through the earth. She peered into the camera and spoke, not in Spanish, but in a Mexican Indian language—Mixtec—a native language spoken with increasing frequency today in the vineyards and vegetable fields of California. From Lucina's village deep in the Mexican state of Oaxaca, she offered greetings and love to her son and her brothers in *el norte*.

With the passing of Cesar Chavez, a new generation of farmworkers inherits the legacy of thirty years of labor reforms shaped by the United Farm Workers. Most of those who now benefit from Chavez's work were not born when his greatest triumphs united farm laborers in *la causa*, and they are strangers to the culture of Chicano farm labor that existed then. An increasing number of farmworkers today are Mexican Indians, like Vicente Domínguez, who have been forced into the migrant stream by extreme poverty. Determined to better their lives in a new country, they are also fiercely proud of their traditions and languages and want to preserve their ties to villages back home. The United Farm Workers union is working with farmworkers like these, attempting to represent their interests at bargaining tables and before government agencies at a time when Mexican immigrant workers are under attack.

Arturo Rodriguez, Cesar Chavez's son-in-law, a union and boycott veteran and now UFW president, has made impressive steps toward involving young farmworkers and Mixtecs in the union. A native of Texas, who himself has the almond-shaped eyes of an Indian, he is the son of a sheet-metal worker and a teacher. Rodriguez was first inspired to get involved in the union as a teenager when he witnessed the 1966 pilgrimage through the Rio Grande Valley. He was studying for a graduate degree in social work at the University

■ *Crystal Gonzalez, with a megaphone, demonstrating against Proposition 187 with other children at the Parlier Middle School in the San Joaquin Valley. The measure proposed forbidding all undocumented children from attending public schools.*

of Michigan in 1973 when he met Linda Chavez, Cesar's daughter, in Detroit. Both were working on the grape boycott. About a year later, they decided to get married, and Rodriguez was startled when Linda told him he would have to ask her father for her hand in marriage. But from that point on, he became part of the Chavez family, learning to organize under the tutelage of Cesar and Richard Chavez and Dolores Huerta, who is still a union officer.

In 1993, Rodriguez signed an agreement with a coalition of Mexican Indian organizations in California, inviting them to join the union and work toward contracts that would protect them. By the mid-1990s, estimates were that Mixtec Indians had become 10 percent of the farm-labor force in California; they often face terrible exploitation because many are undocumented and unaware of their rights. But the Indians have a strong tradition of self-help—the most prosperous here have taken video cameras and fax machines to their villages so they can communicate across borders—and Rodriguez was quick to seize on their capabilities and bring them into the union fold. Mixtec-speaking activists are among the new generation of UFW organizers Rodriguez is cultivating.

In the San Joaquin Valley city of Fresno, farmworker Filemon López, a leader in a Mixtec network called the Benito Juarez Civic Association, agrees. He carried the association's banner at Cesar Chavez's funeral, honoring a man who had blazed the activist trail before him. Among Mixtecs in Fresno, López is a "village leader," a point man for Indian farmworkers who need a place to sleep or directions on how to get to a job. One hot summer night in 1993, a group of women who spoke only limited Spanish showed up at López's house in Fresno, bound for the grape harvest, bringing with them little but the clothes on their backs. Most had been cheated out of wages before. "Some who haven't come here still have the illusion that this is paradise," said López, as the women talked with each other in Mixtec, while heating tortillas and beans and making glasses of powered orange drink for dinner. "Then they get here and work for the contractors. And they exploit them. It's a chain. The reality of the fields here is sad. . . . You learn how to cope once here, but it is not the dream you thought it would be."

Since being named as UFW president following Chavez's death, Rodriguez has proved to be an accomplished leader. He wears cowboy boots, jeans, and union T-shirts—the look of an activist—but businessmen who've gotten to know him laud his pragmatic approach to negotiations. The *Wall Street Journal* has featured him favorably on its front page, and the president of the AFL-CIO calls him one of the most dynamic leaders in labor today. The UFW is now so entrenched in the mainstream of American labor unions that the AFL-CIO voted Rodriguez onto its national labor council—the policy-making group for the nation's largest employee federation.

Before Chavez died, Rodriguez says, he made it clear he wanted to revitalize the UFW's grassroots organizing, which had declined in the 1980s. As Rodriguez takes the UFW toward the twenty-first century, he is making good on that desire. After a round of vigorous and sometimes stealthy campaigns, the UFW won fifteen new contracts by the end of 1997 and brought in five thousand new members. It had the best record for organizing new members of any AFL-CIO union that year.

At least one of those victories in 1995, however, was bittersweet. When four hundred strawberry workers at VCNM Farms in Salinas struck and then voted overwhelmingly for the UFW in August 1995, the company plowed under a quarter of its crop and laid off the workers. California's Agricultural Labor Relations Board found that VCNM had retaliated against union workers by destroying its crop and reducing the workload. The company went out of business a month after the election; the union cried foul again, accusing VCNM of shutting down so its owners could reshuffle assets and open a new business. A legal review of charges like these can drag on for months, if not years—too much time to wait for many mobile strawberry workers, who must rush to the next harvest or make their way back home to Mexico for the winter holidays.

Rodriguez and the UFW learned an important lesson from that experience. Rather than trying to organize workers on hundreds of other strawberry ranches on California's central coast one by one, the union decided to pursue "master contracts" covering all the big shipping companies—known as coolers—for which growers produce their berries. The AFL-CIO was so impressed by Rodriguez, it donated money and summer interns and helped make the UFW's strawberry-organizing campaign the largest union drive in America in 1996. The campaign was aimed at more than fifteen thousand laborers on the central coast who tend and harvest more than half the strawberries grown in the United States. Rodriguez and Dolores Huerta spent days at a time in Watsonville and Salinas during the height of the campaign, running marathon meetings, leading marches, and advising young organizers on how to deal with supervisors who were threatening, even punching, activists when they entered fields at legally prescribed times to talk to employees.

During that summer, as the union's presence escalated, a strawberry "growers and workers alliance" was formed that was managed by the Dolphin Group, the high-powered pro-grower Los Angeles public relations company. The alliance and other anti-union groups produced buttons and stickers and supported a march of thousands of workers shouting antiunion slogans and demanding that the UFW leave them alone. Growers who felt they paid their workers decently—up to seven or eight dollars an hour for the fastest pickers—believed they were being tarred unfairly by the union, and they were fighting back. But the UFW, who charged that many workers had been coerced to attend the antiunion march, sponsored its own parades, including one that drew seven thousand strawberry pickers and supporters into the streets of Watsonville on September 15, near the end of the harvest season.

Huerta, Rodriguez, and Richard Chavez led marchers through Watsonville, shouting "Long live the strawberry pickers!" "Long live Cesar Chavez!" A gray-haired union veteran waved at young workers who stared and smiled from the front porch of their dilapidated house. "The union is our strength, boys," he told them. The observers silently watched the parade go past; there were sturdy *lechugeros* from the desert oasis of Blythe; winery workers from Napa, Sonoma, and Monterey Counties; and tomato pickers who had union contracts south of Salinas.

Rodriguez, buoyed by the AFL-CIO's support, vowed to continue the strawberry campaign by spreading word throughout the country that the strawberry pickers were fighting for contracts and for basic needs—an echo of days past when Cesar Chavez brought the plight of the grape pickers to public consciousness. Some estimates held that in 1995 strawberry pickers earned an average of $8,500 during a six-to-eight-month harvest, hardly enough to pay the rent in an area that is among the most expensive places to live in the country. Many workers said they just can't live on those wages, especially without health insurance. "Those from the union have taught us our rights," said Isabel Rendon, a twenty-nine-year-old Mexican originally from Oaxaca. "We'd like to earn at least seven dollars an hour and to have a medical plan for our children. If I'm sick or if my son is sick, sometimes we have to decide that we can't go to a doctor. How can we? With the *sueldito*—the meager wage—we make? Sometimes we only get checks of two hundred dollars a week, and it just doesn't stretch. We're paying seven hundred dollars in rent, plus the bills."

In between trips to Watsonville in 1996, Rodriguez spent time traveling the rest of California and taking trips to Florida and Alabama, where he stood on highways with signs asking the public to support the rights of mushroom workers to negotiate a contract. The company had tried to fire pro-UFW workers. Rodriguez was overjoyed when he attended a protest in deeply conservative Tallahassee, and people drove by offering thumbs-up and honking horns. "That just blew me away," he said. "That's a foundation [of public support] that took thirty years to build."

On the eve of the Watsonville-Salinas strawberry campaign, Rodriguez bolstered the union's presence in the area by ending a seventeen-year-old feud with one of the country's largest lettuce growers, Bruce Church Incorporated, which also markets fancy prepackaged salads. The company was on the losing end of a long legal battle, for which both sides had spent much money. But more important, a new generation of corporate and union leadership also cleansed bad blood and opened the way to a contract that gave 450 workers an automatic 4 percent pay increase and set minimum pay rates at about seven dollars an hour.

■ *Arturo Rodriguez and Dolores Huerta flank John Sweeney, then president of the Service Employees International Union, during a march through Salinas in 1995. Sweeney, a strong UFW supporter, is now president of the AFL-CIO.*

In a long overdue ceremony that was steeped in symbolism, Dolores Huerta, Arturo Rodriguez, and Steve Taylor sat beneath a portrait of Cesar Chavez and publicly signed the contract between Bruce Church and the UFW. When Chavez began the UFW, the thought of unionization was heretical. Steve Taylor, a University of California at Berkeley graduate in his early forties, has a different attitude, and he believes he can work well with Rodriguez. "I don't know why an agricultural worker should be treated differently from other people," he says. Avoiding strikes is critical, to be sure, he says, because while General Motors can resume making a car easily enough after a delay, a ripe crop will perish if it's not picked on time. "But you have to weigh that," Taylor says, "against the right of people to organize to improve their working conditions."

"It was very good dealing with Steve," Rodriguez says. "He looked at this thing not only in a businesslike fashion, but I would like to think he looked at it as a human being, as well, and had some sense of respect for the dignity of the workforce. . . . You can sit down and talk with folks like that. They look at farmworkers as equals. And a lot of growers just don't."

The strawberry industry, for example, has had its share of sensational cases of abuse. In 1991, more than one hundred pickers, mostly Mixtecs, were found living on an isolated strawberry ranch. They were sleeping in cardboard crates and drinking irrigation water, not far from affluent suburbs north of Salinas and a half hour's drive from the seaside resort of Carmel, one of the wealthiest communities in the United States. After a community social worker gained their trust, the workers revealed that a Chicano sharecropper had invited them to live on the edge of the ranch and pick berries for him to sell to a processing plant. An army of social workers descended on the workers' encampment, and some of the local growers—embarrassed by the bad press—sent over blankets and food. Some of the migrants were as young as thirteen, and none were being paid minimum wage.

Some of the fruit they picked was shipped just across town, to a J. M. Smucker processing plant in Salinas. The giant company professed ignorance and shock at the conditions on the ranch, where Smucker boxes were doubling as berry containers and material for building shelters. Considering how far removed the company is from life in the fields, the claim was plausible. But Rodriguez says distance is no excuse for big companies, especially the coolers—the shipping companies that the UFW is targeting in its strawberry campaign. Even if they don't hire workers directly, the companies that ship the berries sometimes supply investment money to unscrupulous growers, tell them what pesticides to use, and dictate the price and quality of the fruit. "They've developed a nice system for not being responsible," he says.

■ *Mixtec Indians from Mexico began coming to California to work as farmworkers in greater numbers in the 1980s, forming a new class of immigrant laborers.*

Those who have it worst are the undocumented farmworkers. No one knows how many farmworkers are illegal immigrants, but estimates were the figure was no higher than 30 percent in the mid-1990s. In 1986, Congress passed the Immigration Reform and Control Act, making it, for the first time, a federal offense to hire illegal immigrants. Agribusiness leaders had seen the law coming. Alarmed at the prospect of losing cheap labor, they pressured ranking California congressman Leon Panetta, who later became President Clinton's chief of staff, and then-senator Pete Wilson, a Republican whose top campaign donors were growers, to help shape special provisions for farmers. Wilson, who later campaigned against public education for illegal immigrants in his second race for governor in 1994, vigorously defended the use of undocumented farmworkers in the decade before. "There's no question our economy depends very heavily on Mexican nationals," he said in 1982. "I deplore the INS raids on farms here in the roundup of illegal aliens," Wilson said in Orange County that same year.

Panetta, who represented the agricultural Salinas Valley in Congress at the time, worked with Wilson to get a provision included in the new immigration law that would give farmers the right to import temporary workers in the event of labor shortages. Growers also were exclusive beneficiaries of a special amnesty program that granted legal residency to their undocumented farmworkers. And they were among the chief lobbyists for provisions that eased the burden on employers for determining the legitimacy of *micas* and other documents that can be falsified.

Some of the farmhands legalized during the amnesty program were still working in California's fields a decade later. But harvesting and hoeing are jobs for young spines, and ten years later a new generation, deeply hurt by another round of economic disaster in Mexico, continued flowing north to seek farmwork, despite official claims to the contrary.

During Wilson's 1994 campaign, he held a press conference on the edge of a broccoli field, confidently declaring that California's farmers were now employing a legal workforce. After Wilson left, the workers cutting the broccoli were amused when the governor's words were translated into Spanish. More than one harvester in the crew, which had served as a backdrop for Wilson, admitted he was undocumented. In the spring of 1996, just as the UFW was starting its campaign to organize strawberry workers, three undocumented brothers, Jaime, Salvador, and Benjamin Chavez-Muñoz, left the village of Cherán in the Mexican state of Michoacán, bound for the strawberry fields

■ *Steve Taylor, of Bruce Church Inc., signed a historic contract with the UFW on May 30, 1996, in Salinas. Co-signing the agreement at the Cesar Chavez Library were Arturo Rodriguez and Dolores Huerta.*

■ *Young immigrants, one from Mexico, the other from Guatemala, pick wine grapes in the Napa Valley, California's celebrated "Wine Country."*

of Watsonville. The brothers paid $1,200 to a coyote to smuggle them north, where they had hoped to work at jobs that few American workers had wanted for decades. They never made it to Watsonville; all three were killed when the truck they were riding in crashed not far from the border, with the U.S. Border Patrol in hot pursuit.

When California voters passed Proposition 187 in 1994, California's growers were paralyzed by the contradictions of the proposal that would have kicked all the children of illegal immigrants out of California's public schools. "I think it's a crock," said Ed Angstadt, the director of the Salinas Valley Grower-Shipper Association. "The workforce that's being targeted is our workforce. And we'd be crazy to come out against our workforce." Yet not one growers' association took a stand against Proposition 187. In the end, voters in every agricultural region of California voted for Proposition 187, in some cases by more than 70 percent. A federal court judge declared the law unconstitutional, but supporters hope for a review by the U.S. Supreme Court.

Arturo Rodriguez shakes his head at the subject of illegal immigrants. "The sad thing for us is that that workforce is really exploited by the industry. They purposely, intentionally, clearly know that they can bring in a workforce that they can exploit and pit against the local worker," he says. "But we make no distinction. I couldn't care less. We don't ever ask, we don't ever require that of individuals. We're not the immigration [service] nor will we ever be put in that position. That's clearly not within our framework."

In California's wine country north of San Francisco, undocumented migrants occupy the lowest rungs of farm work, performing the least-paid and least-regulated jobs that sustain the area's vineyards and boutique wineries. Tourists sip wine in chic tasting rooms and bicycle across the picturesque landscape, unaware that Sonoma County, like Watsonville, has become a battlefront for new UFW organizing. Union supporters at the E & J Gallo Winery in the town of Healdsburg are in the thick of it. In July 1994, eighty-one Gallo workers voted for the UFW to cover them in collective bargaining. Only twenty-one voted against the union; most were Anglos. More than two years later, the workers still have no contract with the Modesto-based Gallo, the world's biggest wine company and one of the biggest political donors in American politics.

The habitually antiunion Gallo fought the election so hard with various challenges, it took a year to get the UFW election certified as valid by California's Agricultural Labor Relations Board. Then the company filed suit to reverse the decision. Gallo lost, much to the satisfaction of Salvador Mendoza, a veteran farmworker in his forties, who worked in wineries under union contract in neighboring Napa County. "They brought in an army of lawyers," he says in Spanish, "and they still lost."

Mendoza was the principal organizer of the Gallo workers, whose chief complaint was abuse by supervisors. Little by little, he won the quiet confidence of younger workers who were scared to take on the system. The company would demote anyone to a lower-paying job if supervisors felt like it, work hours were erratic, and pay scales fluctuated. When he asked the workers if they wanted to unionize, they told him: *"Tal vezcito,"* a very meek version of the Spanish word for "maybe." But after months of almost clandestine meetings at workers' homes, after patiently explaining their rights under the law, Mendoza led the workers in a "mini-strike" of one day as a show of protest. The company was stunned.

"It was a big surprise for the company. And we did it all legally, with no violence at all," says a proud Serafín Pérez, a twenty-eight-year-old Mexican immigrant, who was standing on a barren patch of soil in a low-income apartment complex in Healdsburg, north of Santa Rosa. Within months, that soil, behind an ornate wrought-iron gate, would become the Cesar E. Chavez Garden, planted by a group of Gallo workers who live in the complex.

"It just goes to show, that you can be a millionaire, but with facts and truth, you can fall," says Pérez, who learned the early history of the UFW from union films, books, and from old-timers. Rogelio Pérez, another young worker, says he noticed after the workers took their vote, Gallo supervisors began to treat them with more respect. The mere fact that the workers have not been fired for voting pro-union has emboldened other farmworkers in the area, Rogelio Pérez says. "We've kept our jobs, and we have more dignity now."

The Gallo workers complain that labor contractors exploit migrants more vulnerable than themselves. Some who are brought into Sonoma as extra hands during peak harvests are jammed into run-down motel rooms, forced to pay for rides to the fields, and not allowed to talk to union organizers. "If God gives us license, and we can win our fight, we can go out try to talk to these people," says Mendoza, who learned the virtue of patience from Chavez. He keeps a photo of his hero displayed prominently in his living room. "Cesar," he adds, "always wanted the idea behind the union to be respect, and for the *patrón* to share the riches he is able to earn."

INTRODUCTION

xviii "Come on, men": Int. and trip to ranch with Nelly Pruneda and Lauro Barajas, July 2, 1996.

3 "basically a scoundrel": *Los Angeles Times,* July 23, 1995.

3 "refuse to accept our heritage": *Los Angeles Times,* July 23, 1995.

4 "Cesar cared a lot": Int. Mary Magaña, July 3, 1996.

4 "factories in the field": McWilliams, *Factories in the Field*, title.

4 Farmworkers . . . excluded from the National Labor Relations Act: Meister and Loftis, *A Long Time Coming*, p. xi.

5 "farmworker is an outsider": Stan Steiner, *La Raza: The Mexican Americans,* p. 250.

5 earliest days of California's agricultural economy: *Fight in the Fields,* documentary film.

5 "in the hands of people": Tómas Almaguer, *Racial Fault Lines: The Historical Origins of White Supremacy in California,* p. 193.

6 Today California agribusiness, which at $22 billion a year: *California Agricultural Directory,* 1996–1997.

6 "little brown guys": Miriam J. Wells, *Strawberry Fields: Politics, Class, and Work in California Agriculture,* p. 165.

6 Golden State produces: *California Agricultural Directory,* 1996–1997.

6 federal subsidies: *Los Angeles Times,* June 25, 1995.

7 90 percent are thought to be foreign-born: *San Francisco Examiner, Image* magazine, July 18, 1993, p. 9.

7 "can't be bought": Int. Delfina Corcoles, June 22, 1996.

7 "united with the union": Int. Jesus Lopez Avalos, Sept. 15, 1996.

9 "People don't realize it": Int. Arturo Rodriguez, Sept. 15, 1996.

ONE: THE LAST FAMILY FARM

10 "We had never worked for anybody else": Int. Rita Chavez Medina, Jan. 23, 1996.

11 "My grandfather came": KMEX-TV footage, compilation tape, April–June 1973.

12 Cesario Chavez . . . born a peon: Jacques Levy. *Cesar Chavez: Autobiography of La Causa,* p. 7.

12 Dorotea "Mama Tella" learned to write . . . in Mexican convent: Ibid., p. 26

12 thumping her cane to demand attention: Ibid., p. 26.

12 an amazed priest in Yuma: Ibid., pp. 26–27.

13 Juana was also the original inspiration of her son's . . . nonviolence: Levy. *Cesar Chavez,* pp. 17–19. Also, Griswold del Castillo and Garcia: *Cesar Chavez,* p. 5.

13 brandishing a shotgun at a bullying older cousin: Levy, *Cesar Chavez,* p. 19.

13 "didn't know the gun was unloaded": Ibid., p. 19.

14 didn't spare hugs or kisses: Ibid., p. 18.

14 fond of the barbecues: Ibid., pp. 32–33.

14 were always very close: Int. Rita Chavez Medina, Jan. 23, 1996; Levy, *Cesar Chavez,* pp. 20–22.

14 sometimes wearied of being a second mother: Int. Rita Chavez Medina, Jan. 23, 1996.

14 "you feed those two horses and I'll feed this cow": Ibid.

14 refused to sit anywhere except with his beloved sister: Ibid.

15 Cesar and Richard were also inseparable: Int. Richard Chavez, June 8, 1995. Chavez's early nickname was "Manzi,"

short for the manzanilla tea he liked to drink; and Richard's was "Rookie."

15 "We didn't suffer the way people in the city . . . suffered": Levy, *Cesar Chavez,* p. 14. Cesar's grandparents' and parents' farms together totaled 160 acres.

15 Juana Chavez held to the teachings of her patron Saint Eduvigis: Griswold del Castillo and Garcia, *Cesar Chavez,* p. 5.

15–16 "We wanted this one bad. . . . were very respectful": Peter Matthiessen, *Sal Si Puedes,* pp. 219–20.

16 debt of $3,600 was "just nothing. . . . loan was blocked": Levy, unpublished interview with Chavez, 1969, p. 54.

16 "Suddenly two cars bore down on us": Levy, *Cesar Chavez,* pp. 36–37.

17 " 'If he's Mexican, don't trust him' ": Ibid., p. 37.

17 wild mustard greens: London and Anderson, *So Shall Ye Reap,* p. 142.

17 Juana Chavez sold crocheting in the street: John Gregory Dunne, *Delano,* p. 5.

17 "He had the papers that told us we had to leave": Levy, unpublished interview with Chavez, 1969, p. 54; Taylor, *Chavez,* p. 59.

17 "Just a monstrous thing": Levy, *Cesar Chavez,* p. 54.

17 gradually stopped talking about it: Ibid., p. 54.

18 "Maybe that is when the rebellion started": Taylor, *Chavez,* p. 61.

18 Chavez property . . . fell into the hands of the company that would later become Bruce Church: *Los Angeles Times,* April 24, 1993, p. A-23.

18 "If we'd stayed there, possibly I would have been a grower": Levy, *Cesar Chavez,* p. 42.

18 "Some had been born into the migrant stream . . . has no freedom": Taylor, *Chavez,* p. 61.

18 "like a wild duck with its wings clipped": Levy, *Cesar Chavez,* p. 38.

18 "I bitterly missed the ranch": Taylor, *Chavez,* p. 61.

18 suspected that his father was suffering more: Levy, *Cesar Chavez,* p. 48.

19 "He would notice how fertile it was": Ibid., p. 48.

19 "We had never worked for anybody else": Int. Rita Chavez Medina, Jan. 23, 1996.

19 "unemployed people . . . like flocks of starlings": Levy, *Cesar Chavez,* p. 48.

19 "came back with not one, but two": Int. Rita Chavez Medina, Jan. 23, 1996.

20 "just like being nailed to a cross": Griswold del Castillo and Garcia, *Cesar Chavez,* p. 12; Levy, *Cesar Chavez,* p. 74.

20 "Anything you can do is fine with me": Levy, *Cesar Chavez,* p. 71.

21 cheated by unscrupulous labor contractors: de Ruiz and Larios, *La Causa,* pp. 4–10; Matthiessen, *Sal Si Puedes,* pp. 224–25.

21 scoured the roads for tinfoil: Matthiessen, *Sal Si Puedes,* p. 227; Dunne, *Delano,* p. 6.

21 harshest work the family encountered: Int. Richard Chavez, June 8, 1995.

21 a daily pass to . . . *The Lone Ranger:* Levy, *Cesar Chavez,* p. 59.

21 absentee owners . . . distance themselves from starvation wages: John Steinbeck, *The Harvest Gypsies,* pp. 36–37.

21 "when we've picked their crop, we're bums": Ibid., pp. 23–24.

21 "when we joined the big league": Int. Richard Chavez, June 8, 1995.

21 "my father never did put up with it": Ibid.

21 "the strikingest family in California": Levy, *Cesar Chavez,* p. 78.

23 "the first ones to leave the fields": Levy, *Cesar Chavez,* p. 78.

23 not uncommon for migrant children to die of malnutrition: Steinbeck, *The Harvest Gypsies,* pp. 28, 50.

23 on assignment from the *San Francisco News:* Ibid., pp. vi–vii. The Steinbeck series provided the background for his 1939 novel *The Grapes of Wrath,* which caused an uproar among California growers. In *The Other California,* Gerald Haslam describes the publication of Carey McWilliams's *Factories in the Field* later the same year: "Conservative outrage exploded. There had to be a conspiracy. . . . In a futile attempt to counter Steinbeck, pro-grower books, films, and pamphlets were rushed into

print, including *Of Human Kindness, Grapes of Gladness, Plums of Plenty,* and other forgettable epics."

23 "He will die in a very short time": Steinbeck, *The Harvest Gypsies,* p. 30.

23 in a single day in 1937, more than three thousand of them: Dunne, *Delano,* p. 45.

23 dozens of strikes in California during depression: McWilliams, *Factories in the Field,* pp. 211–63.

24 Many were led by Communist Party–sponsored Cannery and Agricultural Workers Industrial Union: Taylor, *Chavez,* pp. 48–58.

24 one of the few labor organizations willing to represent . . . field workers: T. H. Watkins, *The Great Depression,* p. 204: Watkins summarizes the American Federation of Labor's attitude toward migrants by quoting a California AFL leader's declaration that "only fanatics are willing to live in shacks or tents and get their heads broken in the interest of migratory labor." Kevin Starr also writes in *Endangered Dreams: The Great Depression in California*: "Today, when Communism has lost its power to frighten and enrage vast sectors of the American public, it is easier to focus upon the skills and human qualities of the [courageous] CAWIU organizers."

24 terrorized and beat organizers, ran them out of town, and even killed them: McWilliams, *Factories in the Field,* pp. 211–63.

24 Title of Carey McWilliams's 1939 book on migrant labor and the abuses of corporate agriculture in California. After its publication, Associated Farmers called McWilliams "Agricultural Pest No.1 in California, outranking pear blight and boll weevil."

24 Two U.S. congressional investigations: Starr, *Endangered Dreams,* pp. 267–69; McWilliams, *Factories in the Fields,* pp. 230–82.

24 reprisals . . . during Corcoran cotton strike . . . Pixley: Taylor, *Chavez,* pp. 49–58. Discussing the Pixley ambush,

Taylor quotes one of the strike leaders, W. D. Hemmett, who met up with the first armed farmer to reach the union hall: "I told him not to shoot into the hall, into the women and children in there. He shoved the rifle toward me and I grabbed the barrel and shoved it toward the ground. Just then a Mexican fella grabbed the rifle. . . . I heard a shot and I heard Delfino Davila shout: 'Bill, run, they've got me.' I look around in time to see him fall. . . ." Delfino Davila, fifty-eight, was a part-time representative of the Mexican consulate in Visalia.

25 Chambers . . . imprisoned: Starr, *Endangered Dreams,* pp. 170–73; Taylor, *Chavez,* pp. 49–57.

25 "I didn't want to get you in trouble": Int. Marc Grossman, Oct. 10, 1996.

25 "his method is the threat of the deputies' guns": Steinbeck, *The Harvest Gypsies,* p. 35.

25 "No one complains at the necessity of feeding a horse while he is not working": Steinbeck, as quoted by Matthiessen in *Sal Si Puedes,* pp. 225–26.

25 farm lobby helped bankroll opposition to . . . Upton Sinclair's run: Advertisement by the California League Against Sinclairism, Internet: http://www.SFmuseum org/hist1/sinclair.html.

26 thousands of Mexican immigrants . . . forcibly deported: Guerin-Gonzales, *Mexican Workers and American Dreams,* pp. 80–94.

26 "We protect our farmers": Department of Labor Bulletin No. 836 (1945).

26 "herd them like pigs": Ibid.

26 "talking either about an election or strike": Levy, unpublished interview with Chavez, 1969, p. 93.

26 enjoy two 10-minute breaks: Levy, *Cesar Chavez,* p. 80.

26 rapped hard on knuckles and scolded for speaking Spanish: Levy, *Cesar Chavez,* p. 24. In an interview with Levy, Chavez recalled: "When we spoke Spanish, the teacher swooped down on us. I can re-

member the ruler whistling through the air as its edge came down sharply on my knuckles."

26 principal would grab the Mexican children: Ibid., p. 24.

26 "I am a clown; I speak Spanish": Taylor, *Chavez*, p. 64.

26 "Some teachers were really cruel": Levy, *Cesar Chavez*, p. 24.

27 "said something about 'dirty Mexicans' ": Ibid., p. 29.

28 denied the vote to black and Indian residents: *Report [on] the Formation of the State Constitution in September and October, 1849* (Washington, D.C.: John T. Towers, 1850), cited by Almagauer, *Racial Fault Lines*, p. 55.

28 enrolled in at least thirty-six schools: Levy, *Cesar Chavez*, p. 65.

28 dozens of masters' theses promoted the view that Mexican students . . . inferior: McWilliams, *North from Mexico*, pp. 206–7. Field interviews by labor economist Paul Taylor researchers further documented the prevalence of these damaging stereotypes: In one such interview, Edward Tilton, assistant principal of schools in San Diego, declared: "[The Mexican student] is inferior, an inferior race, no doubt. . . . The Mexicans are slow to learn." (Guerin-Gonzales, *Mexican Workers & American Dreams*, pp. 69–70.)

28 "like monkeys in a cage": Levy, *Cesar Chavez*, p. 65.

28 "Well, we've got these kids": Ibid., pp. 65–66.

29 single-handedly steered some young strikers into higher education: Int. Clementina Carmona Olloqui, July 6, 1996.

29 segregated . . . just as in the Deep South: Almaguer, *Racial Fault Lines*, pp. 133, 163; Guerin-Gonzales, *Mexican Workers, American Dreams*, p. 67; Gómez-Quinoñes, *Chicano Politics*, pp. 53, 98.

30 "That laugh . . . seemed to cut us out of the human race": Matthiessen, *Sal Si Puedes*, pp. 223–24.

30 Chavez and his family thrown out of a diner: Ibid., p. 224.

30 expression on his father's face: Matthiessen, Peter, *The New Yorker*, May 17, 1993, p. 82; Taylor, *Chavez*, p. 62. "I still feel the prejudice; whenever I go through a door, I expect to be rejected, even when I know there is no prejudice there," Chavez told Taylor.

30 "nothing to eat": Int. Jessie De La Cruz, July 25, 1995.

30 "wouldn't even talk to us": Ibid. In *The Other California*, Gerald Haslam notes that Okies faced the discrimination and prejudice similar to that encountered by blacks and Mexicans; a sign on a Bakersfield movie theater in 1939 said "Negroes and Okies upstairs," and Associated Farmers helped lead an anti-Okie campaign that charged them with immorality, sexual aberrations, and low intelligence.

31 "I can remember the biggest impression": Int. Dorothy Healy, March 27, 1996.

31 Chavez became enamored of . . . *pachuco* look: Levy, *Cesar Chavez*, p. 81.

31 "little old ladies were afraid of us": Ibid., p. 82.

31 "didn't want to be a square": Ibid., p. 82.

32 strict miscegenation laws: Carlos Bulosan, *America Is in the Heart*, pp. 143–44.

32 "She loved to dance": Int. Richard Chavez, June 8, 1995.

32 "had flowers in her hair all the time": Levy, unpublished interview with Chavez, 1969, p. 89.

33 "He was just like an ordinary": Int. Johnny Serda, July 3, 1996.

33 Eddie Rodriguez . . . visiting back home: Int. Butter Torres Gonzalez, July 3, 1996.

33 "sugar-beet thinning . . . backbreaking job": Levy, *Cesar Chavez*, p. 84.

33 worst two years of his life: Levy, *Cesar Chavez*, p. 84.

33 "saw that others suffered, too": Griswold del Castillo and Garcia, *Cesar Chavez*, p. 19.

33 "got serious about each other": Int. Richard Chavez, June 8, 1995.

34 "stepped right into thick, black clay": Levy, *Cesar Chavez*, p. 87.

34 offered jobs as sharecroppers: Ibid., p. 88.

34 "had to chop wood all the time": Taylor, *Chavez*, p. 82.

35 "hadn't planned anything . . . wanted a free choice of where I wanted to be": Levy, unpublished interview, 1969, p. 86. Taylor, Matthiessen, and other biographers also recount this story.

35 "well, he's not drunk": Ibid., p. 87.

35 "tried to scare me": Ibid,. pp. 87–88.

TWO: SAL SI PUEDES

36 "If you talked about civil rights . . . people to vote": Levy, unpublished interview with Chavez, 1969, pp. 511, 616.

37 trying vainly for days: Int. Fred Ross Jr., Aug. 28, 1995; Ross, *Conquering Goliath*, p. 1.

37 roads were unpaved: Int. Ernie Abeytia. Aug. 27, 1996; and Librado "Lenny" Chavez, Aug. 28, 1996.

38 heard about Helen and Cesar Chavez from . . . nurse and . . . priest: Int. Librado "Lenny" Chavez, Aug. 28, 1996.

38 hide out at his brother Richard's: Int. Richard Chavez, June 8, 1996.

38 gringo seemed friendly enough: Levy, *Cesar Chavez*, p. 97.

39 "Somehow I knew that that gringo had impressed me": p. 98.

39 "real tough-like": Int. Lenny "Librado" Chavez, Aug. 28, 1996.

39 "chiseled forever somewhere deep in my retina": Ross, *Conquering Goliath*, pp. 1–3.

39 pronounce the little Spanish: Matthiessen, *Sal Si Puedes*, p. 44.

39 creek that overflowed: Int. Lenny "Librado" Chavez, Aug. 28, 1996.

39 "He took on the politicians": Levy, *Cesar Chavez*, p. 98.

39 "never heard anything from whites unless it was the police": Matthiessen, *Sal Si Puedes*, p. 43.

39 "I told Cesar and his buddies": Ross, *Conquering Goliath*, p. 3.

42 The CSO was formed: Gómez Quiñones, *Chicano Politics*, pp. 53–54.

42 "Never before in the whole history of Los Angeles": Ross, *Conquering Goliath*, p. 3.

42 "Come on, man, the signal": Int. Librado "Lenny" Chavez, Aug. 28, 1996.

43 "I didn't know what CSO was . . . nothing complicated about it": Levy, *Cesar Chavez*, pp. 98–99.

43 "I'd never been in a group before": Matthiessen, *Sal Si Puedes*, pp. 46–47.

43 Cesar found himself volunteering: Levy, *Cesar Chavez*, pp. 101–3.

45 Cesar probably needed to learn more: Int. Fred Ross Jr., Aug. 28, 1995.

45 "For whatever reasons": Levy, *Cesar Chavez*, p. 102.

45 "a rather anemic handshake": Int. Herman Gallegos, Aug. 28, 1995.

45 "So here I am in charge . . . except when they got sent up someplace": Matthiessen, *Sal Si Puedes*, p. 48.

45 Ross would become Chavez's mentor: Int. Fred Ross Jr., Aug. 28, 1995.

46 "What about the farmworkers?": Matthiessen, *Sal Si Puedes*, p. 47.

46 "I think I've found the guy": Dunne, *Delano*, p. 67.

46 congressional hearings in Bakersfield: Taylor, *Chavez*, pp. 71–75.

46 "mission band," Father Donald McDonnell: Int. Librado "Lenny" Chavez, Aug. 28, 1996; Taylor, *Chavez*, p. 81.

46 Chavez asked to accompany him: Int. Librado "Lenny" Chavez, Aug. 28, 1996.

46 "He told me about social justice": Taylor, *Chavez*, p. 81.

47 most influential books: Levy, *Cesar Chavez*, pp. 91, 108; *The Fight in the Fields* documentary film.

47 "I knew very little about him": Levy, *Cesar Chavez*, pp. 91–92.

47 "I guess we were ripe for it": Int. Richard Chavez, June 8, 1995.

47 "observing the things Fred did": Taylor, *Chavez*, p. 84.

47 hire Chavez at thirty-five dollars a week: Matthiessen, *Sal Si Puedes,* p. 48.

48 Cesar's initial confrontation: Levy, *Cesar Chavez,* pp. 104–5.

48 "agents started asking me a lot of questions": Ibid., p. 106.

49 "Republicans were told they couldn't intimidate people at polls": Ibid., p. 106.

49 "damned naive": Levy, unpublished interview with Chavez, 1969, p. 512.

49 "Another FBI agent came to investigate me at home": Levy, unpublished interview with Chavez, 1969, p. 616.

49 "The Chicanos wouldn't talk to me": Levy, *Cesar Chavez,* p. 107.

49 help immigrants to become U.S. citizens: Int. Beatriz Bedoya, Aug. 28, 1996; Levy, *Cesar Chavez,* p. 110.

50 Chavez collected food donations: Int. Ernie Abeytia, Aug. 27, 1996.

50 perceived as charity: Levy, *Cesar Chavez,* p. 111.

50 suggested opening a small office: Ibid., p. 111.

50 experience working on unfamiliar turf: Int. Beatriz Bedoya, Aug. 28, 1996; Levy, *Cesar Chavez,* p. 113.

50 "Well, I'm the organizer": Levy, *Cesar Chavez,* p. 113.

51 "After the meeting, the first thing I did was sprint to the phone": Ibid., p. 114.

51 a priest declared publicly: Matthiessen, *Sal Si Puedes,* p. 48; Levy, *Cesar Chavez,* p. 116.

51 "Everybody went running for the woods": Levy, *Cesar Chavez,* p. 118.

52 "doing a swell job": Ross letter to Chavez, Sept. 15, 1954.

52 "registrants don't get flushed down the drain . . . hate to keep moving you around": Ross letter to Chavez, Sept. 15, 1954.

52 CSO members sided with him: Levy, *Cesar Chavez,* pp. 123–24.

52 "Some of them thought the CSO was a fly-by-night deal . . . and tequila": Chavez letter to Ross, Dec. 22, 1954.

52 challenging new project in Oxnard: Ross, *Conquering Goliath,* pp. 4–5.

53 "I was deathly afraid I would take that money": Ibid., p. 6.

55 took their seven kids to a beach: Levy, *Cesar Chavez,* p. 157.

55 what irked people most in Oxnard's barrio: Chavez letters to Ross, Sept. 25, 1958; Oct. 10, 1958; Oct. 24, 1958.

55 the system was corrupt: Ernesto Galarza, *Merchants of Labor,* is an exhaustive investigation of abuses in bracero system.

55 "The grower has little expense of recruitment": Ross, *Conquering Goliath,* pp. 10–11.

55 "Helen was hurt, but she took it pretty well . . . Even if you stay with us, you're not really going to be here": Ibid., p. 12.

56 Mexican Americans who were seeking help: Ibid., p. 22.

56 "We have a long talk": Ross, *Conquering Goliath,* p. 22.

56 "Can't get over how these old-timers": Chavez letter to Ross, Nov. 4, 1958.

56 "there's an old *dicho* . . . conditions local people wouldn't tolerate": Levy, *Cesar Chavez,* p. 129.

56 tripling the turnout of voters: Ibid., p. 21.

56 Al Rojas . . . recalls meeting the thirty-two-year-old Chavez: Int. Al Rojas, June 30, 1996.

58 crucial piece to the puzzle: Ross, *Conquering Goliath,* pp. 27–30.

58 accompany him to the farm placement office: Ibid., pp. 36–37.

58 each day, locals forced to fill out new job applications: Ibid., pp. 38–39.

58 managed to interest . . . federal labor department investigators: Levy, *Cesar Chavez,* p. 135.

59 "Surely, it is your right": Ross, *Conquering Goliath,* p. 82.

59 "He can keep his damn tomatoes . . . We're through": Ibid., p. 122.

59 "Get all the publicity you can": Ibid., p. 122.

59 "As soon as it happens, everyone—even the growers—get real quiet": Ibid., p. 128.

60 firing the regional director of the Farm Placement Service: Levy, *Cesar Chavez,* p. 142.

60 "We could have built a union there, but the CSO wouldn't approve": Ibid., p. 143.

60 "This has been a wonderful experience . . . never dreamed that so much hell could be raised": Chavez letter to Ross, May 7, 1959.

60 blossomed into twenty-two chapters: Gómez Quiñones, *Chicano Politics*, p. 55.

60 "I mean, we saw him for about five seconds": Int. Dolores Huerta, June 8, 1995.

61 "You were not supposed to talk": Ibid.

61 "I was raised with two brothers": Griswold del Castillo and Garcia, *Cesar Chavez*, p. 64.

61 "seemed uncomfortable that Chicano, black, white, and Filipino kids were mixing": Int. Dolores Huerta, June 8, 1995.

61 "rich kids always got special treatment": Ibid.

61 Huerta wondered if he was a Communist: Griswold del Castillo and Garcia, *Cesar Chavez*, p. 65.

61 "This was, of course, something I had been looking for all my life": Ibid., p. 64.

62 it was turned down: Ibid.; Levy, *Cesar Chavez*, pp. 146–47.

62 "I resign": Levy, *Cesar Chavez*, p. 147.

62 "It took me six months to get over leaving CSO": Ibid., p. 157

62 "she chose Delano because she had family there": Int. Linda Chavez Rodriguez, June 8, 1995.

63 "What I didn't know was that we would go through hell": Levy, *Cesar Chavez*, p. 148.

THREE: DELANO

64 "Anyone who comes in with the idea": Dunne, *Delano*, p. 71.

65 "For the first time": Levy, unpublished interview with Chavez, 1969, p. 19.

65 1223 Kensington Street: Chavez letters to Ross, April 26 and Aug. 17, 1962.

65 *"el movimiento"*: Chavez letter to Ross, Aug. 7, 1962.

65 avoid calling their association a union: Chavez letter to Ross, May 16, 1962; Taylor, *Chavez*, p. 107.

66 "They come and they go": Chavez letter to Ross, July 3, 1962.

66 San Joaquin workforce was a mix: Chavez letter to Ross, April 26, 1962.

67 "marvelous secret": Levy, *Cesar Chavez*, p. xxi.

67 "Everyone but everyone": Chavez letter to Ross, May 10, 1962.

67 "heck of a lot more": Chavez letter to Ross, May 10, 1962.

67 1960 survey: Dunne, *Delano*, p. 62.

67 map of the valley: Dunne, *Delano*, p. 72; Int. Gilbert Padilla, July 25, 1995.

67 Julio Hernandez . . . remembers: Int. Julio Hernandez, Sept. 6, 1996.

68 "We found out that the harder a guy is to convince": Dunne, *Delano*, p. 70.

68 Angie Hernandez Herrera . . . was another early recruit: Int. Hernandez Herrera, Sept. 6, 1996.

68 Migrant Ministry . . . early supporters: Levy, *Cesar Chavez*, p. 162.

69 Drake had to think the man was crazy: Ibid., p. 162.

69 "He was organizing us": Int. Chris Hartmire, July 1, 1996.

69 Huerta carved out time and Padilla started organizing: Int. Dolores Huerta, June 8, 1995; Int. Padilla, July 25, 1995.

69 "I left Stockton at 2:30 A.M.": Chavez letter to Ross, May 22, 1962.

69 treating them to ice cream: Int. Linda Chavez Rodriguez, June 8, 1995.

69 "We are now really in the swing": Chavez letter to Ross, May 16, 1962.

70 "The local office-supply store": Chavez letter to Ross, June 19, 1962.

70 "When I missed the fourth paycheck": Levy, unpublished interview with Chavez, Feb. 3, 1969, p. 19.

70 contacts tended to drop out: Chavez letter to Ross, July 11, 1962.

70 "They are working every day": Chavez letter to Ross, July 11, 1962.

70 "Anyone who comes in": Dunne, *Delano*, p. 71.

70 conflict with AWOC and Dolores Huerta participation in founding of AWOC:

Chavez letters to Ross in 1962; Int. Dolores Huerta, June 8, 1995; Profile of Huerta in Rose, "Women in the United Farm Workers," pp. 35–38.

70 "no place for a woman": Rose, "Women in the United Farm Workers," p. 36.

71 "very threatened and jealous": Int. Dolores Huerta, June 8, 1995.

71 office in his brother Richard's garage: Int. Richard Chavez, June 8, 1995; Levy, *Cesar Chavez*, p. 162.

71 "I went down and got my own boys": Chavez letter to Ross, May 22, 1962.

71 "that Mexican": Meister and Loftis, *A Long Time Coming*, p. 130.

72 "I told this guy Goldberger": Chavez letter to Ross, Aug. 28, 1962.

72 went in search of a hall: Chavez letter to Ross, Aug. 28, 1962; Levy, *Cesar Chavez*, pp. 173–74.

72 official stationery: Chavez letter to Ross, Sept. 18, 1962.

72 design of the union flag: Int. Richard Chavez, June 8, 1995; Int. Andy Zermeño, Sept. 24, 1996.

73 plan of action: Levy, *Cesar Chavez*, p. 176.

73 unveil the association's banner: Levy, *Cesar Chavez*, p. 170.

73 "When that damn eagle flies": Ibid., p. 175.

73 *"Viva La Causa"*: Ibid., p. 175.

74 Rosa Gloria had written a *corrido:* Ibid., p. 170.

74 "Who finances the organization": undated union leaflet.

74 "I told him this might be many, many months": Huerta letter to Chavez, Oct. 3, 1962.

74 Chavez told Ross: Chavez letter to Ross, Oct. 18, 1962.

74 "chasing the rat": Chavez letter to Ross, March 11, 1963.

74 Getting the credit union started: Int. Richard Chavez, June 8, 1996.

75 loaning out more than $5.5 million: Rose, "Women in the United Farm Workers," p. 147.

75 "Your mother and your father": Levy, *Cesar Chavez*, p. 172.

75 "He came home and gave Mama": Chavez letter to Ross, July 3, 1962.

75 "people who give you their food": Matthiessen, *Sal Si Puedes,* p. 57.

75 "met this fairly young Negro": Chavez letter to Ross, May 16, 1962.

75 "Both Helen and I can't quite accept": Chavez letter to Ross, May 16, 1962.

76 "Found out that I can still pull the vines": Chavez letter to Ross, June 1963.

76 "Oh, boy, what a lot of food": Levy, *Cesar Chavez*, p. 168.

76 "I don't know if she cried": Dunne, *Delano*, p. 73.

76 Dolores Huerta . . . mother of seven: Int. Dolores Huerta, June 8, 1995; Rose, "Women in the United Farm Workers," p. 31.

76 "experienced lobbyist": Huerta letter to Chavez, undated.

77 "I had been working as a teacher": Int. Dolores Huerta, June 8, 1995.

77 "not the quiet long-suffering": Huerta letter to Chavez cited in Rose, "Women in the United Farm Workers," p. 47.

77 "If I can make it through August": Huerta letter to Chavez, undated.

77 tried to get her a job working for the state: Huerta letter to Chavez, Nov. 1963.

77 "big box of groceries": Int. Dolores Huerta, June 8, 1995.

77 "In the eyes of the public it was slow": Int. Padilla, July 25, 1995.

78 "wanted to pay us less": Int. Padilla, July 25, 1995.

78 tried to offer Padilla: Taylor, *Chavez*, p. 89.

79 Doug Adair, who happened to be in the valley: Int. Doug Adair, July 6, 1996.

80 "took the glasses off my nose and—*pow!*": Int. Doug Adair, March 10, 1995.

80 *El Malcriado* means "ill bred," Chavez letter to Ross, March 25, 1965.

81 "The cartoons were very important": Int. Andy Zermeño, Sept. 24, 1996.

81 Don Sotaco: Northridge int. Doug Adair, March 10, 1995, pp. 73–74.

82 "Every Saturday we'd have to get up": Int. Eliseo Medina, June 14, 1996.

82 Communist Infiltration: FBI file, Oct. 8, 1965.

82 "Go see Cesar Chavez": Int. Epifanio Camacho, July 3, 1996.

83 promised $9.00 per thousand plants: Levy, *Cesar Chavez*, p. 197.

83 crucifix: Ibid., p. 179.

83 "Things are getting exciting": Chavez letter to Ross, March 25, 1965.

83 she moved her truck: Levy, *Cesar Chavez*, p. 180.

83 Strike fever was sweeping the state: Meister and Loftis, *A Long Time Coming*, p. 127.

86 "We who are picking": Northridge int. Doug Adair, March 10, 1995, p. 80.

86 caught off guard: Levy, *Cesar Chavez*, p. 182.

86 treasury: Int. Dolores Huerta, June 8, 1995.

87 "you're going to suffer a lot of hardship": Archival int. Larry Itliong from *The Fight in the Fields* documentary film.

87 "We knew that we had to support it": Int. Dolores Huerta, June 8, 1995.

87 "Aren't we a union?": Ibid.

88 "Cesar's genius": Int. Doug Adair, July 6, 1996.

88 "Now is when every worker": Northridge int. Doug Adair, March 10, 1995, pp. 81–82.

88 *"Estámos en huelga!"*: Int. Eliseo Medina, June 14, 1996.

88 growers received a boost: Meister and Loftis, *A Long Time Coming*, p. 132.

88 "in the state of Guanajuato": Levy, *Cesar Chavez*, p. 184.

89 Camacho complied: Int. Doug Adair, July 6, 1996.

89 "I saw two strikers murdered": Eugene Nelson, *Huelga*, p. 27.

89 union counted some twenty-seven hundred authorization cards: Levy, *Cesar Chavez*, p. 185.

FOUR: HUELGA!

90 "The experience of working in the civil rights movement . . . Mississippi": Int. Marshall Ganz,. Jan. 6, 1996.

92 showering the crowd with fine Delano dirt: Eugene Nelson, *Huelga*, pp. 77–79.

92 rushed off registered letters: Levy, *Cesar Chavez*, pp. 185–85.

92 "dangerous potential for violence": *Los Angeles Times*, Sept. 29, 1965.

93 studied civil rights movement and nonviolence: Ibid., p. 196.

93 tried to get word of strike to people outside Delano: Levy, *Cesar Chavez*, p. 193.

93 "He wasn't a good speaker at all": Int. Marion Moses, April 23, 1996.

93 sought immediate court injunctions: Matthiessen, *Sal Si Puedes*, p. 85.

96 "They were like members of our family": Nelson, *Huelga*, p. 33.

96 menace erupted at the C. J. Lyons Ranch: Ibid., pp. 10–11.

96 sprayed pickets with pesticides: Matthiessen, *Sal Si Puedes*, p. 86.

96 backed his pickup truck into Eugene Nelson: Nelson, *Huelga*, pp. 36–38.

96 drawing blood with his punches: Ibid., pp. 91–92.

96 "Oh, pardon me, Mex": Ibid., p. 55.

96 knock a Mexican striker to the ground: Matthiessen, *Sal Si Puedes*, p. 85.

96 protective custody: Ibid., p. 56.

97 "Those early days were really mean": Levy, *Cesar Chavez*, p. 189.

97 "We took every case of violence and publicized what they were doing": Ibid., p. 194.

97 "Love is the most important ingredient": Ibid., p. 196.

97 Delano forms ad hoc groups: Int. Richard Chavez, June 8, 1996.

97 "I don't like the people that are here in our town": Archival footage from *The Fight in the Fields* documentary film.

97 "little United Nations": *California Farmer*, March 5, 1966.

98 moonlighted as labor contractor: Dunne, *Delano*, p. 117.

98 doctoral thesis: Ibid., p. 63.

98 "These men are extremely happy": Archival footage from *The Fight in the Fields* documentary film.

98 Grape pickers in Delano earned about $2,400 a year: Matthiessen, *Sal Si Puedes,* p. 54.

98 poverty line in 1965: U.S. Department of Labor.

98 "We just cannot in our own mind see this complete power of destruction": *California Farmer,* Jan. 1966.

99 "They had developed this mythology": Int. Chris Hartmire, July 1, 1996.

99 "I could not talk to my family about this back then": Int. George Zaninovich, Sept. 3, 1996.

99 "only too willing to accept socialist Big Government subsidies": Nelson, *Huelga,* p. 63.

100 "an anachronism: the last-ditch stand for a feudalistic way of life": Ibid., p. 72.

100 "It became like a brotherhood": Int. Gilbert Padilla, July 25, 1995.

100 "Hello, brother": Int. Pete Velasco, June 8, 1995.

100 "I was convinced they wanted to shoot us": Int. Eliseo Medina, June 14, 1996.

101 chanted in Tagalog: photo insert in Nelson, *Huelga.*

101 "You—you with the stringy hair—come over here": Ibid., p. 8.

101 "hard to organize those that believed that women should stay home": Int. Jessie De La Cruz, July 25, 1995.

101 Cesar washing dishes: Int. Doug Adair, July 6, 1996.

102 "I was warned not to take the volunteers": Levy, *Cesar Chavez,* p. 197.

102 "it's beautiful to work with other groups": Ibid., p. 197.

102 short-wave radios: Dunne, *Delano,* p. 24.

103 "I hadn't really made the connection between what was happening in Mississippi and where I'd grown up . . . often a very constructive tension": Int. Marshall Ganz, Jan. 6, 1996.

103 "Some of the volunteers were for ending the Vietnam War above all else": Levy, *Cesar Chavez,* p. 197.

103 blurb offering a "cash reward": FBI documents, Jan. 21, 1996.

104 "I—a Republican—could be picketing next to a Communist": Int. Doug Adair, July 6, 1996.

104 "in the hot summer of California in 1965, the movement of the farmworkers began": Northridge int. Doug Adair, March 10, 1995.

104 placed a call to the FBI in early October: FBI document, Oct. 8, 1965.

104 "The general public in Delano": FBI document, Oct. 20, 1965.

105 "clean background": Oct. 11, 1965.

105 "We have conducted limited inquiries . . . members of this organization have extremely liberal views": FBI document, April 14, 1966.

105 "driving force on the picket lines": FBI documents, Oct. 15 and 20, 1965, and April 14, 1966.

105 evidence enough to the FBI: FBI documents, Oct. 20, 1965 and April 14, 1966.

105 the word *huelga*—"strike"—seemed suspicious: FBI document, Oct. 15, 1965.

106 dossiers on pickets: Levy, *Cesar Chavez,* p. 190; Taylor, *Chavez,* p. 165.

106 "More than once, the authorities threatened to close our office": Levy, *Cesar Chavez,* p. 189.

106 Roy Galyen issued a special directive: Nelson, *Huelga,* p. 96.

106 Dodd motioned for a squad car: Ibid., pp. 97–99.

107 forbidding the use of the word *"huelga"*: Ibid., p. 101.

107 "More than a victory issue was the morale issue": Int. Chris Hartmire, Jan. 24, 1996.

107 Forty-four pickets were transported: Nelson, *Huelga,* pp. 101–3.

107 "Being in jail didn't scare me": Rose, "Women in the United Farm Workers," p. 141.

107 $6,700 in cash donations: Levy, *Cesar Chavez,* p. 193; Matthiessen, *Sal Si Puedes,* p. 86.

107 350 people gathered to picket: Nelson, *Huelga,* p. 103.

107 "urgent" teletype: FBI document, Oct. 22, 1965.

108 Valdez was eager to celebrate: *San Francisco Chronicle*, May 2, 1966.

108 FBI informant claimed that Valdez: FBI document, Oct. 11, 1966.

108 penned a strident letter: FBI copy of letter dated Oct. 1965 and published in Progressive Labor Party newspaper *Spark*.

109 "Cesar came to San Francisco—and Dolores, too—in the first weeks of the strike . . . the urge to express . . .was just waiting for the 'huelga'": Int. Luis Valdez, Jan. 24, 1996.

111 "So I said, 'Well, you're the show'": Ibid.

111 the *teatro* developed characters: Ibid.

112 "What's *dignidad?*": Ibid.

112 "The farmworker . . . is now speaking": *San Francisco Chronicle*, May 2, 1966.

112 "There were no bathrooms out there": Int. Luis Valdez, Jan. 24, 1996.

112 "We call that Mexican golf": Levy, unpublished interview with Chavez, 1969, p. 346.

112 "Down with the Grape Society!": Matthiessen, *Sal Si Puedes*, p. 236.

112 "From Selma, to Watts, to Berkeley, to Delano": Allen, "Communist Wrath in Delano," p. 1.

113 "Southern law enforcement authorities": Allen, *Communist Revolution in the Streets*, p. 760.

113 "Up until the late 1920s and 1930s, people lived on farms in great numbers": Int. Luis Valdez, Jan. 24, 1996.

113 "Here was Cesar, burning with a patient fire": Matthiessen, *Sal Si Puedes*, p. 148.

113 Cesar directed volunteers to follow grapes: Levy, *Cesar Chavez*, p. 201.

113 "You're not big enough to threaten me": Ibid., p. 202.

113–14 boycott of Schenley Industries: Meister and Loftis, *A Long Time Coming*, p. 140.

114 DiGiorgio Fruit Corporation: Ibid, pp. 221–22.

114 "There is no power in the world like the power of free men": Nelson, *Huelga*, p. 121.

114 "Sooner or later these guys are going to win": Levy, *Cesar Chavez*, p. 203.

114 "This is not your strike, this is our strike": Nelson, *Huelga*, p. 121.

114 Teatro Campesino performed: Int. Luis Valdez, Jan. 24, 1996.

115 "preliminary investigation": FBI document, March 9, 1966.

115 Murphy aide had called the FBI: FBI document, March 7, 1966.

115 Chavez offered convincing examples of police bias: Testimony included in FBI file #161-4719.

115 "All these bills do is say people who work on farms should have the same human rights": Testimony in FBI file #161-4719.

116 Catholic Bishops of California: Taylor, *Chavez*, p. 167.

116 "Do you take pictures of everyone": Taylor, *Chavez*, p. 166; Archival footage from *The Fight in the Fields* film documentary.

117 "I'd never seen any Democrat . . . do that before": Int. Paul Schrade, July 15, 1995.

117 idea of a march to Sacramento: Int. Marshall Ganz, Jan. 6, 1996.

117 "long, long struggle . . . wanted to be fit not only physically but also spiritually": Levy, *Cesar Chavez*, p. 207.

117 "We were right there on Albany Street": Int. Angie Hernandez Herrera, Sept. 6, 1996.

119 few minutes of public negotiations: Levy, *Cesar Chavez*, p. 208.; Int. Angie Hernandez Herrera, Sept. 6, 1996.

119 "March group began with about one hundred persons": FBI document, March 18, 1966. At least two dozen FBI documents detail various events or details about the march to Sacramento, including lists of every city passed through over the course of twenty-five days.

119 farmworkers trudged north: Archival footage in *The Fight in the Fields* film documentary; Int. Angie Hernandez Herrera, Sept. 6, 1996.

119 Epifanio Camacho . . . refused to carry any images: Int. Camacho, June 3, 1996.

119 "Some people had bloody feet": Int. Angie Hernandez Herrera, Sept. 6, 1996.

120 subscription list of *El Malcriado*: Int. Doug Adair, July 6, 1996.

120 "We walked into Parlier": Int. Adair, July 6, 1996.

120 "We are the sons of the Mexican Revolution": Reprinted as Appendix C in the Fourteenth Report of the Senate Factfinding Subcommittee on Un-American Activities, 1967.

120 250 Chicagoans: FBI document, April 4, 1966.

121 ask Chavez a few questions: Int. Hijinio Rangel, Sept. 6, 1996.

121 "The marchers have a field kitchen for emergencies": FBI document, March 28, 1966.

121 card tables with bottles of Schenley liquors: Levy, *Cesar Chavez*, p. 210.

121 "White people would honk": Int. Angie Hernandez Herrera, Sept. 6, 1996.

121 FBI put out an alert: FBI document, April 5, 1966.

121 "Oh, the hell with him": Levy, *Cesar Chavez*, pp. 215–16.

121 false rumor: Taylor, *Chavez*, pp. 175–76.

122 "a nice little guy": Ibid., p. 175.

122 Chris Hartmire drove Cesar: Levy, *Cesar Chavez*, p. 216.

122 "When we fought for it, bled for it": Ibid., p. 216.

122 preliminary agreement was signed: Ibid., pp. 216–17. Note: after the agreement was signed, FBI director J. Edgar Hoover requested background summary on the union and on Schenley attorney Sidney Korshak, FBI document, April 14, 1966.

122 "On behalf of all the farmworkers of this state": Archival KQED television footage, April 10, 1966, from *The Fight in the Fields* documentary film.

123 "It is well to remember there must be courage": Archival KQED television footage, April 10, 1966, from *The Fight in the Fields* documentary film.

FIVE: BOYCOTT

124 "I saw these two young white men walking toward me": Int. Jessica Govea, April 25, 1996.

125 Adelina Gurola was at her job: Int. Adelina Gurola, July 5, 1996.

126 floated a hint over the radio: Levy, *Cesar Chavez*, p. 223; Archival footage from *Fight in the Fields* documentary film.

126 "We go back to continue the strike that's lasted now almost seven months": Archival footage from *The Fight in the Fields* documentary film.

126 Chavez broke them off: Levy, *Cesar Chavez*, p. 223.

126 taking on DiGiorgio was like wrestling with a titan: Meister and Loftis, *A Long Time Coming*, p. 140; Taylor, *Chavez*, p. 174; Levy, *Cesar Chavez*, pp. 221–22.

127 *The Grapes of Wrath*: Matthiessen, *Sal Si Puedes*, p. 131.

127 broke a strike in Yuba City: Levy, *Cesar Chavez*, p. 221.

127 company pitted Mexicans: Int. Adelina Gurola, July 5, 1996.

127 Sierra Vista ranch: Levy, *Cesar Chavez*, p. 222.

127 "We were facing a giant": Ibid., p. 222.

127 strike from within the ranch: Levy, *Cesar Chavez*, p. 222; Int. Marshall Ganz, Jan. 6, 1996.

128 "anything that was legal and moral": Levy, *Cesar Chavez*, p. 222.

128 *submarinos*: Int. Johnny Serda, July 3, 1996.

128 Dolores Huerta was dispatched to Texas: Levy, *Cesar Chavez*, pp. 224–25.

128 "My mind just flashed to all the possibilities": Ibid., pp. 226–27.

128 blocked an S&W Foods distribution center: FBI documents, May 24, 1966.

129 stinging news: Taylor, *Cesar Chavez*, pp. 191–92.

129 Teamsters union, notorious for its corruption: Brill, *The Teamsters,* is an exhaustive investigation into the union's history.

129 "The decision was made to test Chavez's resolve": Int. Bill Grami, Aug. 28, 1995.

129 sweetheart agreement: Meister and Loftis, *A Long Time Coming*, p. 102.

129 DiGiorgio's June 24 elections: Levy, *Cesar Chavez*, p. 230; Int. Fred Ross Jr., Aug. 28, 1995.

130 the winner of elections at both ranches: Meister and Loftis, *A Long Time Coming*, p. 146.

130 Dolores Huerta won the support of Mexican-American Political Association: Taylor, *Chavez*, p. 194.

130 "This meeting is two months overdue": Ibid., p. 194.

130 "We couldn't find *huelga* headquarters": Int. Luis Valdez, Jan. 24, 1996.

131 shackled the men together: Ibid.; Levy, *Cesar Chavez*, pp. 231–33.

131 "In the backseat I fell asleep right away": Ibid., p. 233.

131 convicted later of criminal trespassing: Ibid., p. 233.

131 membership voted to merge: Meister and Loftis, *A Long Time Coming*, p. 149.

131 "Don't leave *anything* to chance": Int. Eliseo Medina, June 14, 1996.

132 "I was looking for people I had no addresses for": Int. Gilbert Padilla, July 25, 1995.

132 "We had mounted the sound system on top": Int. Eliseo Medina, June 14, 1996.

132 rushed to San Francisco for the count: Levy, *Cesar Chavez*, p. 245.

132 "That was an emotional day": Ibid., p. 245.

132 "We were all worried as hell": Int. Eliseo Medina, June 14, 1996.

132 "merchants began to close down . . . day of mourning": Levy, *Cesar Chavez*, p. 246.

133 sent to San Francisco with Marshall Ganz: Int. Adelina Gurola and Teofilo Garcia, July 5, 1996.

133 "They didn't see why they should be required": Levy, *Cesar Chavez*, p. 254.

133 shop steward had to produce a copy of the rules: Int. Teofilo Garcia, July 5, 1996.

133 "hicks from the sticks": Int. Teofilo Garcia, July 5, 1996.

133 "everyone became brave": Int. Adelina Gurola, July 5, 1996.

133 DiGiorgio contract broke new ground: Meister and Loftis, *A Long Time Coming*, p. 150.

136 boycott against Perelli-Minetti's labels: Meister and Loftis, *A Long Time Coming*, p. 150; Taylor, *Chavez*, p. 204.

136 Gallo . . . Christian Brothers, Almaden, and Paul Masson: Taylor, *Chavez*, p. 184.

136 admitted to a hospital: Levy, *Cesar Chavez*, p. 262.

136 *perigrinación* through Texas: FBI documents, July through Oct. 1966.

136 "I was in jail the second day I was in Texas": Int. Gilbert Padilla, July 25, 1995.

136 Texas' tough right-to-work laws: Int. Gilbert Padilla, July 25, 1995.

137 Chavez lured him away from CRLA: Int. Jerry Cohen, June 8, 1996.

137 lack of labor laws protecting farmworkers: Ibid.

137 injunctions prohibiting . . . "secondary boycotts": Ibid.

137 transferred the nine warehouse workers out of UFWOC: Ibid.

138 contracts, mostly in a handful of wineries: Meister and Loftis, *A Long Time Coming*, p. 151.

138 children as young as six to work as strikebreakers: Nelson, *Huelga*, photo; Matthiessen, *Sal Si Puedes*, pp. 252–55.

138 wages were lower: Levy, *Cesar Chavez*, p. 263.

138 "If we can crack Giumarra": Meister and Loftis, *A Long Time Coming*, p. 151.

138 Giumarra workers voted to strike: Levy, *Cesar Chavez*, p. 264; Int. Jerry Cohen, June 8, 1996.

138 "The first day we went out on the picket line": Levy, *Cesar Chavez*, p. 267.

139 "Where's Chicago?": Int. Eliseo Medina, June 14, 1996.

139 boycott against all California table grapes: Meister and Loftis, *A Long Time Coming*, p. 151.

139 send him to his grandparents: Int. Marc Grossman, Oct. 10, 1996.

139 "All his kids go to private schools in Switzerland": Int. Linda Chavez Rodriguez, June 8, 1996.

139 ran over Rivera: Matthiessen, *Sal Si Puedes*, pp. 87–88.

140 Pereira got behind the wheel: Ibid., p. 89.

140 cells of two or three men: Int. Epifanio Camacho, July 3, 1996.

140 "It's very tough. I don't know if I can continue": Int. Paul Chavez, April 25, 1996.

140 Chavez would become furious: Int. Al Rojas, June 30, 1996.

140 dissidents who had lost faith in nonviolence: Matthiessen, *Sal Si Puedes*, p. 177.

141 "I despise exploitation and I want change": Levy, *Cesar Chavez*, p. 270.

141 confidential meeting: Int. Al Rojas, June 30, 1996; Jerry Cohen, June 8, 1996.

141 "moral jujitsu": Levy, *Cesar Chavez*, p. 93.

142 Chavez began his fast: Int. Leroy Chatfield, Jan. 24, 1996.

142 drinking only Diet-Rite cola: Levy, *Cesar Chavez*, p. 272; Int. Leroy Chatfield, Jan. 24, 1996.

142 meeting Filipino Hall: Int. Al Rojas, June 30, 1996; Matthiessen, *Sal Si Puedes*, p. 178.

142 "He's just going to kill himself": Int. Al Rojas, June 30, 1996.

143 Tony Orendain . . . sat with his back toward Chavez: Levy, *Cesar Chavez*, p. 282.

143 "The irony of the fast": Int. Leroy Chatfield, Jan. 24, 1996.

143 "He had tacos": Matthiessen, *Sal Si Puedes*, p. 190.

144 Cesar's fast was no publicity stunt: Int. Jerry Cohen, June 8, 1996.

144 "No union movement": Meister and Loftis, *A Long Time Coming*, p. 152.

144 "Kern County courthouse . . . was enemy territory": Int. Jerry Cohen, Aug. 28, 1995.

144 Cesar looked up and winked: Int. Jerry Cohen, June 8, 1996.

145 ruin the planned celebration: Int. Al Rojas, June 30, 1996.

145 "Well, how goes the boycott": Int. Ethel Kennedy, March 27, 1996.

146 Dolores Huerta accompanied Kennedy to the stage: Archival footage from *The Fight in the Fields* documentary film; Int. Dolores Huerta, June 8, 1995.

146 Reagan . . . showily eating grapes: Archival footage from *The Fight in the Fields* documentary film.

146 Al Rojas . . . left the Forty Acres for Pittsburgh, Pennsylvania: Int. Al Rojas, June 30, 1996.

148 "Rojas distributed to each participant one grape": FBI documents, Nov. 21, 23, 26, 1968.

148 Whitaker & Baxter public relations firm: Meister and Loftis, *A Long Time Coming*, p. 155.

148 sending [grapes] to troops in Vietnam: Ibid., p. 157.

148 Govea volunteered to go to Toronto: Int. Jessica Govea, April 25, 1996; Rose, "Women in the United Farm Workers," p. 210.

149 "liberating and wonderful experience": Int. Jessica Govea, April 25, 1996.

149 "the girl from the south": *Toronto Star*, Oct. 5, 1968.

152 dazzled the Ontario Federation of Labor: Rose, "Women in the United Farm Workers," p. 214.

152 mistaken for a Canadian Indian: Int. Jessica Govea, April 25, 1996.

152 one of the largest chain stores in Montreal agreed to stop selling grapes: Rose, "Women in the United Farm Workers," p. 227.

152 "We all worked for five dollars a week": Int. Jessica Govea, April 25, 1996.

153 "We just totally disrupted": Int. Eliseo Medina, June 14, 1996.

153 "Cesar told me he needed me to go to Detroit": Int. Hijinio Rangel, Sept. 6, 1966.

153 "red squads": Int. Manny Lopez, Oct. 23, 1996.

153 boycott group in San Diego: Int. Linda and Carlos LeGerrette, July 7, 1996; Int. Manny Lopez, Oct. 23, 1996.

154 Chicano Moratorium march: Gómez Quiñones, *Chicano Politics*, p. 127.

155 pesticides were "trade secrets": Int. Jerry Cohen, June 8, 1996.

155 $25 million in losses: Meister and Loftis, *A Long Time Coming*, p. 160.

155 "immediate advantage over our competitors": Int. Lionel Steinberg, Jan. 26, 1996.

155 Both [Dispoto and Roberts] signed union contracts: Meister and Loftis, *A Long Time Coming*, p. 162.

155 "I realized something important": Int. Jerry Cohen, Aug. 28, 1995.

156 grape pickers would have a hiring hall: Meister and Loftis, *A Long Time Coming*, p. 163.

157 "If it works well here, if this experiment in social justice": Archival footage from *The Fight in the Fields* documentary film.

157 "The strikers and people involved in this struggle sacrificed a lot": Archival footage from *The Fight in the Fields* documentary film.

157 "It was like a boa constrictor swallowing a big pig": Int. Jerry Cohen, Aug. 28, 1995.

SIX: BLOOD IN THE FIELDS

158 "The opposition was fighting very hard": Int. Richard Chavez, June 8, 1996.

159 "Who sent you": Int. Sabino Lopez, June 8, 1996.

160 secretly signed contracts: *Salinas Californian*, Aug. 1, 1970.

161 as many as eleven thousand workers: Ibid., Aug. 4, 1970.

161 operatives in positions as high as the Nixon White House: Levy, *Cesar Chavez*, pp. 472–73.

161 Rumors were flying that a back-room contract had been struck: Int. Jerry Cohen, Aug. 28, 1995; *Salinas Californian*, July 27, 1970.

161 "The grape boycott scared the heck out of the farmers": Int. Daryl Arnold, Jan. 26, 1996.

162 Chavez drove to Salinas to speak to supporters: *Salinas Californian*, July 25, 1970.

162 Eric Brazil, covered Chavez's appearance: Int. Eric Brazil, June 18, 1996.

162 "A prompt reply will avoid the bitter conflict experienced in Delano": *Salinas Californian*, July 25, 1970.

162 found out that rumors of the Teamster takeover were true: *Salinas Californian*, July 28, 1970.

162 "They can't get away with this": Ibid.

162 "rage of workers was just palpable": Int. Eric Brazil, June 18, 1996.

163 began a four-day pilgrimage: *Salinas Californian*, Aug. 1, 1970.

163 prospered using cheap bracero labor: Int. Eric Brazil, June 18, 1996.

163 "crazies" from Delano: Int. Jerry Cohen, June 8, 1996.

163 "Some of them didn't like the notion of dealing equally with people of a different race": Ibid.

163 faceless conglomerates: Levy, *Cesar Chavez*, pp. 327–28.

163 "Mama United": Stephen Schlesinger and Stephen Kinzer's *Bitter Fruit* offers a detailed investigation into United Fruit's involvement in overthrowing Guatemala's democratically elected government in 1954.

163 Pic N Pac employed up to two thousand berry pickers: Wells, *Strawberry Fields*, p. 83.

163 "Of late [Governor Ronald Reagan] seems very concerned about farmworker rights": *Salinas Californian*, Aug. 1, 1970; Levy, *Cesar Chavez*, p. 329.

164 more than three thousand farmworkers responded: *Salinas Californian*, Aug. 3, 1970.

164 "It's tragic that these men have not yet come to understand": Ibid.

165 "There was so much anger among people": Int. Sabino Lopez, Aug. 28, 1995.

165 Teamsters still had only a fraction: *Salinas Californian*, Aug. 6, 1970.

165 pouring into the little union office: Int. Marshall Ganz, Jan. 6, 1996; Int. Jerry Cohen, Aug. 28, 1995.

165 most militant were the lettuce cutters: Int. Marshall Ganz, Jan. 6, 1996.

165 "There was more work being done on more fronts in that period": Levy, *Cesar Chavez*, p. 332.

166 strike-fund assistance: Ibid., p. 332.

166 "I knew that if we ever started a public boycott": Ibid., p. 332.

166 Leroy Chatfield's brainstorm: Ibid., p. 332.

166 executives' college-age children had boycotted grapes: Ibid., p. 334.

166 daughter, Eloise, was getting married: Ibid., pp. 336.

166 workers voted to strike: *Salinas Californian,* Aug. 10 and 11, 1970.

167 "I am here to be served the order": *Salinas Californian,* Aug. 12, 1970.

167 "no raid" pact: *Salinas Californian,* Aug. 10, 1970.

167 "They've got the money, we've got the people": *Salinas Californian,* Aug. 12, 1970.

168 "The field was something you didn't even think about, it was so low": Int. Eric Brazil, June 18, 1996.

168 local ranchers were thrown into crisis: Int. Eric Brazil, June 18, 1996.

168 Will Lauer flew into Salinas: Levy, *Cesar Chavez,* p. 334.

168 "Come in tougher than hell": Ibid., p. 345.

169 "We picketed his office in Los Angeles": Ibid., p. 347.

169 "There was a lot of hatred building up": Ibid., p. 340.

169 retreated to San Juan Bautista: Ibid., p. 351.

169 "The talk is quiet, friendly": Ibid., p. 386.

169 "You know, I think we've got you guys": Ibid., p. 388.

169 lettuce workers exploded: Int. Marshall Ganz, Jan. 6, 1996; Int. Jerry Cohen, Aug. 28, 1995; Levy, *Cesar Chavez,* p. 348.

169 "The only thing we have to lose by waiting is our chains": *Salinas Californian,* Aug. 19, 1970.

169 sew new banners: Wells, *Strawberry Fields,* p. 84.

169 verifying the legitimacy of all 856 union cards: Levy, *Cesar Chavez,* pp. 373–74.

170 growers insisted on keeping their contracts: *Salinas Californian,* Aug. 21 and 22, 1970.

170 "Everything we have done, we have done in good faith": *Salinas Californian,* Aug. 24, 1970.

170 "It looked like a revolution": Int. Jerry Cohen, Aug. 28, 1995.

170 injunctions against picketing: *Salinas Californian,* Aug. 25, 1970.

170 belligerent thugs: Levy, *Cesar Chavez,* pp. 380, 412.

170 later suspended and indicted: *Salinas Californian,* Sept. 13 and 14, 1970.

170 Seafarers' Union: Levy, *Cesar Chavez,* p. 411.

170 "Get 'em, boys!": Int. Jerry Cohen, Aug. 28, 1995; Levy, *Cesar Chavez,* pp. 383–84.

171–72 "The people don't understand how dangerous this is": Levy, *Cesar Chavez,* p. 419.

172 *Chavistas* were arrested for shooting a Teamster: Ibid., p. 423.

172 "If we mean nonviolence": Int. Marc Grossman, Oct. 10, 1996.

172 driven a hard bargain with InterHarvest: Levy, *Cesar Chavez,* p. 398; *Salinas Californian,* Aug. 31, 1970.

172 growers' wives: Int. Eric Brazil, June 18, 1996.

172 "social bones of Salinas stood out": Int. Eric Brazil, June 18, 1996.

172 "Purex had been called communistic because we signed with Mr. Chavez": Meister and Loftis, *A Long Time Coming,* p. 170.

172 unknown assailants planted dynamite: Levy, *Cesar Chavez,* p. 424.

173 judge jailed Chavez: Levy, *Cesar Chavez,* pp. 426–28.

173 Chavez made the most of his solitude: Ibid., p. 430.

173 Ethel Kennedy came calling: *Salinas Californian,* December 7, 1970; Archival footage from *The Fight in the Fields* documentary film; Int. Eric Brazil, June 18, 1996.

173 "You really throw some weird parties": Int. Paul Schrade, July 15, 1995.

173 California Supreme Court: Levy, *Cesar Chavez,* pp. 433, 476.

174 "Everyone wants to be acknowledged": Int. Sabino Lopez, Aug. 28, 1995.

174 official member of the AFL-CIO: Meister and Loftis, *A Long Time Coming,* p. 174.

174 reluctant to move back there: Int. Marc Grossman, Oct. 15, 1996.

175 "There were people who thought he was removing himself from the workers": Int. Paul Chavez, April 25, 1996.

175 "What I really would like is to be alone somewhere": Nelson, *Huelga,* p. 52.

175 private rendezvous: Int. Frank Denison, June 14, 1996.

175 "Play it, Sam": Int. Mike Ybarra, June 8, 1995.

176 "picture of a guy who an informant said had been hired to shoot Cesar": Int. Jerry Cohen, June 8, 1996.

176 person who reported the plot was Larry Shears: Levy, *Cesar Chavez,* pp. 444–46.

176 "all steps of the investigation": Ibid., p. 446.

177 "For a long time I didn't believe there was a plot": Ibid., p. 443.

177 "we did have something to worry about": Int. Marc Grossman, Oct. 10, 1996.

177 Jerome Joseph Ducote, a veteran John Bircher: *San Jose Mercury News* obituary, Oct. 13, 1987.

177 Ducote confessed: *San Francisco Chronicle,* February 17, 1976; FBI document, March 31, 1974.

177 Ducote said that Kenneth Wilhelm: *San Francisco Chronicle,* Feb. 17, 1976.

177 used as a source: FBI document, Jan. 20, 1976.

177 Ducote was an authentic spy: Int. former agent with California Department of Justice, Oct. 24, 1996.

177 irreplaceable items that were never recovered: Int. Marc Grossman, Oct. 15, 1996.

177 first knew this man as "Fred Schwartz": Int. Jerry Cohen, June 21, 1996; FBI document, March 28, 1974.

177 filled an ashtray with Camel cigarettes: FBI document, March 31, 1975.

178 "I was pulled from case": Int. former agent with California Department of Justice, Oct. 24, 1996.

178 died of a heart attack: *San Jose Mercury News* obituary, Oct. 13, 1987.

178 "Those were very awful days for everybody": Int. Richard Chavez, June 8, 1996.

178 abbreviated the names: Levy, *Cesar Chavez,* pp. 359–60.

178 "There was more confusion to add to the confusion": Ibid., p. 361.

178 first legal tools: Levy, *Cesar Chavez,* p. 361; Int. Epifanio Camacho, July 3, 1996.

179 "there was a need for it": Int. Al Rojas, June 30, 1996.

179 Rojas's office in Poplar was shot up: FBI documents, Jan. 31, 1973; Int. Al Rojas, June 30, 1996.

179 "one-man show": Int. Ray Huerta, July 6, 1996.

180 "hard-core, arrogant people": Int. Lionel Steinberg, Jan. 26, 1996.

180 "doing its job beautifully": Int. Doug Adair, July 6, 1996.

180 Nixon courted Teamsters president Frank Fitzsimmons: Brill, *The Teamsters,* pp. 96–100; Levy, *Cesar Chavez,* pp. 472–73; Meister and Loftis, *A Long Time Coming,* p. 185.

180 "some way to work against the Chavez union": Levy, *Cesar Chavez,* p. 473.

181 Colson became an attorney for the Teamsters: Brill, *The Teamsters,* p. 97.

181 "revolutionary movement": Meister and Loftis, *A Long Time Coming,* p. 184.

181 wage war against the UFW: Levy, *Cesar Chavez,* pp. 470–71.

181 "He was stretching the meaning": Ibid., p. 471.

181 picketed Republican headquarters: FBI documents, March 16, 17, 1972.

181 "have got us cold": Levy, *Cesar Chavez,* p. 475.

182 "there's people coming out and signing up Filipinos": Int. Ray Huerta, July 6, 1996.

182 talks collapsed: Levy, *Cesar Chavez,* p. 477.

182 new contracts that eliminated the hiring hall: Ibid., p. 478.

182 "It was like being in a war. They arrested farmworkers": Int. Clementina Carmona Olloqui, July 6, 1996.

182 Teamster enforcers: Meister and Loftis, *A Long Time Coming,* p. 186; Levy, *Cesar Chavez,* p. 485; *Fighting for Our Lives,* UFW documentary film on Coachella strike in 1973.

183 Metheny indignantly expunged: Levy, *Cesar Chavez,* p. 483.

183 "We had standing orders": Int. Ray Huerta, July 6, 1996.

183 "Cesar understood . . . favorable press": Int. Jerry Cohen, June 8, 1996.

183 shots crackled all around them: Int. Ray Huerta, July 6, 1996; Int. Carlos LeGerrette, July 7, 1996.

183 others were injured seriously: Levy, *Cesar Chavez*, pp. 491–92; FBI documents, July 9, 1973.

183 Mike Falco attacked John Bank: Levy, *Cesar Chavez*, p. 488; Meister and Loftis, *A Long Time Coming*, p. 196.

184 "escalate the violence": FBI document, July 9, 1973.

184 contracts . . . were probably lost: Levy, *Cesar Chavez*, p. 483.

184 $1.6 million strike fund: Ibid., pp. 486–87.

184 "has no authority to conduct such an investigation": FBI document, July 5, 1973.

184 "you're a lousy bunch of Commies": *Fighting for Our Lives*, UFW documentary film on Coachella strike.

184 "This may be a blessing in disguise": Levy, *Cesar Chavez*, p. 484.

185 Chavez began to lose faith in the power of the strike: Int. Jerry Cohen, Aug. 28, 1995; Int. Luis Valdez, Jan. 24, 1996.

185 UFW contracts tumbled: Levy, *Cesar Chavez*, p. 495.

185 Marta Rodriguez was picketing: Int. Marta Rodriguez, April 25, 1996; *Fighting for Our Lives*, UFW documentary film; Levy, *Cesar Chavez*, pp. 497–98.

186 "greasers": Archival footage in *The Fight in the Fields* documentary film; Levy, *Cesar Chavez*, p. 496.

186 "I'm not sure how effective": Meister and Loftis, *A Long Time Coming*, p. 203.

186 Paddy wagons in the valley were filled: Int. Jerry Cohen, Aug. 28, 1995.

186 Chavez roamed the picket lines: Levy, *Cesar Chavez*, p. 501.

187 "It's like a resurrection": Ibid., p. 505.

187 Nagi Daifullah: Levy, *Cesar Chavez*, pp. 505–7; Meister and Loftis, *A Long Time Coming*, p. 191.

187 Juan de la Cruz: Int. Jose de la Cruz, July 5, 1996; Levy, *Cesar Chavez*, p. 578; Meister and Loftis, *A Long Time Coming*, p. 191.

188 "until the federal law enforcement agencies": Meister and Loftis, *A Long Time Coming*, p. 191.

188 90 percent of the UFW's grape contracts were gone: *The Fight in the Fields* documentary film.

188 "After people started to get killed there was a visible change in Cesar": Int. Luis Valdez, Jan. 24, 1996.

188 "The opposition was fighting very hard": Int. Richard Chavez, June 8, 1996.

188 "We weren't all angels": Int. Jerry Cohen, June 8, 1996.

189 "She didn't last much longer after that": Int. Jose de la Cruz, July 5, 1996.

189 jury acquitted: Meister and Loftis, *A Long Time Coming*, p. 191.

189 UFW needed a special set of laws: Int. Jerry Cohen, June 8, 1996.

SEVEN: WITH THE LAW ON OUR SIDE

190 "Whether it's Sacramento": Int. Jerry Brown, Aug. 28, 1995.

191 More than 17 million Americans had stopped: *Dollars and Sense*, Jan., 1976, p. 13, among other sources.

191 The *New York Times* shocked the union: Int. Fred Ross Jr., Sept. 3, 1996.

191 quoted cynics who predicted: *New York Times Sunday Magazine*, Jan. 4, 1975.

191 "It said the Teamsters were too rich": Int. Fred Ross Jr., Sept. 3, 1996.

191 Major growers like John Giumarra: Meister and Loftis, *A Long Time Coming*, p. 224.

191 Chavez had offered to put up a million-dollar "performance bond": Ibid., p. 213.

192 $1.6 million strike loan: Griswold del Castillo and Garcia, *Cesar Chavez*, p. 123.

192 many UFW allies in politics and organized labor: Int. Jerry Cohen, Sept. 3, 1996.

192 The union's school for members: Rose, "Women in the United Farm Workers," p. 246.

192 attorneys were drawing up drafts of an ideal farm labor law: Int. Jerry Cohen, Sept. 3, 1996.

192 The UFW's Agbayani Village had opened its doors: *El Malcriado,* June 24, 1973.

192 "We needed to do something really public and dramatic": Int. Fred Ross Jr., Sept. 3, 1996.

193 Was the nation's largest winemaker: *San Francisco Examiner,* Feb. 23, 1975. (By 1996, federal campaign contribution records indicate Gallo's owners were among the leading benefactors of both major political parties. In 1974, the *Examiner* article reported that Gallo produced over a third of the nation's wine, and almost half of California's total output.)

193 Operation Clean Sweep: *El Malcriado,* Feb. 22, 1974.

193 others within the union were leery: Int. Fred Ross Jr., Sept. 3, 1996.

193 intense public relations campaign that Gallo was underwriting: *San Francisco Examiner,* March 2, 1974, among other sources.

193 Cohen backed the march: Int. Jerry Cohen, Sept. 3, 1996.

193 Brown's victory in 1974 had signaled a power shift in Sacramento: Ibid.

193 A liberal with a background as a labor law specialist: Ibid.

193 But Brown had avoided any face-to-face meetings since he took office: Ibid.

193 Brown's neglect seemed odd for a man who had been thrilled: Int. Jerry Brown, Aug. 28, 1995.

193–94 Chavez, whose reading list included works on Asian religions: Int. Marc Grossman, Oct. 15, 1996.

194 it was the growers who actually gave Brown his final push: Int. Jerry Cohen, Sept. 3, 1996; Int. Fred Ross Jr., Sept. 3, 1996.

194 On February 22, several hundred people gathered on Union Square: *San Francisco Examiner,* Feb. 23, 1975; Int. Fred Ross Jr., Sept. 3, 1996.

194 Gallo was a formidable foe: *New York Times,* Feb. 3, 1975.

194 "Gallo's 500 Union Farm Workers Best Paid in U.S.": Photo accompanying *San Francisco Examiner* article, Feb. 23, 1975; Int. Jerry Cohen, Sept. 3, 1996; Int. Fred Ross Jr., Sept. 3, 1996.

194 Gallo's Teamster contract did call for good hourly rates: Guest editorial by Cesar Chavez, *San Francisco Examiner,* March 2, 1975.

195 About 130 workers and union volunteers walked the whole week: Int. Fred Ross Jr., Sept. 3, 1996.

195 there were upwards of twenty thousand: Estimate by Fred Ross Jr., Sept. 3, 1996, who coordinated the march. Other sources, in particular newspapers, put the crowd count at no more than fifteen thousand. The newspapers relied mostly on police estimates.

195 *"Chavez sí! Teamsters no!":* San Francisco Examiner, March 2, 1975.

195 "73 more miles to go.": Ibid.

195 "It meant the growers were now telling Brown that there should be a law": Int. Jerry Cohen, Sept. 3, 1996.

197 Arizona had passed a law that outlawed secondary boycotts: Griswold del Castillo and Garcia, *Cesar Chavez,* p. 120, among other sources.

197 "It was May, and all these labor leaders": Int. Marc Grossman, Oct. 15, 1996.

197–98 But when doctors monitoring Chavez noticed an erratic heartbeat: Griswold del Castillo and Garcia, *Cesar Chavez,* p. 121, among other sources.

198 Raul Castro, Arizona's first Mexican-American governor: Ibid.

198 Hoping to replicate the industry's success in Arizona: Int. Leroy Chatfield, Jan. 24, 1996.

198 "Brown was a great politician back then": Int. Jerry Cohen, Sept. 3, 1996.

198 Once the group met at Brown's house in Laurel Canyon: Int. Carlos LeGerrette, Sept. 17, 1996; Int. Jerry Cohen, Sept. 19, 1996.

198 Cohen and Chavez had planned a unique version: Int. Jerry Cohen, Sept. 3, 1996.

198 Brown appeared aloof, but he, too, was traveling: Meister and Loftis, *A Long Time Coming*, p. 216.

198 the Teamsters, who angrily roamed the halls: Int. Jerry Cohen, June 21, 1996.

199 There, all the players were gathered: Ibid.

199 Once again Rose Bird laid out new proposals: Ibid., Int. Tom Dalzell, former UFW lawyer, Nov. 4, 1996.

200 "Jerry's right. Let's move on.": Ibid.

200 Rural legislators tried to change the law's wording: United Press International, *San Diego Union*, May 23, 1975.

200 The UFW prepared for another major march: Ibid.

200 "I think today marks a victory": Jerry Brown remarks at a ceremony, press conference, Sacramento, June 5, 1975, archival news footage. Chavez wanted Brown and the legislators to have their day in the spotlight. And, although he rarely said much before the law was passed, upon its signing he did issue a statement, reprinted in the United Farm Workers, *The Sabotage and Subversion of the ALRA,* December, 1975, p. 1: "In the long and sordid history of American agribusiness, there has never been a law enforced for farmworkers against growers. Child labor, health and safety, pesticide controls [—] these laws have been disregarded or forgotten. The ALRA is a good law and we want it enforced. If nothing else, it is a law the growers must be made to respect."

200 "Certainly I liked the law": Int. Jerry Brown, Aug. 28, 1995.

200 Growers were upset that Brown, despite his vow: Int. Jerry Cohen, Sept. 3, 1996, among other sources.

201 Field organizers were already reporting that Gallo supervisors: United Farm Workers, *The Sabotage and Subversion of the ALRA,* Dec., 1975, p. 53.

201 On July 1, 1975, he embarked on his least-noticed, but toughest *perigrinación:* Int. Mike Ybarra, June 8, 1995; Int. Marc Grossman, Oct. 15, 1996.

201 "In some places, there was no one around.": Int. Marc Grossman, Oct. 15, 1996.

201 It was on this march, say Cesar's friends and aides: Ibid.

201 Luis Valdez remembers many meetings: Int. with Luis Valdez, Jan. 24, 1996; Int. Marc Grossman, Oct. 15, 1996.

201 "After a while we were so tough.": Int. Mike Ybarra, June 8, 1995.

202 With a $2.8 million budget for its first year: ALRB, First Annual Report, June 30, 1976, Appendix E, rounded amount.

202 In only the first two months of the state's new farm-labor law: Int. Leroy Chatfield, Jan. 24, 1996, among other sources. Chatfield's is an estimate that seems to fall midpoint among the various estimates from different sources. Even the state figures vary, depending on which periods are covered in the first year of the law's existence.

202 Cohen and his band of legal monkey wrenchers: Int. Jerry Cohen, June 8 and Sept. 3, 1996; Int. Tom Dalzell, Nov. 2, 1996. Dalzell explains that the lawyers were young, virtual volunteers, and were improvising their negotiating and organizing tactics. Both Cohen and Dalzell point to a particular ploy the UFW legal staff perfected: During his gubernatorial campaign, Jerry Brown was still avoiding talks with the union over a farm labor bill. To get his attention, Cohen instructed Dalzell and other union lawyers to accompany a group of picket-wielding farmworkers to Brown's San Francisco campaign headquarters, which were in the same building as bureaus of the *New York Times,* the *Wall Street Journal,* and news wire services. There, the lawyers measured the dimensions of the office, and told the staff they were looking to see how many farmworkers would fit into the office. Faced with the impending sit-down, Brown's aide, Leroy Chatfield, called Cohen. Within days, Brown called as well.

202 UFW organizers had tried and failed to enter the Ballantine grape vineyards: *Dollars and Sense* magazine, Jan., 1976, p. 11, among other sources.

203 the UFW filed more than a thousand complaints: Meister and Loftis, *A Long Time Coming*, p. 222; Estimate checked against ALRB annual report for 1975–76, Appendix C.

203 one of the more daring challenges: Rose, "Women in the United Farm Workers," p. 142.

203 The Teamsters, in turn, charged that the UFW: Griswold del Castillo and Garza, *Cesar Chavez*, p. 129.

203 In the fall of 1975, Teamsters stormed: Meister and Loftis, *A Long Time Coming*, p. 222.

203 By the end of the ALRB's first fiscal year: ALRB annual report, June, 1976, p. 77.

203 Workers at the M. Caratan Ranch: Int. Ben Maddock, July 25, 1995.

203 "A month turned into twenty years.": Ibid.

203 "We wanted to prove that in 1973.": Ibid.

204 "As it went on, I knew we had made up the fifteen votes.": Ibid.

204 leaving the agency with insufficient operating funds: The ALRB ceased operations on April 1, 1976. *Sacramento Bee*, July 30, 1976, among other sources.

204 Bud Antle Company of Salinas, threatened to sue: *Sacramento Bee*, June 9, 1976, p. C1, among other sources.

205 A platoon of assistants and farmworkers-turned-union activists: Int. Clementina Carmona Olloqui, July 6, 1996.

205 Within a short time, more than 700,000 signatures were recorded: UFW voter registration flyer, summer, 1976.

205 "an impossible, no-win situation": Int. Leroy Chatfield, Jan. 24, 1996.

205 The ALRB's near-death spurred the UFW: *Sacramento Bee*, Sept. 23, 1976, among other sources.

208 "We're going to teach the growers a lesson": Cesar Chavez, speaking to delegates at the AFL-CIO California Labor Federation meeting, reported by the *Sacramento Bee*, Sept. 23, 1976.

208 Looking back to the fall of 1976: Int. Jerry Cohen, Sept. 3, 1996; Int. Marshall Ganz, Oct. 10, 1996; and other sources.

208 Clementina Carmona Olloqui was attacked: Int. Clementina Carmona Olloqui, July 6, 1996.

208 anti–Proposition 14 forces invested nearly $2 million: Griswold del Castillo and Garcia, *Cesar Chavez*, p. 130.

208 The No on 14 campaign—run by an influential lobbying firm: Int. Jerry Cohen, Sept. 3, 1996.

208 "They played the race card": Ibid.

208 Although Chavez had started a fast: Int. Ben Maddock, Sept. 5, 1996.

209 friends and union workers noticed a change: Int. Cohen, Sept. 3, 1996; Ganz, Oct. 10, 1996; and others; Griswold del Castillo and Garcia, *Cesar Chavez*, p. 132, Chavez quoted after Proposition 14 defeat: "The world was really changing. Now we had to start planning."

209 "We've been asking the workers to be patient": Associated Press, *Woodland Daily Democrat*, Jan. 28, 1977.

209 "Seventeen months after the farm-labor law": Associated Press, *Sacramento Bee*, Jan. 26, 1977.

209 Regulators failed to verify months-old elections: Associated Press, *Woodland Daily Democrat*, Jan. 28, 1977.

209 The bureaucratic foot-dragging: *San Francisco Examiner, Image* magazine, July 18, 1993, p. 10.

210 "The law says it is the policy": Ibid.

210 participated in nineteen new elections in 1976: *Sacramento Bee*, March 22, 1977.

210 the board announced its first substantial settlement: *Sacramento Bee*, Dec. 18, 1976.

210 The UFW finally won an enormous victory: *Sacramento Bee*, Jan. 22, 1977.

210 "Cesar's not the underdog": Press conference in Burlingame at signing of Teamster-UFW peace pact. Archival news footage.

210 "You know, it was one of the most difficult": Ibid.

211 "All the money we've thrown into the fight": Associated Press, *Imperial Valley Press*, April 25, 1977.

211 Chavez had said publicly that the union was looking toward a future: Griswold del Castillo and Garcia, *Cesar Chavez*, p. 133.

211 "We in the top echelon of the organization": *New Republic*, May 22, 1976, p. 15.

211 He was obsessed with finding the right management technique: Int. Marc Grossman, Oct. 15, 1996.

212 It was the drive to deal with these contradictions, some say, that pushed Chavez: Griswold del Castillo and Garcia, *Cesar Chavez*, p. 132; *Los Angeles Times*, March 19, 1979, p. A1.

212 "He didn't want the UFW to become a traditional labor union": Int. Jerry Cohen, June 21, 1996.

212 Chavez became intrigued with the psychological techniques used in a drug-rehabilitation program known as Synanon: Int. Marc Grossman, Oct. 15, 1996; Int. Jerry Cohen, June 8, 1996; Int. Marshall Ganz, Oct. 10, 1996; published accounts such as the *Los Angeles Times*, March 19, 1979, and the *Village Voice*, Aug. 21, 1984.

212 Nick Jones, had been asked to leave: *Village Voice*, Aug. 21, 1984, p. 9; Int. Cohen, June 21, 1996; Int. Eliseo Medina, June 14, 1996.

212 that same year he was convicted: *Village Voice*, Aug. 21, 1984.

213 Unexpected criticism: Int. Grossman, Oct. 15, 1996.

213 "It was just a blip on the screen": Int. Jerry Cohen, June 8, 1996.

213 In 1978 the union laid off more than half of Cohen's legal team: *Los Angeles Times*, March 19, 1979.

214 Loyal organizer Eliseo Medina: Int. Eliseo Medina, June 14, 1996.

214 "Lax bookkeeping": *Sacramento Bee*, Nov. 9, 1983.

214 strong pro-union sentiment in the fields: Int. Marshall Ganz, Sept. 5, 1996; Int. Marc Grossman, Oct. 15, 1996.

214 "wage competition": Ibid.

215 "The growers just freaked out": Int. Marshall Ganz, Oct. 10, 1996.

215 "He told me he never wanted to be friends with any grower": Int. Marc Grossman, Oct. 15, 1996.

215 In 1978, in the fertile farm belt: Int. Joe Livernois, Sept. 9, 1996.

215 "And then, just like that": Ibid.

215 The protracted conflict tore apart communities: Int. Joe Livernois, Sept. 9, 1996; Int. Marshall Ganz, Sept. 5, 1996.

216 "I feel the way I have never felt before": Cesar Chavez address to UFW rally, *Imperial Valley Press*, Feb. 2, 1979.

216 "coming of age": Int. Marshall Ganz, Oct. 10, 1996.

216 A cross was burned in one lettuce field: Int. Joe Livernois, Sept. 9, 1996, among other sources.

216 his "primary interest is the safety of Anglo": *Imperial Valley Press*, Feb. 8, 1979.

216 did not want to sit down: Int. Marc Grossman, Oct. 15, 1996.

217 The Dolphin Group, Chavez's nemesis: Int. Joe Livernois, Sept. 9, 1996, among other sources.

217 At the schools, confrontations over recruiting handbills: *Los Angeles Times*, Feb. 16, 1979.

217 The tension in the valley snapped violently: *Los Angeles Times*, Feb. 11, 1979, among other published accounts.

218 "lack of evidence": *Los Angeles Times*, Feb. 14, 1979; Int. Joe Livernois, Sept. 9, 1996.

218 "Look at whose blood is being shed": Int. Marc Grossman, Oct. 15, 1996.

218 Grower Carl Maggio: *Los Angeles Times*, July 30, 1991.

218 Mario Saikhon would never fully recover: *Los Angeles Times*, Sept. 9, 1993.

218 "I think I was a little skeptical": Int. Joe Livernois, Sept. 9, 1996.

218 "It's pretty poor": *Los Angeles Times*, Feb. 4, 1979.

219 When the strike arrived with the harvest: Int. Marshall Ganz, Sept. 5, 1996; Int. Jerry Cohen, Sept. 3, 1996.

EIGHT: THE POISONED EAGLE

220 "In the old days": Wayne State archives, Cesar Chavez speech at Pacific Lutheran College, Tacoma, Wash., March 19, 1989.

221 membership in the UFW: Various sources differ on the union's peak membership. Some news accounts put it as high as 120,000, while others listed it as no higher than 70,000. Union literature often had the mark at 107,000. The listed figure is therefore rounded off.

221 fastest workers could earn: Int. Marshall Ganz, Sept. 5, 1996.

221 medical plan: Ibid.

221 On forty-six ranches in California: Agricultural Labor Relations Board statistics, reported by William H. Friedland, *From Social Movement to Trade Union*, Table 2.

221 "The UFW made us, as an industry": Int. Larry Galper, presented in *The Golden Cage: A Story of California's Farm Workers*, film documentary, University of California, Berkeley, 1990.

222 "There was no question that the farmers learned a lot.": Int. Daryl Arnold, Jan. 26, 1996.

222 Third Way: William H. Friedland, *From Social Movement to Trade Union*, p. 5.

222 "Take Preventative Action!": *Western Grower and Shipper*, July, 1984, p. 24.

222 "Spanish for Farmers": Regular ad appearing in *Ag Alert*, California Farm Bureau Federation, 1984.

223 "I said, 'What about' ": Int. Bill Grami, Aug. 1, 1995.

223 "We've been working more on": *Sacramento Bee*, May 20, 1982.

223 He was especially patient with young staff members: Int. Marc Grossman, Oct. 15, 1996.

223 In one instance Cesar personally "handled": Int. Al Rojas, Jan. 24, 1996.

223 "There were some who thought we should.": Int. Marc Grossman, Oct. 15, 1996.

224 Chavez's position reflected a goal he had written into the union's first constitution: Jacques Levy, *Cesar Chavez*, p. 176.

224 Since the defeat of Proposition 14: California Fair Political Practices Commission figures. *Sacramento Bee*, Nov. 9, 1983.

224 "Chicano lobby" to push the interests: *Sacramento Bee*, Sept. 9, 1983.

224 "We must continue to keep our membership involved": Jacques Levy, *Cesar Chavez*, p. 537.

225 Marshall Ganz remembers that Cesar tried to convince: Int. Marshall Ganz, Oct. 10, 1996.

225 "The energy seemed to go out of the strike": Int. Joe Livernois, Sept. 9, 1996.

225 Chavez urged a return to the boycott: Int. Marshall Ganz, Oct. 10, 1996; Int. Jerry Cohen, Nov. 3, 1996.

225 As far back as the Delano grape-strike days: Int. Doug Adair, July 6, 1996.

225 Adair traces some of the problems: Northridge int. Doug Adair, March 10, 1995.

226 The idea, gleaned from autoworkers' contracts in Detroit: Int. Marshall Ganz, Oct. 10, 1996.

226 The tumult was underscored by the departures: Int. Marshall Ganz, Oct. 10, 1996; Int. Jerry Cohen, June 21, 1996.

226 Growing personal and philosophical differences: Int. Gilbert Padilla, July 27, 1995; among other sources.

226 he admitted that he had given up his life for the UFW: *Imperial Valley Press*, April 25, 1977. "His tactics and zeal are unchanged. He has sacrificed his private life for the union. He can't remember when he last saw a movie. His credo is self-discipline and denial."

226 Former union supporters complained: William H. Friedland, *From Social Movement to Trade Union*, p. 8.

226 "He so believed in the law": Int. Marc Grossman, Oct. 15, 1996.

227 "I think we are all responsible for what happened": Int. Marshall Ganz, Jan. 7, 1996.

227 In 1979, Chavez had quietly endured: Int. Marc Grossman, Oct. 15, 1996; *New York Times*, Feb. 8, 1979.

227 "mirthless grin": *Village Voice,* Aug. 14, 1984.

227 a column by muckraker Jack Anderson: Anderson's Merry Go-Round column, *Bakersfield Californian,* March 7, 1980.

228 he turned to famed San Francisco lawyer Melvin Belli: Int. Marc Grossman, Oct. 15, 1996.

228 Belli had sent a letter: letter from Belli firm to the *Escondido Times Advocate,* April 18, 1980: "We wish to use this opportunity to inform you about [the UFW's] concerns and to impart to you how and why many of the statements contained in Mr. Anderson's article are in error. We will also summarize very briefly the evidence that is available to document these errors."

228 "Flanked by three lawyers": Anderson's Merry Go-Round column, *Salinas Californian,* April, 1989.

228 "collected money from a consortium of powerful growers": Ibid.

228 In late 1979, Newman's research had tipped: Int. Marc Grossman, Oct. 15, 1996.

228 Salinas field representatives claimed: *Village Voice,* Aug. 21, 1984.

229 In an attempt to have their own voice: *Sacramento Bee,* Sept. 6, 1981.

229 There was clearly not enough support: Int. Marc Grossman, Oct. 15, 1996.

229 Chavez responded by blaming the attempted rebellion on "external forces": *Sacramento Bee,* Sept. 6, 1981.

229 A saddened Ganz: Int. Marshall Ganz, Oct. 5, 1996.

229 Cesar insisted that the representatives had actually been appointed: *New York Times,* Oct. 28, 1982.

229 One of the fired representatives: Int. Sabino Lopez, June 21, 1996.

229 In 1982 the number of new elections at ranches fell to twenty-one: Agricultural Labor Relations Board statistics, reported by William H. Friedland, *From Social Movement to Trade Union,* Table 2.

229 The union also succeeded in getting an affirmative ruling: *Sacramento Bee,* Nov. 1, 1983.

230 Not all court decisions were so sweet, however: *San Jose Mercury News,* May 21, 1978.

230 In another particularly bitter fight: Wire services report, *Sacramento Bee,* Feb. 22, 1984.

230 Stirling, an unsuccessful candidate for state attorney general: *San Jose Mercury News,* Nov. 30, 1985.

230 "There's no question that the standard.": Ibid.

230 Chavez complained bitterly that the labor board was preventing: *San Jose Mercury News,* Sept. 9, 1985.

230 But Dolores Huerta believes that politics: Int. Dolores Huerta, June 9, 1995.

231 Chavez was developing plans to work with the Mexican government: Int. Marc Grossman, Oct. 10, 1996. In 1991, that early work finally created a benefits plan that covered union members who lived in Mexico part of the year. Participation in the program, however, has diminished as the premiums charged to workers have increased along with the cost of medical care.

231 Chavez remained in high regard among farmworkers: Barger and Reza, *The Farm Labor Movement in the Midwest,* tables, p. 48.

231 In September 1983, Rene Lopez: *Los Angeles Times,* Sept. 27, 1983; among other sources, including Int. Marc Grossman, Oct. 10, 1996, and Tanis Ybarra, Parlier, CA, UFW director, Nov. 3, 1996. The story of the Lopez shooting is further detailed in wire service and *Los Angeles Times* articles in Sept., 1983.

232 "How long can I go to the well": *Los Angeles Times,* Sept. 28, 1983, p. A3.

232 "Rene is gone because he dared to hope": *Los Angeles Times,* Sept. 27, 1983, p. A3.

232 Daryl Arnold, then head of the Western Growers Association, demurred: *Los Angeles Times,* Oct. 24, 1983, p. D1.

232 "too little, too late": *Los Angeles Times,* Sept. 28, 1983.

232 In the end, it was determined that Estrada had shot Lopez: Int. Marc Grossman, Oct. 10, 1996; Int. Tanis Ybarra, Nov. 3, 1996.

233 Pablo Romero was dealing with another kind of tragedy: Int. with Pablo Romero, Oct. 24, 1996.

233 "One of the things that happens": Ibid.

234 In the mid-1980s, Monterey County would enact some of the toughest: Ibid., among other sources.

234 Everyone would become acutely aware of pesticides: Federal Environmental Protection Agency statistics reported by the Institute for Food & Development Policy, *Farmworkers in the 1990s, Where Do We Stand?*, 1994.

234 He and Dolores Huerta had negotiated away the use: *San Jose Mercury News*, Sept. 9, 1985; among other sources.

235 Chavez, in the summer of 1984, called on American consumers: *San Jose Mercury News*, July 12, 1984.

235 Richie Ross, a union activist: Int. Marc Grossman, Oct. 15, 1996.

235 Marion Moses, once a UFW nurse: Int. Marion Moses, April 23, 1996.

235 The accounts were shocking: Ibid., among other sources.

237 "If food safety is the issue": *San Jose Mercury News*, Jan. 20, 1988.

237 Perhaps Obbink hadn't noticed: *San Jose Mercury News*, Oct. 4, 1985.

237 To supplement the media campaign: Setterberg and Shavelson, *Toxic Nation*, p. 202.

237 The Dolphin Group was quietly directing the show: In 1993, an ALRB claim filed by the UFW against the coalition had Adam Ortega on the stand during a hearing. Ortega, often quoted as the head of the group, testified that he was actually being paid by the Dolphin Group, and that the coalition's office is located in the Dolphin Group's Los Angeles headquarters. Source: May 5, 1993 memo from UFW lawyer Rebecca Harrington to Rees Lloyd of the Robin Hood Foundation.

237 In 1985, the nation was jolted: Mott and Snyder, Natural Resources Defense Council, *Pesticide Alert*, p. 131.

238 In 1986, Moses got a call: *Hippocrates* magazine, Nov.–Dec., 1996, p. 57.

238 enlisted volunteers from Hollywood to produce: *Sacramento Bee*, Feb. 2, 1987.

238 Roughly fifty thousand copies of the short film: Setterberg and Shavelson, *Toxic Nation*, p. 200.

238 Some McFarland mothers sued the union: Ibid.

239 In 1988, she founded the Pesticide Education Center: *Hippocrates* magazine, Nov.–Dec., 1996.

239 "And then, in 1989": Int. Marion Moses, April 23, 1996.

239 As late as the summer of 1996: *Los Angeles Times*, Aug. 29, 1996.

239 While McFarland became: Ferriss, *The Golden Cage: A Story of California's Farm Workers*, film documentary, 1990, among other sources.

240 A woman's voice interrupted: *San Jose Mercury News*, Sept. 24, 1985.

241 Ballin would eventually be convicted: *San Jose Mercury News*, Nov. 26, 1985.

241 Ballin's ranch was similar to some of the strawberry farms: Wells, *Strawberry Fields*, p. 212.

241 Because Ballin's ranch was so isolated: Unpublished notes from Ferriss, *The Golden Cage: A Story of California's Farm Workers*, film documentary, 1990; Int. Hector de la Rosa, Oct. 25, 1996.

242 Near an affluent area north of San Diego: *Los Angeles Times* Sunday magazine, Nov. 24, 1994.

242 In Stockton, a tireless Mexican immigrant: *Stockton Record*, April 2, 1988.

242 "Some farm workers now earn.": *Philadelphia Inquirer*, July 10, 1985.

242 They set up "Ag Help": *Monterey County Herald*, July 1, 1987.

243 The UFW vehemently opposed the new immigration law's provision for possible guest workers: Griswold del Castillo and Garcia, *Cesar Chavez*, p. 167.

243 "We never worked out a solution": Int. Jerry Cohen, June 21, 1996.

244 In 1973, Cesar had sent his cousin Manuel: Griswold del Castillo and Garcia, *Cesar Chavez*, p. 164.

244 Some accounts of the Arizona strike: Ibid., among other sources, including FBI surveillance memos, 1973 and 1974, obtained through the Freedom of Information Act.

244 Still, the citrus campaign was a disaster: Griswold del Castillo and Garcia, *Cesar Chavez*, p. 164.

244 Earlier that year, Chavez had insisted: Griswold del Castillo and Garcia, *Cesar Chavez*, p. 124.

244 industry kept reporting increased sales: Associated Press, *San Jose Mercury News*, Aug. 21, 1988.

244 Firms like Abatti and SunHarvest used the ploy: *Los Angeles Times*, Dec. 16, 1983; *Monterey County Herald*, Dec. 13, 1987; Ferriss, *The Golden Cage: A Story of California's Farm Workers*, film documentary.

244 In one celebrated 1983 case, however: *Los Angeles Times*, Oct. 18, 1983.

245 A medical team, which included: Int. Marion Moses, April 23, 1996; Int. Pablo Romero, Oct. 25, 1996.

245 "It was scary": Int. Marion Moses, April 23, 1996.

245 "What are you doing here?": Ibid.

246 "It was very hard on him": Int. Chris Hartmire, July 2, 1996.

246 "Today, I pass on the fast": *San Jose Mercury News*, Aug. 21, 1988.

247 Stepping out of his quarters: Int. Marc Grossman, Oct. 15, 1996.

247 "Health experts think the high rate": Wayne State Archives, Cesar Chavez address at Pacific Lutheran College, Tacoma, Wash., March, 1989.

247 "Last year California's Republican Governor.": Ibid.

NINE: A PINE COFFIN

248 "Regardless of what": Wayne State Archives, Commonwealth Club speech, Nov. 9, 1984, San Francisco.

249 Chavez walked onto a stage: Int. Clementina Carmona Olloqui, July 6, 1996.

249 rest uneasy in death if monuments: Int. Marc Grossman, Oct. 15, 1996.

249 "I just want to stop those grapes, Al": Int. Al Rojas, Jan. 24, 1996.

249 "It was odd to see him": Int. Clementina Carmona Olloqui, July 6, 1996.

250 "I do not believe that the people meant": Wayne State Archives, Chavez speech at school dedication, Oct. 19, 1990.

250 Cesar found that campaign exhilarating: Int. Marc Grossman, Aug. 14, 1996.

250 noisy, anti-pesticide rallies: McClatchy News Service, *San Jose Mercury News*, Feb. 8, 1990.

250 "Direct confrontation": Wayne State Archives, Address to Nader environmental conference, Oct. 27, 1991.

250 Three hundred family members: *San Jose Mercury News*, Dec. 19, 1991.

250 "Juana Estrada Chavez does not": Wayne State Archives, Chavez's address during services, Dec. 18, 1991.

250 Juana Chavez's death: Int. Marc Grossman, Oct. 10, 1996; Mike Ybarra, June 8, 1995.

250 "We were the strikingest family": Chavez's address during services, Dec. 18, 1991.

251 George Deukmejian had passed the Republican Party's torch: McClatchy News Service, *San Jose Mercury News*, April 30, 1991.

251 Chavez estimated that if the state would only force growers: Arturo Rodriguez address to Sacramento rally, April 24, 1994, text supplied by Marc Grossman/UFW.

251 Arturo Rodriguez remembers: Ibid.

251 Rodriguez muses: Ibid.

251 In 1992, Cesar marched again: *Monterey County Herald*, July 25, 1992, p. A1.

252 "feels a little strange": *Monterey County Herald*, July 25, 1992.

252 died at eighty-two: *San Jose Mercury News*, Sept. 30, 1992.

252 "He loved to sing": Ibid.

252 Paying tribute to Ross: Int. Marc Grossman, Oct. 15, 1996.

252 Chavez, slowed by a severe flu: Archival film footage of Ross services, Oct. 17, 1992; Int. Marc Grossman, Oct. 15, 1996.

252 Grossman recalls that just before he was to speak: Int. Marc Grossman, Oct. 15, 1996.

252 "Father of all Chicanos": Oscar Zeta Acosta, *Revolt of the Cockroach People*, p. 45; among other sources.

253 It is ironic that the land upon which the Chavez family: *Los Angeles Times*, April 24, 1993, among other sources.

253 Bruce Church was one of the companies: *Salinas Californian*, Aug. 24, 1970.

253 In 1984, the company sued the UFW: *San Jose Mercury News*, June 17, 1993.

254 But in 1992, the Bruce Church challenge loomed: Ibid.

254 There were only twenty-two thousand farmworkers officially in the union's: Ibid.

254 only in 1989 and 1990 was there a serious dip in sales: Griswold del Castillo and Garcia, *Cesar Chavez*, p. 137.

254 he began one of his now-routine fasts: Griswold del Castillo and Garcia, *Cesar Chavez*, p. 173.

255 His high spirits, it seemed, had returned: *Los Angeles Times*, April 24, 1993.

255 teachers hit him for speaking Spanish: Levy, *Cesar Chavez*, p. 24.

255 Later, Cesar called Arturo Rodriguez: Int. Arturo Rodriguez, June 9, 1995.

255 He broke his fast with a simple vegetarian meal: *Los Angeles Times*, April 25, 1993.

255 It was Martinez, a man who'd grown up in service of Chavez: Ibid.

255 After calling for paramedics: Ibid.

255 Sixty-six-year-old Dona Maria Hau: Ibid.

255 As word spread among union field offices: Int. Sabino Lopez, June 21, 1996.

256 Arturo Rodriguez was driving on a lonely stretch of road near Yuma: Int. Arturo Rodriguez, June 9, 1995.

256 "Regardless of what": Commonwealth Club speech, Nov. 9, 1984, San Francisco.

256 Richard Chavez was among the first in the family: Int. Richard Chavez, June 8, 1995.

257 Dolores Huerta also believes Chavez had sensed he might be nearing the end: Int. Dolores Huerta, June 9, 1995.

257 She was also still on the mend from a clubbing: *San Jose Mercury News*, Sept. 20, 1988.

257 In 1991, the city settled with Huerta: *San Jose Mercury News*, March 4, 1991.

257 Chavez's wish for a simple burial: Int. Richard Chavez, June 8, 1995.

260 just a few feet from where Boycott: Int. Carlos LeGerrette, Sept. 21, 1996.

260 Mike Ybarra, Cesar's longtime driver and bodyguard: Int. Mike Ybarra, June 8, 1995.

260 Ybarra last saw Cesar alive: Ibid.

260 Lucio Gonzalez and his wife, Butter: Int. Lucio Gonzalez, July 3, 1996.

261 Comedian Paul Rodriguez agreed that Chavez: *San Jose Mercury News*, April 30, 1993.

261 seventy-seven-year-old former farmworker Richard Lopez: *Los Angeles Times*, April 30, 1993.

263 Employing the tactics he learned from Chavez, Medina had scored impressive gains: Int. Eliseo Medina, June 14, 1996.

263 Al Rojas was at the funeral, too: Int. Al Rojas, July 1, 1996.

263 Jerry Cohen was there, pushing aside criticism: Int. Jerry Cohen, June 21, 1996.

263 "Cesar wasn't afraid of anybody": Ibid.

263 "I remember the first time I saw him": Ibid.

265 Brown had reconnected with Cesar: Int. Jerry Brown, Aug. 28, 1995.

265 "An authentic hero": *Los Angeles Times*, April 24, 1993, p. A1.

265 "He was the spring, the root, where it all started": Ibid.

265 "For all the workers,": Ibid.

265 "God has taken the strongest arm we have": Ibid.

265 Baldemar Velásquez, president of the Farm Labor Organizing Committee: Int. Baldemar Velásquez, Nov. 1, 1996.

265 Velásquez had emulated UFW tactics: Barger and Reza, *The Farm Labor Movement in the Midwest*, p. 46.

267 To this day, the UFW and FLOC remain close: Int. Arturo Rodriguez, Sept. 15, 1996.

267 "Cesar opened the world of possibility for us": Int. Baldemar Velásquez, Nov. 1, 1996.

268 Filemon López was singing in his second language: Ferriss, "Bitter Harvest," *Image* magazine, *San Francisco Examiner*, p. 11.

268 Some were holding the memorial program: Griswold del Castillo and Garcia, *Cesar Chavez*, p. 178.

268 "My dad says, 'Let me see if I can explain it to you'": Int. Paul Chavez, Aug. 25, 1996.

268 "Cesar was a gift to the farmworkers": Int. Pete Velasco, June 8, 1995.

269 "When a pine tree falls": Int. José Padilla, Nov. 3, 1996, in San Francisco at the showing of *The Fight in the Fields* documentary.

EPILOGUE

270 "Cesar always wanted the idea behind the union to be respect": Int. Salvador Mendoza, May 19, 1996.

271 Vicente Dóminguez pushed: Ferriss, *San Francisco Examiner, Image* magazine, July 18, 1993.

271 increasing number of farmworkers today are Mexican Indians: *San Francisco Examiner, Image* magazine, July 18, 1993.

271 son of a sheet-metal worker and a teacher: *Los Angeles Times*, February 17, 1996.

272 Rodriguez was startled: Int. Arturo Rodriguez, Sept. 15, 1996; Int. Linda Chavez Rodriguez, June 8, 1995.

272 In 1993, Rodriguez signed an agreement: UFW press advisory issued in Los Angeles on Sept. 24, 1993.

272 10 percent of the farm-labor force: *San Francisco Examiner, Image* magazine, July 18, 1993.

272 Filemon López, a leader in a Mixtec network: *San Francisco Examiner, Image* magazine, July 18, 1993.

273 *Wall Street Journal* has featured him: *Wall Street Journal*, Dec. 19, 1995.

273 most dynamic leaders: *Wall Street Journal*, Dec. 19, 1995.

273 wanted to revitalize the UFW's grassroots organizing: Rodriguez quoted in *Los Angeles Times*, Feb. 17, 1996.

273 UFW won thirteen elections by end of 1995: California Institute for Rural Studies, Fall 1996 report.

273 strawberry workers at VCNM Farms: *Wall Street Journal*, Dec. 19, 1995; UFW reports.

273 legal review of charges like these can drag on: *San Francisco Examiner, Image* magazine, July 18.

274 strawberry "growers' and workers alliance" was formed: *San Francisco Examiner*, Nov. 24, 1996.

274 believed they were being tarred unfairly: *New York Times*, July 3, 1996.

274 Watsonville on September 15: authors' observations.

274 estimates are that strawberry pickers earn an average $8,500 during a year: California Institute for Rural Studies, based on surveys of pay stubs and Employment Development Department statistics in Monterey and Santa Cruz Counties.

274 "Those from the union have taught us our rights": *San Francisco Examiner*, Nov. 24, 1996.

274 in 1996, Rodriguez spent time: Int. Arturo Rodriguez, Sept. 15, 1996.

275 ending a seventeen-year-old feud: *New York Times*, May 30, 1996.

275 "I don't know why an agricultural worker should be treated differently": Int. Steve Taylor, June 10, 1996.

275 "It was very good dealing with Steve": Int. Arturo Rodriguez, Sept. 15, 1996.

275 In 1991, more than one hundred pickers: *Monterey County Herald*, Sept. 8, 1991.

276 giant company professed ignorance: *Monterey County Herald*, Sept. 8, 1991.

276 supply investment money: *Monterey County Herald*, Sept. 8, 1991; *San Francisco Examiner, Image* magazine, July 18, 1993.

276 no higher than 30 percent: Int. Don Villarejo, director of California Institute for Rural Studies, Oct. 28, 1996.

276 Immigration Reform and Control Act: *San Francisco Examiner, Image* magazine, July 18, 1996; *Monterey County Herald*, July 1, 1987, March 15, 1988; *Los Angeles Times*, Sept. 25, 1995.

276 "our economy depends very heavily on Mexican nationals": *San Francisco Examiner*, Aug. 15, 1993.

276 "I deplore the INS raids": Ibid., Aug. 15, 1993.

277 held a press conference on the edge of a broccoli field: *San Francisco Examiner,* Oct. 30, 1994.

277 three undocumented brothers, Jaime, Salvador, and Benjamin Chavez-Muñoz: *Pacific News Service,* June 19, 1996.

278 "I think it's a crock": *San Francisco Examiner,* Oct. 30, 1994.

278 every agricultural region of California voted for Proposition 187: California Secretary of State's Office.

278 "that workforce is really exploited": Int. Arturo Rodriguez, Sept. 15, 1996.

278 Gallo workers voted for the UFW: *San Francisco Chronicle,* May 4, 1996.

279 one of the biggest political donors: Federal campaign records.

279 "They brought in an army of lawyers": Int. Salvador Mendoza, May 19, 1996.

279 "It was a big surprise for the company": Int. Serafin Perez, May 19, 1996.

279 "we have more dignity now": Int. Rogelio Perez, May 19, 1996.

279 "If God gives us license": Int. Salvador Mendoza, May 19, 1996.

THE FIGHT IN THE FIELDS INTERVIEWS

Fred Abad, July 25, 1995, Delano, CA

Ernie Abeytia, Aug. 27, 1996, telephone interview

Doug Adair, July 6, 1996, Coachella, CA

Daryl Arnold, Jan. 26, 1996, Corona Del Mar, CA

Jesus Lopez Avalos, Sept. 15, 1996, Watsonville, CA

Lauro Barajas, July 2, 1996, Watsonville, CA

Beatriz Bedoya, Aug. 28, 1996, telephone interview

Eric Brazil, June 18, 1996, San Francisco, CA

Jerry Brown, Aug. 28, 1995, Oakland, CA

Juanita Brown, April 23, 1996, Mill Valley, CA

Epifanio Camacho, July 3, 1996, McFarland, CA

Leroy Chatfield, Jan. 24, 1996, Sacramento, CA

Librado "Lenny" Chavez, Aug. 28, 1996, telephone interview

Paul Chavez, April 25, 1996, Keene, CA

Richard Chavez, June 8, 1995, and April 25, 1996, Keene, CA

Jerry Cohen, Aug. 28, 1995; June 8, 1996, Carmel, CA; Sept. 3, 1996, telephone interview Sept. 19, 1996

Delfina Corcoles, June 22, 1996, Watsonville, CA

Bert Corona, July 16, 1995, Los Angeles, CA

Dorothy Coyle, April 23, 1996, San Francisco, CA

Tom Dalzell, Nov. 4, 1996, telephone interview

Jessie De La Cruz, July 26, 1995, Kingsburg, CA

Jose de la Cruz, July 5, 1996, Arvin, CA

Hector de la Rosa, July 30, 1996, telephone interview

Frank Denison, June 14, 1996, San Diego, CA

Herman Gallegos, Aug. 30, 1995, Brisbane, CA

Marshall Ganz, Jan. 6, 1996, Oakland, CA; Sept. 5 and 10, 1996; and Oct. 10, 1996, telephone interview

Teofilo Garcia, Aug. 2, 1995, and July 5, 1996, Arvin, CA

Butter Torres Gonzales, July 3, 1996, Delano, CA

Lucio Gonzalez, July 3, 1996, Delano, CA

Jessica Govea, April 25, 1996, San Francisco, CA

Bill Grami, Sept. 1, 1995

Marc Grossman, Oct. 10, 1996, telephone interview

Adelina Gurola, July 25, 1995, and July 5, 1996, Arvin, CA

Chris Hartmire, Jan. 24, 1996, and July 1, 1996, Sacramento, CA

Dorothy Healy, March 30, 1996, Washington, DC

Julio Hernandez, Sept. 6, 1996, telephone interview

Angie Hernandez Herrera, Sept. 6, 1996, telephone interview

Monsignor George Higgins, March 27, 1996, Washington, DC

Dolores Huerta, June 9, 1995, Keene, CA

Ray Huerta, July 6, 1996, Coachella, CA

Mo Jourdane, July 30, 1996, telephone interview

Ethel Kennedy, March 28, 1996, McLean, VA

Carlos and Linda LeGerrette, July 7, 1996, San Diego, CA; Sept. 14, San Francisco, CA; Sept. 17, 1996

Joe Livernois, Oct. 28, 1996, telephone interview

Manny Lopez, Oct. 23, 1996, telephone interview

Sabino Lopez, Aug. 28, 1995, and June 8, 1996, Salinas, CA

Ben Maddock, July 25, 1995, Sanger, CA, and Sept. 11, 1996, telephone interview

Mary Magaña, July 3, 1996, McFarland, CA

Roger Mahony, July 14, 1995, Los Angeles, CA

Pete Maturino, August 28, 1995, Salinas, CA

Eliseo Medina, June 14, 1996, San Diego, CA

Rita (Chavez) Medina, Jan. 23, 1996, San Jose, CA

Salvador Mendoza, May 19, 1996, Santa Rosa, CA

Marion Moses, April 23, 1996, San Francisco, CA

Lupe and Cathy Murguia, July 24, 1995, Tehachapi, CA

Clementina Carmona Olloqui, July 6, 1996, Coachella, CA

Gilbert Padilla, July 27, 1995, Fresno, CA

Rogelio Perez, May 19, 1996, Healdsburg, CA

Serafin Perez, May 19, 1996, Healdsburg, CA

Nelly Pruneda, July 2, 1996, Watsonville, CA

Hijinio Rangel, Sept. 6, 1996, telephone interview

Isabel Rendon, Sept. 15, 1996, Watsonville, CA

Arturo "Artie" Rodriguez, June 9, 1995, Keene, CA, and Sept. 15, 1996, Watsonville, CA

Linda Chavez Rodriguez, June 8, 1996, La Paz

Marta Rodriguez, April 25, 1996, Arvin, CA

Al Rojas, Jan. 24, 1996, and June 30, 1996, Sacramento, CA

Pablo Romero, Oct. 24, 1996, telephone interview

David Ronquillo, July 14, 1995, Los Angeles, CA

Fred Ross Jr., Aug. 28, 1995, San Francisco, CA; and Sept. 3, 1996 and Oct. 10, 1996, telephone interview

Paul Schrade, July 15, 1995, Los Angeles, CA

Johnny Serda, July 3, 1996, Delano, CA

Lionel Steinberg, Jan. 26, 1996, Thermal, CA

Steve Taylor, June 10, 1996, Salinas, CA

Luis Valdez, Jan. 26, 1996, San Juan Bautista, CA

Pete Velasco, June 8, 1995, Keene, CA

Baldemar Velasquez, Nov. 1, 1996, telephone interview

Don Villarejo, Oct. 28, 1996, telephone interview

Tanis Ybarra, Nov. 3, 1996

Mike Ybarra, June 8, 1995, Keene, CA

Andy Zermeño, Sept. 24, 1996, telephone interview

George Zaninovich, Sept. 3, 1996, telephone interview

BOOKS, FILMS, AND ARCHIVES

Adair, Doug. "Oral History Transcript of Interview." California State University, Northridge: Provost's Committee on Chicano Labor, March 10, 1995.

Allen, Gary (John Birch Society). *Communist Revolution in the Streets.* Boston and Los Angeles: Western Islands, 1967.

Almaguer, Tomás. *Racial Fault Lines: The Historical Origins of White Supremacy in California.* Berkeley, CA: University of California Press, 1994.

Balderrama, Francisco E., and Raymond Rodriguez. *Decade of Betrayal: Mexican Repatriation in the 1930s.* Albuquerque: University of New Mexico Press, 1995.

Barger, W. K., and Ernesto M. Reza. *The Farm Labor Movement in the Midwest: Social Change and Adaptation among Migrant Farmworkers.* Austin: University of Texas Press, 1994.

Brill, Steven. *The Teamsters.* New York: Simon and Schuster, 1978.

Chavez Letters. Dolores Huerta Collection. Detroit: Wayne State University, Special Collections.

Chavez Letters. Fred Ross Collection. Detroit: Wayne State University, Special Collections.

De Ruiz, Dana Catherine, and Richard Larios. *La Causa: The Migrant Farmworker Story.* Austin: Raintree Steck-Vaugh, 1993.

Dunne, John Gregory. *Delano: The Story of the California Grape Strike.* New York: Farrar, Straus & Giroux, 1967.

El Malcriado papers, private collection.

Farm Security Administration archives, U.S. Library of Congress.

Ferriss, Susan. *The Golden Cage: A Story of California's Farm Workers.* Documentary film. Distributed by Filmmakers Library, New York, 1990.

Fred Ross Collection, Stanford University.

Galarza, Ernesto. *Barrio Boy.* London: University of Notre Dame Press, 1971.

Griswold del Castillo, Richard, and Richard A. Garcia. *Cesar Chavez: A Triumph of Spirit.* Norman and London: University of Oklahoma Press, 1995.

Herrera, Albert S. *Union Member's Handbook.* Washington, D.C.: Public Affairs Press, 1963.

Kiser, George C., and Martha Woody Kiser, eds. *Mexican Workers in the United States: Historical and Political Perspectives.* Albuquerque: University of New Mexico Press, 1979.

Levy, Jacques E. *Cesar Chavez: Autobiography of La Causa.* New York: W. W. Norton & Company, 1975.

———. Unpublished interviews with Cesar Chavez. Private collection.

London, Joan, and Henry Anderson. *So Shall Ye Reap: The Story of Cesar Chavez and the Farm Workers' Movement.* New York: Thomas Crowell Company, 1970.

Martinez, Elizabeth, ed. *500 Años del Pueblo Chicano/500 Years of Chicano History in Pictures.* Albuquerque: Southwest Organizing Project, 1991.

Matthiessen, Peter. *Sal Si Puedes: Cesar Chavez and the New American Revolution.* New York: Dell Publishing Company, Inc., 1969.

McWilliams, Carey. *Factories in the Field.* 1935. Reprint, Santa Barbara and Salt Lake City: Peregrine Publishers, 1971.

Meister, Dick, and Anne Loftis. *A Long Time Coming: The Struggle to Unionize America's Farm Workers.* New York: Macmillan Publishing Co., Inc., 1977.

Mott, Lawrie, and Karen Snyder. *Pesticide Alert: A Guide to Pesticides in Fruits and Vegetables.* San Francisco: Sierra Club Books and the Natural Resources Defense Council, 1987.

National Advisory Committee on Farm Labor. *Farm Labor Organizing, 1905–1967.* New York: NACLD, 1967.

Nelson, Eugene. *Huelga: The First Hundred Days of the Great Delano Grape Strike.* Delano, Calif.: Farm Worker Press, 1966.

Quiñones, Juan Gómez. *Chicano Politics, Reality and Promise, 1940–1990.* Albuquerque: University of New Mexico Press, 1990.

Rose, Margaret Eleanor. "Women in the United Farm Workers: A Study of Chicana and Mexicana Participation in a Labor Union, 1950–1980." Ph.D. diss., University of California at Los Angeles, History, 1988.

Rosenberg, Howard R., Valerie J. Horwitz, and Daniel L. Egan. *Labor Management Laws in California Agriculture.* Oakland: University of California/Division of Agriculture and Natural Resources, 1995.

Ross, Fred. *Conquering Goliath: Cesar Chavez at the Beginning.* Keene, Calif.: El Taller Grafico Press/United Farm Workers, 1989.

Samora, Julian. *Los Mojados: The Wetback Story.* Notre Dame: University of Notre Dame Press, 1971.

Setterberg, Fred, and Lonny Shavelson. *Toxic Nation.* New York: John Wiley & Sons, 1993.

Smith, Sydney D. *Grapes of Conflict.* Pasadena, Calif.: Hope, 1987.

Starr, Kevin. *Endangered Dreams: The Great Depression in California.* New York: Oxford University Press, 1996.

Steinbeck, John. *The Harvest Gypsies: On the Road to the Grapes of Wrath.* 1936. Reprint, Berkeley: Heyday Books, 1988.

Steiner, Stan. *La Raza: The Mexican Americans.* New York: Harper & Row Publishers, Inc., 1970.

Taylor, Ronald. *Chavez and the Farm Workers.* Boston: Beacon Press, 1975.

Valdez, Luis, and Teatro Campesino. *Actos.* San Juan Bautista, Calif.: Menyah Productions, 1971.

UFW archives. La Paz, California.

UFW. *Fighting for Our Lives.* Documentary film, 1974.

Watkins, T. H. *The Great Depression: America in the 1930s.* Boston: Little, Brown & Co., 1993.

Wells, Miriam, J. *Strawberry Fields: Politics, Class, and Work in California Agriculture.* Ithaca: Cornell University Press, 1996.

Williams, Juan, ed. *Eyes on the Prize: America's Civil Rights Years, 1954–1965.* New York: Penguin Books, 1987.

Zeta, Oscar Acosta. *The Revolt of the Cockroach People*. New York: Vintage Books, 1989.

ADDITIONAL RESOURCES

Ballis, George. *Basta! The Tale of Our Struggle*. Delano, Calif.: Farm Workers Press, 1966.

Braconi, Joan Marie, Alan Nicholas Kopke, and the Center for Labor Research and Education. *California Workers Rights: A Manual of Job Rights, Protections and Remedies*. Berkeley: Center for Labor Research and Education, 1994.

Bulosan, Carlos. *America Is in the Heart*. Seattle: University of Washington Press, 1973.

Day, Mark. *Forty Acres: Cesar Chavez and the Farm Workers*. New York: Praeger Publishers, 1971.

Fusco, Paul, and George D. Horowitz. *La Causa: The California Grape Strike*. New York: Collier, 1970.

Galarza, Ernesto. *Merchants of Labor: The Mexican Bracero Story*. Charlotte and Santa Barbara: McNally & Loftin, 1964.

Guerin-Gonzales, Camille. *Mexican Workers and American Dreams: Immigration, Repatriation, and California Farm Labor, 1900–1939*. New Brunswick, N.J.: Rutgers University Press, 1994.

Haslam, Gerald. *The Other California: The Great Central Valley in Life and Letters*. Reno: University of Nevada Press, 1994.

Institute for Food & Development Policy. *Farmworkers in the 1990s, Where Do We Stand*, 1994.

Kushner, Sam. *Long Road to Delano*. New York: International Publishers, 1975.

Light, Ken, Roger Minick, and Reesa Tansey. *In These Fields*. Oakland: Harvest Press, 1982.

McWilliams, Carey. *North From Mexico: The Spanish-Speaking People of the United States*. 1948. Reprinted, New York: Greenwood Press, 1968.

Moore, Truman. *The Slaves We Rent*. New York: Random House, 1965.

Parini, Jay. *John Steinbeck: A Biography*. New York: Henry Holt and Company, 1995.

Scharlin, Craig, and Lilia V. Villanueva. *Philip Vera Cruz: A Personal History of Filipino Immigrants and the Farmworkers' Movement*. Los Angeles: UCLA Labor Center, Institute of Industrial Relations and UCLA Asian American Studies Center, 1992.

Schlesinger, Stephen, and Stephen Kinzer. *Bitter Fruit: The Untold Story of the American Coup in Guatemala*. New York: Doubleday, 1982.

NEWSPAPERS AND MAGAZINES

Ag Alert/California Farm Bureau Federation
American Opinion
Bakersfield Californian
California Farmer
Delano Record
Dollars and Sense
Imperial Valley Press
Los Angeles Times
Monterey County Herald
New Republic
New York Times
Oxnard Courier
Philadelphia Inquirer
Sacramento Bee
Salinas Californian
San Diego Union
San Francisco Chronicle
San Francisco Examiner
San Jose Mercury News
Stockton Record
Toronto Daily Star
Village Voice
Wall Street Journal
Woodland Daily Democrat

STUDIES AND GOVERNMENT REPORTS

Agricultural Labor Relations Board. *Annual Reports*. 1976–1978.

California Assembly, Joint Hearings. Senate Committee on Industrial Relations. Assembly Committee on Labor Relations. *Review of the Implementation of the Alatorre-Zenovich-Dunlap-Berman Agricultural Labor Relations Act of 1975, Fresno, Nov. 25 & 26, 1975*. Sacramento, Dec. 19, 1975.

California Assembly. Senate Committee on Industrial Relations: Senate Bill 1—Third Extraordinary Session Relating to Farm Labor Relations, Sacramento, May 21, 1975.

California Assembly. Committee on Labor Relations. *Non Payment of Wages to Farm Workers.* Modesto, Aug. 16, 1975.

California Legislature. *Fourteenth Report of the Senate Factfinding Subcommittee on Un-American Activities.* 1967.

Friedland, William. *From Social Movement to Trade Union: The United Farm Workers in 1984.* Report to the Society for the Study of Social Problems. San Antonio, Texas, 1984.

Perry, Herbert A. "The Agricultural Labor Relations Board at the Crossroads." Report to the Social Science Association, 26th Annual Meeting. San Diego, CA: April 25–28, 1984.

United Farm Workers. *The Sabotage and Subversion of the Agricultural Labor Relations Act.* Keene, Calif.: El Taller Grafico Press/United Farm Workers, December 1975.

U.S. Congress. Senate Committee on Labor and Human Resources. *Hearings on Examination of the Agricultural Labor Dispute and to Encourage Both Parties to Seek a New and More Productive Relationship.* 96th Cong., 1st sess., 1979. Committee Print.

FEDERAL BUREAU OF INVESTIGATION DOCUMENTS (obtained under freedom of information act)

FBI #100-478197 Sept. 1973–Aug. 1974; "UFW Activity with Other Groups"; "Phoenix Memo on UFW planned demonstration to protest INS green card policy and Sen. Barry Goldwater"; and other files.

FBI #139-2387 Nov. 1965; "Los Angeles Memo."
157-30054 1973. "Memo from Delano, Tulare and Kern County Pickets"; "Daifullah Killing"; "Sacramento and Tulare Counties"; "Congressman Ed Roybal"; and other files.

FBI #44-66143 1975. "UFW Activity in Pecos and Reeves Counties, TX."
376-815 1973. "Memo: Confrontation between UFW and Teamsters Union"; and other files.

FBI #58-10898 1982. "National Farm Workers Service Center, UFWU, La Paz, Keene, CA. Washington DC."

FBI #159-3729 1973. "Cesar Chavez telegraph alleged Delano offices dynamited"; "Terra Bella office destroyed and Poplar office recently attacked"; and other files.

FBI #92-9831 1967. "Farm Worker strike in Rio Grande City, TX." "Investigation as to possible Communist infiltration of UFW." "Inquiries into allegations of mistreatment by law enforcement officials. Texas Rangers"; and other files.

FBI #63337 "Phoenix." "Yuma." "Civil Rights Lawsuit"; and other files.

FBI #176-69 1968. "Chavez and UFW protest Delano Police Department." "Chavez Requests US Marshals to Protect UFW During Labor Day Parade"; and other files.

FBI #100-444762 1965. "Communist infiltration of NFWA." "Grape Strike." "Larry Itliong." "Dolores Huerta." "Collected Biographies on Investigated NFWA Members and Strike Supporters." "Office of Economic Development grant to NFWA." "David Ward Havens." "*People's World*: 'Eyewitness Account of Big Farm Strike'"; and other files.

FBI #157-27530 "UFW." "Homestead Farm Owners, FLA."

FBI #100-444762-114 1967. 1968. "San Antonio, TX." "San Francisco, CA." "Strike." "Rio Grande City, TX."; and other files.

FBI #100-444762-189 1969. "US Dept. of Defense." "San Antonio." "Grape Boycott at Meijer Inc. Supermarkets, Grand Rapids, MI." "UFWOC March from Indio to Calexico, CA." "Pittsburgh March Led by Albert Rojas." "Krogers Supermarket, St. Louis and Kirkwood, MO." "Mather Air Force Base." "Jewel Food Co., Melrose Park, Ill. Picket." "Fort McPherson, Atlanta, GA UFWOC Protest"; and other files.

FBI #444762-187 1970 "Chicago." "UFWOC FDA Protest Pesticide Use." "March on Dow Chemical Pittsburgh"; and other files.

FBI #44-5908 1973. "Kern County." "Congressman Don Edwards (D-CA)." "Civil Rights Investigations, UFW and Teamsters

Union Confrontations." "Guimarra Ranch, Kern County"; and other files.

FBI #44-43004 1969. "Rio Grande Valley Growers Call for Texas Rangers." "UFW Charges Lack of Protection from Delano Police Dept." "UFW"; and other files.

FBI #100-444-762 1966. "Picket S & W Food Co., Chicago." "DiGiorgio Products Boycott." "Students for Demo Society and Committee on Racial Equality, Chicago." "Catholic Church." "Political Association of Spanish-Speaking Organizations." "Texas News Stories March to Austin." "*El Malcriado,* NFWA Newspaper." "Sam Kushner, editor, *People's World*." "Dolores Huerta." "Detailed Itinerary of Farm Worker March from Rio Grande City to Austin, 7/18/66." "Hiram Moon, United Auto Workers, Area Director, Dallas." "Paul Montemayor, United States Steel Workers, Representative." "Farmworker Blockade of Roma International Bridge, US/Mexico Border." "Antonio Orendain." "Delano March to Sacramento 3/25/67." "UFWOC

Members Magdaleno Divas and Ben Rodriguez Beaten and Jailed by Starr County Texas Rangers, 6/1/67"; and other files.

FBI #100-478197 1973. "Boston." "Leftist Political Activities"; and other files.

FBI #173-9333914. 100-42662 1973. "Operational Bread Basket, Chicago." "Cesar Chavez." "Jesse Jackson." "Delano Grape Strike." "UFWU and Teamsters Dispute Union, Fresno, Kern, Tulare Counties." "Shooting." "Violence against UFW Strike." "AFL-CIO." "Nagi Moshin Daifullah Killing. CA." "Juan Trujillo de la Cruz Killing." "UFW 'Operation Wet Blind' and Brutality Against Mexican Workers"; and other files.

FBI #44-60006 1975. "Jerome Ducote, AKA 'Fred Schwartz.'" "Extortion Plot." "UFW Office Burglaries." Various CA Grape Growers. "UFW Charges FBI with Foot-Dragging on Office Burglaries." "Senator Hughes, Chairman of U.S. Un-American Activities Committee." "Western Research, Oakland, CA"; and other files.

INTRODUCTION

pp. xviii, 2, 8 David Bacon. p. 5 *Arizona Daily Star,* Photo by José Galvez. p. 6 Thor Swift, Impact Visuals.

ONE: THE LAST FAMILY FARM

pp. 10, 20 Dorothea Lange, The Bancroft Library, University of California at Berkeley. p. 12 Rita Medina. pp. 13, 14, 15, 23, 32, 33 Helen Chavez. pp. 18–19 Photography Collection, Harry Ransom Humanities Research Center, University of Texas at Austin. pp. 24, 27 The Bancroft Library, University of California at Berkeley. p. 28 E. E. Mireles and Jovita Gonzalez de Mireles Papers, Special Collections and Archives, Texas A&M University—Corpus Christi Bell Library. p. 29 Library of Congress, Dorothea Lange. p. 34 Courtesy of Houston Metropolitan Research Center, Houston Public Library.

TWO: SAL SI PUEDES

pp. 36, 38 Department of Special Collections, Stanford University Libraries, Fred Ross Collection. p. 40 Department of Special Collections, Stanford University Libraries. p. 41 UFW archives. p. 42 Alexis Gonzales. p. 43 Helen Chavez. p. 44 Corbis-Bettmann Archive. p. 47 Archives of Labor and Urban Affairs, Wayne State University. p. 53 The Dorothea Lange Collection, The Oakland Museum of California, The City of Oakland, Gift of Paul S. Taylor. p. 54 Hearst Collection, USC Libraries.

THREE: DELANO

pp. 64, 66 Archives of Labor and Urban Affairs, Wayne State University. pp. 69, 78, 87 George Ballis, Take Stock. p. 73 Joe Gunterman. p. 81 Ernest Lowe Collection, The Oakland Museum of California, The City of Oakland. p. 82 *El Malcriado,* Wayne State University Collection.

p. 85 Fred Cordova, Filipino American National Historical Society.

FOUR: *HUELGA!*

pp. 90, 93, 95, 101, 102, 115 George Ballis, Take Stock. p. 98 Paul Fusco, Magnum Photos. p. 105 Federal Bureau of Investigation, United States Government. pp. 108, 109 Archives of Labor and Urban Affairs, Wayne State University. p. 110 El Teatro Campesino archives. p. 122 The Ernest Lowe Collection, The Oakland Museum of California, The City of Oakland.

FIVE: BOYCOTT

pp. 124, 142, 156 Archives of Labor and Urban Affairs, Wayne State University. p. 127 Paul Fusco, Magnum Photos. p. 134 *Arizona Daily Star,* Photo by José Galvez. pp. 141, 154 Oscar Castillo. p. 144 George Ballis, Take Stock. p. 145 *San Jose Mercury News,* from the Collection of Scott Montoya and Family. pp. 147, 152, 157 Bob Fitch. p. 149 Jessica Govea Thorbourne. p. 151 UFW archives.

SIX: BLOOD IN THE FIELDS

p. 158 Archives of Labor and Urban Affairs, Wayne State University. pp. 160, 164, 167, 181, 186, 187, 188 Bob Fitch. p. 171 *Salinas Californian.* p. 173 Russ Cain, *Monterey County Herald.* pp. 185, 189 Archives of Labor and Urban Affairs, Wayne State University, Bob Fitch.

SEVEN: WITH THE LAW ON OUR SIDE

p. 190 Carlos LeGerrette. pp. 192, 196 Rick Tejada-Flores. p. 194 Bob Fitch. p. 197 Corbis-Bettmann Archive. p. 199 © Jesús M. Mena Garza p. 205 Associated Press/World Wide. pp. 206, 211 Archives of Labor and Urban Affairs, Wayne State University. pp. 216, 217 Paul Noden, *Imperial Valley Press.* p. 207 Jessie De La Cruz.

EIGHT: THE POISONED EAGLE

p. 220 Archives of Labor and Urban Affairs, Wayne State University. p. 224 Earl Dotter, Impact Visuals. p. 234 Lonny Shavelson, Impact Visuals. p. 238 Lonny Shavelson. p. 240 Jesus Lopez, California Rural Legal Assistance of Salinas. pp. 241, 243 © Ken Light. p. 245 Corbis-Bettmann Archive. p. 246 Victor Aleman © 1997 VICTOR ALEMAN/2 MUN-DOS COMMUNICATIONS.

NINE: A PINE COFFIN

p. 248 Victor Aleman © 1997 VICTOR ALEMAN/2 MUN-DOS COMMUNICATIONS. pp. 251, 253 Alain McLaughlin, Impact Visuals. p. 258 Ester Hernandez. p. 259 El Malcriado (top) and Laurie Coyle (bottom). pp. 261, 267 © José Galvez. p. 262 United Farm Workers.

EPILOGUE

pp. 270, 276, 278 David Bacon. p. 272 Thor Swift, Impact Visuals. p. 275 David Bacon, Impact Visuals. p. 277 Orville Meyers, Monterey County Herald.

Grateful acknowledgment is made to the following to reprint previously published material:

"We're human beings, not dogs!" Excerpted from Rain of Gold by Victor Villaseñor (Dell Publishing, 1991). Copyright © 1991 by Victor Edmundo Villaseñor. Reprinted by permission of the author.

"The Zoot Suit Riots," adapted from North from Mexico by Carey McWilliams (Greenwood Press Publishers, 1968). Copyright © 1948 by Carey McWilliams. Reprinted by permission of the publisher.

"My Crime Is Being Filipino in America," excerpt from America Is in the Heart by Carlos Bulosan. Copyright © 1943, 1945 by Harcourt Brace & Company. Reprinted by permission of the publisher.

"The Braceros," adapted from Merchants of Labor: The Mexican Bracero Story by Ernesto Galarza (McNally and Lofus, 1964). Reprinted by permission of the Galarza estate.

"Zeferino's Fathers," from The Silver Cloud Café by Alfredo Véa, Jr. Copyright © 1996 Alfredo Véa, Jr. Used by permission of Dutton Signet, a division of Penguin Books, USA, Inc.

"Chavez and El Teatro Campesino" and "Chavez's Legacy: He Nurtured Seeds of Art," expanded and adapted by Max Benavidez from an article he wrote for the April 28, 1993, issue of the Los Angeles Times. Copyright © 1993 Max Benavidez. Used by permission of the author.

"Diary of a Strikebreaker," adapted from El Malcriado: The Voice of the Farmworker, 1974. Reprinted by permission of the UFW.

"The Spraying," excerpted from Under the Feet of Jesus by Helena Maria Viramontes. Copyright © 1995 Helena Maria Viramontes. Used by permission of Dutton Signet, a division of Penguin Books, USA, Inc.

"Postscript, 1993: Remembering Cesar Chavez," adapted from Peter Matthiessen's eulogy for Chavez published in the June 17, 1993, issue of the New Yorker. Reprinted with permission of the author.

"An Open Letter to the Grape Industry" by Cesar Chavez, reprinted by permission of the UFW.

"A Migrant Harvester's Letters Home" from an article by Jane Kay published in the December 4, 1977, issue of the Arizona Daily Star. Expanded and adapted by the author. Used with permission of the author.

"The Peasant and the Pauper" by Jose Antonio Burciaga. Reprinted with permission of Pacific News Service.

Grateful acknowledgment is also made to use the following original material:

Excerpt from "Elegy on the Death of César Chávez," copyright © 1996 by Rudolfo Anaya, and published by permission of the author and Susan Bergholz Literary Services, New York. All rights reserved.

BOOK STAFF

Writers
SUSAN FERRISS
RICARDO SANDOVAL

Editor
DIANA HEMBREE

Photo Editor
MICHELE McKENZIE

Research Editor
JUAN AVILA HERNANDEZ

Photo Research
LAURIE COYLE
MICHELE McKENZIE

Editorial Consultant
ROBERT LAVELLE

Consulting Editors
BARBARA JAMISON
CONSTANCE MATTHIESSEN

Additional Research
SOFIA MARQUEZ, OXNARD PUBLIC
   LIBRARY
JAVIER OCOSIO
MARIO OLIVAREZ
DENNIS PRICELER, UNIVERSITY
   LIBRARY, ARIZONA STATE
MAGGIE ROTH
ADRIANA SANDOVAL
CELIA SANDOVAL
CLAUDIA SANDOVAL
MARIA SISON
WAYNE STATE UNIVERSITY STAFF

Photo Research Interns
JAVIER FRANCISCO
SHAKINA HALL
RUTH WYNKOOP

FILM STAFF

Directed by
RAY TELLES & RICK TEJADA-FLORES

Produced by
RICK TEJADA-FLORES & RAY TELLES

Narrator
HENRY DARROW

Associate Producer
LAURIE COYLE

Editor
HERB FERRETTE

Executive in Charge of Production
FREDERICK D. PERRY

Camera
VICENTE FRANCO

Sound
DOUG DUNDERDALE

Original Score
PETE SEARS

Additional Music Composed by
AGUSTÍN LIRA

On-Line Editor
ED RUDOLPH

Audio Post Production
MARK ESCOTT

Assistant Camera
MATTHEW UHRY

Still Photography
EMILIO MERCADO

Additional Camera
RICK BUTLER

Archival Research
LAURIE COYLE

Assistant Archivist
DEBORAH J. HARTLEY

Archival Consultant
KENN RABIN

Additional Archival Research
ANDREW NOREN
KAREN WYATT
MICHELE McKENZIE

Archival Research Assistant
LUIS CRUZ

Post Production Coordinator
DAWN HAWK

Assistant Editors
TAL SKOOT
SHIRLEY GUTIERREZ
KARYN FUKUI
ARTHUR PINES
ALAN MICHELS

Production Assistants
DOMINIC VISSARI
DAVID· TELLES

Assistant Production Manager
JIM SUMMERS

Project Advisors
PHILIP MASON
MARIO BARRERA
MARIO GARCIA
JIM GREEN
RICARDO GRISWOLD DEL CASTILLO
MARGARET ROSE
VICTOR SORELL

Paradigm Productions Inc.
Board of Directors
ROSARIO ANAYA
JUAN GONZALEZ
WENDY LARIVIERE
JULIE MACKAMAN

Special Thanks
CESAR E. CHAVEZ FOUNDATION
UNITED FARMWORKERS OF AMERICA
  AFL-CIO
JIM YEE
LES HOUGH
EDWIN GUTHMAN
RICHARD GONZALES
JOSE PADILLA
RICHARD SAIZ

A Production of
PARADIGM PRODUCTIONS INC.

In Association with the
INDEPENDENT TELEVISION SERVICE

THE FIGHT IN THE FIELDS is a presentation
of the INDEPENDENT TELEVISION SERVICE,
with funds provided by the NATIONAL
ENDOWMENT FOR THE HUMANITIES, the
CORPORATION FOR PUBLIC BROADCASTING,
the SAN FRANCISCO FOUNDATION, the
CALIFORNIA WELLNESS FOUNDATION,
the CALIFORNIA COUNCIL FOR THE
HUMANITIES, the JOHN D. AND CATHERINE
T. MACARTHUR FOUNDATION, the WALLACE
ALEXANDER GERBODE FOUNDATION, the
COLUMBIA FOUNDATION, the CALIFORNIA
COMMUNITY FOUNDATION, and the HAZEN
FOUNDATION.

California

Crescent City

Eureka

Marysville

Healdsburg
Santa Rosa
Sonoma
Sacramento
Stockton
San
Francisco
Oakland
Modesto
San Jose
Merced
Watsonville
Madera
Monterey
Salinas
Fresno

Pixley
Delano
Wasco
McFarland
Bakersfield
Arvin

Oxnard

Los Angeles

Coachella
Blythe
Salton
Sea

El Centro
Calexico
Yuma

San Diego

Tijuana

MEXICO

Arizona

Pacific Ocean

N

0 _____ 250 kilometers
0 _____ 150 miles

Writer SUSAN FERRISS, Mexico City correspondent for Cox Newspapers, has won awards from the Associated Press and other organizations for her coverage of immigration, business fraud, toxic waste hazards, and the agribusiness ties between Mexico and California for the *San Francisco Examiner.* She has also reported extensively from Mexico and Central America. Her documentary film about migrant farmworkers, *The Golden Cage,* won top awards at the National Educational Film and Video Festival, the Columbus International Film Festival, and the Chicago International Film Festival.

Writer RICARDO SANDOVAL is an award-winning reporter with the *San Jose Mercury News,* now based in Mexico City, who has written extensively on agribusiness, the savings and loan industry, and energy as well as labor, immigration, and international trade. He was born in Mexico, and his parents, Leopoldo and Ofelia, worked the lettuce and tomato fields of San Diego County. His grandfather, Manuel Palos, spent most of his life working on farms in Mexico and on the migrant trail throughout the western United States.

Editor DIANA HEMBREE, a former news editor and staff writer at the Center for Investigative Reporting in San Francisco, has received more than a dozen national awards for her investigative stories. Other books and anthologies she has helped edit include *Global Dumping Ground: The International Traffic in Hazardous Waste, The Electronic Sweatshop* and *The Self-Care Advisor.*

Photo Editor MICHELE MCKENZIE has contributed to several major PBS documentaries and their companion volumes, including Blackside Inc.'s *The Great Depression.* Her award-winning photo research earned a News and Documentary Emmy for *Malcolm X: Make It Plain.*